BIOLOGY OF FRESHWATER POLLUTION

FOURTH EDITION

Christopher. F. Mason
Department of Biological Sciences, University of Essex

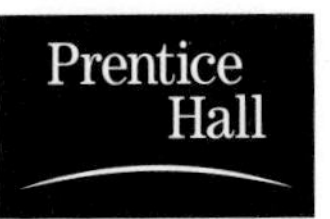

An imprint of Pearson Education

Harlow, England · London · New York · Reading, Massachusetts · San Francisco
Toronto · Don Mills, Ontario · Sydney · Tokyo · Singapore · Hong Kong · Seoul
Taipei · Cape Town · Madrid · Mexico City · Amsterdam · Munich · Paris · Milan

Pearson Education Limited
Edinburgh Gate
Harlow
Essex CM20 2JE
England

and Associated Companies throughout the world

Visit us on the World Wide Web at:
www.pearsoneduc.com

First published 1981
Second edition 1991
Third edition 1996
Fourth edition 2002

ISBN 0130-90639-5

British Library Cataloguing-in-Publication Data
A catalogue record for this book is available from the British Library

Library of Congress Cataloging-in-Publication Data
Mason, C.F.
Biology of freshwater pollution/ C.F. Mason.-4th ed.
p. cm.
Includes bibliographical references (p.).
ISBN 0-13-090639-5 (hbk)
1. Freshwater biology-Great Britain.2. Water-Pollution-Great Britain.

QH137 .M37 2001
628.1'68'09169-dc21

2001036566

10 9 8 7 6 5 4 3
06 05 04 03

Typeset by 3 in Janson Text 10/12pt
Printed and bound in China
GCC/03

CONTENTS

PREFACE

This fourth edition of *Biology of Freshwater Pollution* develops some of the new and pressing themes that have emerged in the field of water pollution over the last decade. In particular, global warming caused by human activities is now accepted as fact by most environmental scientists, if not by the majority of industrialists and politicians. If our use of fossil energy is not curbed, the effect on our environmental resources, including water, will be very severe. It has been known for a long time that some contaminants adversely influence reproduction but, in recent years, a whole host of materials in daily use have been grouped under the heading of endocrine disruptors. Male fish living below discharges of treated sewage effluent exhibit female characters. In some places, alligators, turtles and birds are similarly affected. These wildlife species could be acting as sentinels for possible endocrine disrupting effects on ourselves. In Britain the water industry has been reorganized yet again, the newly emerged Environment Agency having a wider environmental remit based firmly on the management of entire catchments. These new subjects are explored in the following pages, this edition having been reorganized so that more topics are described in fewer chapters.

New themes emerge but old problems do not go away. The contamination of water with our wastes remains the biggest killer of children worldwide. Only a few of our eutrophic lakes have been effectively treated while our attention is turning more to the eutrophication of rivers. Pollution from farmland is a major cause of concern but unlike other industries, which have to pay the costs of their waste treatment, the cost of treating water to remove agricultural chemicals is borne by the consumer in terms of increased charges. Farm run-off, containing nutrients, pesticides and soil, is causing considerable damage to the ecosystems of rivers and adjacent coastal waters. Although there have been great strides made in reducing acidifying gases produced by power stations, the acidification of upland streams remains a major concern. Research priorities change but the solutions to these older problems still require enacting on a large scale by politicians and managers before environmental damage can be reversed. They rarely are.

I would like to thank my wife, Sheila Macdonald, who has read and commented on the contents of this book, as she has on previous editions. The staff of Pearson Education, and especially Alex Seabrook and Pauline Gillett, have been supportive throughout.

Chris Mason
February 2001

ACKNOWLEDGEMENTS

We are grateful to the following for permission to reproduce copyright material:

Figs. 1.9 and 2.10 and the respective authors (Holdgate, 1979), Fig. 5.13 (Turnpenny, 1989), Fig. 5.14 (Brown and Sadler, 1989) and Fig. 7.2 ('Aquatic Invading Species' in Arthington, A. and Mitchell, D.S., *Ecology of Biological Invasions*, 1986) from Cambridge University Press; Fig. 2.2 (*The chemistry and Ectotoxicology of Pollution*, Connell, D.W. and Miller, G.J., 1984) and Fig. 3.2 (Cabelli, 1983), reprinted by permission of John Wiley & Sons, Inc © 1984 John Wiley & Sons, Inc. from the Water Environment Federation; Fig. 2.3 (Herbert, 1961) from The Institution of Water Engineers & Scientists; Fig. 2.4 (Sprague, 1964) Fig. 4.14 (Schindler *et al.*, 1973) and Fig. 4.16 (Dillon and Rigler, 1975) from the *Canadian Journal of Fisheries and Aquatic Science*; Figs 2.5 (McCahon and Pascoe, 1990), Figs 3.16 and 3.19 (Holland and Harding, 1984), Table 5.1 (Gee and Stoner, 1988), Fig. 5.9 (Fryer, 1980) and Fig. 9.17 (Hay, 1996) from Blackwell Scientific Publications Ltd; Fig. 2.6 (Lloyd, 1960) from the Board of Editors, *Annals of Applied Biology*; Fig. 2.7 (Sprague, 1970), Fig. 2.8 (Calamari and Marchetti, 1973), Table 3.2 (Curtis and Curds, 1971), Fig. 6.10 (Vandermeulen, 1987) and Figs 4.11 and 4.12 (Payne, 1975) from Pergamon Press PLC; Fig. 2.9 (Fox, 1993) from the International Association for Great Lakes Research; Fig. 2.14 (Westlake and Van der Schalie, 1977), © 1977 American Society for Testing and Materials from ASTM; Fig. 2.15 (Diamond *et al.*, 1988) from Ellis Horwood; Table 2.3 (Evans *et al.*, 1986) from WRC; Fig. 2.17 from *Environment Agency* 1988, p. 2 ('Environment-disrupting substances in the Environment: what should be done'); Fig. 2.18 (Jobling S. *et al.*, 1998), © 1998 American Chemical Society *Environmental Science & Technology*, Volume 32, 2498–2506; Fig. 2.19 (Becker, D.S. and Bigham, G.N. *Water, Air, Soil Pollution* **80**, 563–571, 1995) and Fig. 4.5 (Hussein and Mason, 1988) from Kluwer Academic Publishers; Fig. 2.21 (Häkkinen and Häsänen, 1980) from Dr I. Häkkinen; Fig. 2.22 (Czarnezki, 1985) from Springer-Verlag and the respective authors; Fig. 3.4 (Wood, 1982) from IOP Publishing Ltd; Fig. 3.7 (Cooper, P.F. *et al.*, *Journal of Water and Environmental Management* **3**, 60–74, 1989) from the Chartered Institution of Water and Environmental Management; Fig. 3.8 from *Wildfowl & Wetlands Trust* Magazine **130**, p. 17; Figs 3.13 and 3.15 (Hynes, 1960) and Fig. 6.5 (Langford, 1983) from Liverpool University Press; Fig. 2.15 from *Biol. Conserv.* **7**, 79–118, Newbold, C., 'Herbicides in aquatic systems', 1975, Fig. 3.17 from *Environmental Pollution* **5**, 1–10, Aston, R.J.,

'Tubificids and water quality: a review', 1973, Fig. 4.3 from 'Concept of stress and recovery in aquatic ecosystems', R.D. Gulati, in *Ecological Assessment of Environmental Degradation, Pollution and Recovery*, O. Ravera, ed., Fig. 4.21 from 'Biomanipulation of aquatic food chains to improve water quality in eutrophic lakes', De Barnardi, R. in *Ecological Assessment of Environmental Degradation, Pollution and Recovery*, Ravera, O., ed., Figs 5.6 and 5.11 from 'Air pollution effects on aquatic ecosystems and their restoration', Henrikson, A., in *Ecological Assessment of Environmental Degradation, Pollution and Recovery*, Rivera, O., ed., Figs 6.3 and 6.4 from Hellawall, J.M., *Biological Indicators of Freshwater Pollution and Environmental Management*, 1986, Fig. 6.8 from Green and Trett, *The Fate and Effects of Oil in Freshwater*, 1989, Fig. 8.8 from *Environmental Pollution* **58**, 55–70, Raven, P.J. and George, J.J., 'Recovery by riffle macroinvertebrates in a river after a major accidental spillage of chlorphyrifos', 1989 and Fig. 8.11 from *Environmental Pollution* **81**, 217–28, Rutt, G.P. *et al.*, 'The impact of livestock farming on Welsh streams: the development and testing of a rapid biological method for use in the assessment and control of organic pollution from farms' from Croom Helm Ltd; Fig. 4.2 (Edmondson, 1969) from the National Academy of Sciences; Fig. 4.7 from 'A Guide to the Restoration of Shallow Lakes' (Moss, B., Magdwick, J. and Phillips, E.C., 1996); Fig. 4.10 (Hartmann, 1977) from Birkhauser Verlag: Fig. 4.15 from Eutrophication of Waters: Monitoring, Assessment and Control, reproduced by permission of the OECD, © 1982 OECD; Fig. 4.17 (Rast and Holland, 1988) from The Royal Swedish Academy of Sciences; Fig. 5.5 (Havas *et al.*, 1984) and Fig. 8.3 (Foster and Bates, 1978) from The American Chemical Society; Fig. 5.7 (Henriksen *et al.*, 1984) from the Norwegian Institute for Water Research; Fig 5.15 (Ormerod, J.J. and Tyler, J.J., 'Birds as indicators of change in water quality', in Furness, R.W. and Greenwood, J.J.D., eds, *Birds as Monitors of Environmental Change*, 1993) and Fig. 9.5 (Edwards, R.W., 'Predicting the environmental impact of a major reservoir development', in Roberts, R.D. and Roberts T.M., eds, *Planning and Ecology*, 1984) from Chapman & Hall, London; Fig. 6.6 from the Atomic Energy Authority; Fig. 6.7 (Preston, 1974) from the International Atomic Energy Agency; Table 7.1 (Moyle, P.B. and Light, T., 'Biological invasions of fresh water; empirical rules and assembly theory', reprinted from *Biological Conservation* **78**, 149–161, 1996) from Excerpta Medica Inc.; Fig. 11.2 (Mouvet, 1985) from E. Schweizerbart'sche; The Saprobic Index of Pantle and Buck (1955); Fig. 8.9 from 'The multimetric approach to bioassessment, as used by the United States of America', printed as Fig. 19.3, p. 289 (adapted from Barbour, M.T. and Yoder, C.O., 2000) reproduced with the permission of the Freshwater Biological Association; Fig. 9.4 from 'River restoration' (Dobbs, A.J. and Sabel, T.F., 1996).

Photographs: Figure 1.1 reproduced by kind permission of Miguel Delibes; Figure 1.2 reproduced by kind permission of Estación Biologica de Doñana; Figure 1.3 reproduced by kind permission of K. Bavinck; Figure 1.7 reproduced by kind permission of The International Commission for the Protection of the Rhine against Pollution; Figure 2.24 reproduced by kind permission of Chris Perrins; Figure 2.27 reproduced by kind permission of Mike Gilbertson; Figure 3.9 reproduced by kind permission of Matthew Millett, Wildfowl & Wetlands Trust; Figure 4.13 reproduced by kind permission of Geoff Phillips, Environment Agency; Figure 7.1 reproduced by kind permission of Chris Gibson; Figure 9.7 reproduced by kind permission of Jonathan Wortley.

Whilst every effort has been made to trace the owners of copyright material, in a few cases this has proved impossible and we take this opportunity to offer our apologies to any copyright holders whose rights we may have unwittingly infringed.

1 CHAPTER INTRODUCTION

Of the Earth's total resource of water, 97 per cent is in the oceans and is too salty to be used for drinking, irrigation or industry. The majority of the remaining 3 per cent is frozen in ice caps and glaciers, or is buried too deeply underground to be utilized. The exploitable volume is only 0.003 per cent of the total, though it is replenished by the hydrological cycle. Not only is fresh water essential to life but it is also a relatively scarce resource, and is likely to become more so with the impacts of global warming and population growth: the human population, currently estimated at 6.24 billion, is predicted to rise to 10 billion by the year 2050. Some 80 per cent of the global population live in developing countries.

Many freshwater resources are contaminated through human activities. More than 20 per cent of the world's population do not have access to safe drinking water and a greater proportion still do not have even basic sanitation. Each day some 25 000 children die from their everyday use of water; 4 million of these deaths every year are simply from diarrhoea. It is not only children who suffer. It is estimated that, at any one time, half the inhabitants of developing countries are ill with diseases caused by dirty water and poor sanitation. Some 80 per cent of all illness and a third of all deaths are due to unwholesome water. As well as causing much suffering, water-borne diseases also result in great economic loss. In India, for example, it is estimated that 73 million working days are wasted each year, costing $600 million in lost production and in healthcare (Lean *et al.*, 1990).

The United Nations launched the International Drinking Water Supply and Sanitation Decade in 1980. Its aim was to provide clean water and adequate sanitation for all by the year 1990. Every day during that programme, about 330 000 people in developing countries were given access to a safe supply of drinking water and some 210 000 were provided with improved sanitation facilities. However, at the same time, the population of developing countries was growing by some 200 000 each day. The problem of unwholesome water in the developing world is not one of science, but of politics and economics. The United Nations 1994 Development Report has estimated that a 12 per cent cut in military spending worldwide could provide safe drinking water and primary healthcare for everyone. Nevertheless, 18 developing countries were spending more on the military than on health and education combined.

Some 40 per cent of the world's population in over 80 countries are affected by serious water shortages. In other countries water is available but too expensive to use because more accessible resources have already been depleted and new sources cost much more to treat to an acceptable standard. Pollution must be seen as a gross misuse of an essential but scarce resource.

NATURAL CAPITAL

Capital is the stock of materials and information that exists at a given point in time. Typically we think of capital in terms of the money we have in the bank, or tied up in buildings, cars and other possessions. Capital stock will generate a flow of services which may be used to transform materials and the way they interact to enhance human welfare. The wellbeing of a population will depend on the value of the services flowing from the total capital stock existing within their community (Pretty, 1998).

We can easily put a value on our houses and cars but we are wholly dependent on the air we breathe, the water we drink, etc., the values of which are extremely difficult to assess. Natural capital refers to the stocks of plants and animals and ecosystems of the Earth and these goods create services, comprising flows of materials, energy and information which we can combine with manufactured and human capital to produce welfare.

It has been estimated that the combined value of the world's ecosystem goods and services is in the range of US$16–54 trillion ($10^{12}$) per year (Constanza *et al.*, 1997). These goods and services include water regulation and supply, climate regulation, nutrient cycling, soil formation, waste treatment, wild food production, biological control of pests and recreation. A best estimate of US$33 trillion within this range values the total ecosystem goods and services at twice the global gross national product (GNP) of US$18 trillion per year. GNP is a measure of the total flow of goods and services produced by the economy over a year and is obtained by adding up, at market prices, the total output of goods and services. Although the precise value of natural capital can be disputed, it is clearly of critical importance to our wellbeing.

As our population increases year on year, the natural capital available per individual diminishes and we get closer to the time when the ecosystem will no longer be able to support us. By polluting that ecosystem we further diminish its capacity to support us. It is therefore an urgent priority that future development uses resources in a sustainable way (Everard, 2000). The sustainable use of resources requires that current use does not reduce the potential for future use. It should not impair the long-term viability of the species or ecosystem being used, and the long-term viability of supporting and dependent ecosystems must be maintained. Pollution can be seen as a threat to sustainability and, rather than disposing of waste materials to the environment, we should be aiming to reuse, recycle and recover them.

WHAT IS POLLUTION?

The dictionary defines pollution as the 'act of making dirty, defiling, contaminating, profaning, corrupting'. For our purposes such definitions are too broad to be useful. We will restrict our definition to include only the effects of substances or energy released by humans themselves on their resources. The definition followed in this book is that given by Holdgate (1979):

> The introduction by man into the environment of substances or energy liable to cause hazards to human health, harm to living resources and ecological systems, damage to structure or amenity, or interference with legitimate uses of the environment.

Pollutants may be derived from *point sources*, often discharges known to the authorities and readily treatable provided resources are available. Examples of point sources are discharges of effluents from sewage treatment works or of wastes from factories. Alternatively, sources of pollution may be *diffuse*, entering watercourses from run-off and land drainage. Fertilizers and pesticides applied to crops, and acid precipitation, provide examples of diffuse pollution.

Much pollution is *chronic* (or steady state), that is, the watercourse receives discharges continuously or regularly. Given the right legal framework and resources, such pollution can often be reduced so its impact on the aquatic ecosystem is acceptable. In much of the developed world, marked improvements in water quality have taken place over the last four decades. A much greater problem now is that of *episodic* (or intermittent) pollution, which is unpredictable in both space and time. Heavy rainfall, releasing large amounts of acid from soils or causing sewerage systems to overflow, may result in pollution episodes. Accidents are a major cause, for example carelessness in handling wastes or the crashing of road tankers close to rivers. Vandalism and the deliberate discharge of wastes also result in episodic events. Some recent examples of pollution events in England which resulted in successful prosecutions include the release of ferric sulphate by a water company to a river, killing 35 000 fish; the release of potentially toxic cyanobacteria from a water treatment works to a canal; the release of butchery waste from the kennels of a fox hunt, turning the receiving stream blood red; and the release of the pesticide cypermethrin from the kennels of a fox hunt (where it was used to treat mange in the hounds) to a stream where it eliminated thousands of the endangered white-clawed crayfish (*Austropotamobius pallipes*).

Episodic pollution is of especial concern to water managers, for a single event can destroy years of careful, patient work in reducing the impact of pollution from known discharges. Of course, as chronic pollution is increasingly brought under control, the impact of single pollution events becomes much more apparent.

THREE EXAMPLES OF WATER POLLUTION

Pollution of the Doñana National Park, Spain

At the mouth of the River Guadalquivir in south-west Spain is one of Europe's most important wetlands. It consists of a vast area of fresh, salt and brackish water habitats totalling some 230 000 ha, part of it forming the Doñana National Park (Fig. 1.1). The whole area is classed as a world heritage site. Not only does it hold internationally important populations of breeding birds, but millions more from northern Europe migrate through or winter in the marshes.

One of the main rivers feeding the wetland is the Rio Guadiamar. On 24 April 1998 at a mineral mine at Aznalcollar, a dam holding back a reservoir of toxic waste burst, releasing some 2 million cubic metres of metal-rich sludge and 4 million cubic metres of acid water (pH 2) into the river. The banks of the river were over-topped and around 400 m of land on either side were inundated with sludge and acid water, a total of some 4286 ha being contaminated (Fig. 1.2). It was estimated that the mud contained 16 000 t of zinc and lead, 10 000 t of arsenic, 4000 t of copper, 1000 t of antimony, 120 t of cobalt, 100 t of thallium and bismuth, 50 t of cadmium and silver, 30 t of mercury and 20 t of selenium and other metals (Grimalt *et al.*, 1999). Some 60 km below the accident site, walls were built to retain the

Fig. 1.1 The marismas (marshes) of the Doñana National Park in Andalucia, Spain (photograph by Miguel Delibes).

Fig. 1.2 The Aznalcollar mine tailings dam and the blackened river valley following its collapse in April 1998 (photograph by Estación Biologica de Doñana).

waste but more than 1000 ha of the national park were inundated, as were several other areas of important wetlands. In this holding area the water was treated and then pumped into the estuary, but this itself has economically important fisheries. A priority was the removal of the mud before autumn rains.

All fish and shellfish in the watercourses were killed, with 26 t of dead fish being removed in the 18 days following the spill (Pain *et al.*, 1998). Many amphibians were also killed. Two hundred birds were found dead in the weeks following the accident and several species were seen feeding in the contaminated area where samples of *Typha* and *Scirpus*, two wetland plant species important as food for birds, revealed levels of cadmium 2 to 40 times higher than controls, and zinc 20 to 100 times higher (Meharg *et al.*, 1999). Analyses of dead birds found in the months after the accident revealed elevated amounts of zinc and copper but apparently not of other metals (Hernández *et al.*, 1999). However, it has been claimed that birds wintering at the site have suffered considerable contamination, with 50 per cent of greylag geese (*Anser anser*), 33 per cent of mallards (*Anas platyrhynchos*), 41 per cent of coots (*Fulica atra*) and 82 per cent of purple gallinules (*Porphyrio porphyrio*) having near lethal levels of heavy metals (Anon., 1999).

This incident was one of episodic pollution, an accident, but one that could have been avoided. A manager at the mine had predicted potential problems with the dam but his warnings had been ignored. With heavy metals now within the food chain, the area is likely to suffer from the chronic effects of pollution for years to come.

There have been some 25 major pollution incidents involving reservoirs holding mine wastes since 1971. The latest took place in Romania on 30 January 2000, when heavy rains caused a storage pond to overflow at a gold mine, releasing 100 000 cubic metres of toxic waste, including cyanide. The waste flowed down the River Szamos into neighbouring Hungary and entered the River Tisza, a tributary of the Danube. Several hundred kilometres of the Szamos and Tisza were declared biologically dead, with more than 100 t of fish being killed and many predators such as otters and sea eagles being poisoned by contaminated prey. We have here an example of pollution crossing international borders between countries.

The decline of the otter

The otter (*Lutra lutra*) is an amphibious, mammalian carnivore, feeding extensively on fish and therefore at the top of the aquatic food chain (Fig. 1.3). The species was once widespread over most of Europe (Fig. 1.4a) but declined over an extensive area, especially since the 1950s, and by 1990 was very restricted in range (Fig. 1.4b). Otters were absent, for example, from much of southern and central England and most of central Europe. Even in Sweden, with its small human population and myriad rivers and apparently pristine lakes, the otter was almost extinct in the south and rare in the north. The best populations in Europe are closest to the Atlantic coast (Norway, Scotland, Ireland, Portugal) and in east and southeast Europe.

Fig. 1.3 European otter (photograph by K. Bavinck, Otterstation *Aqualutra*, the Netherlands).

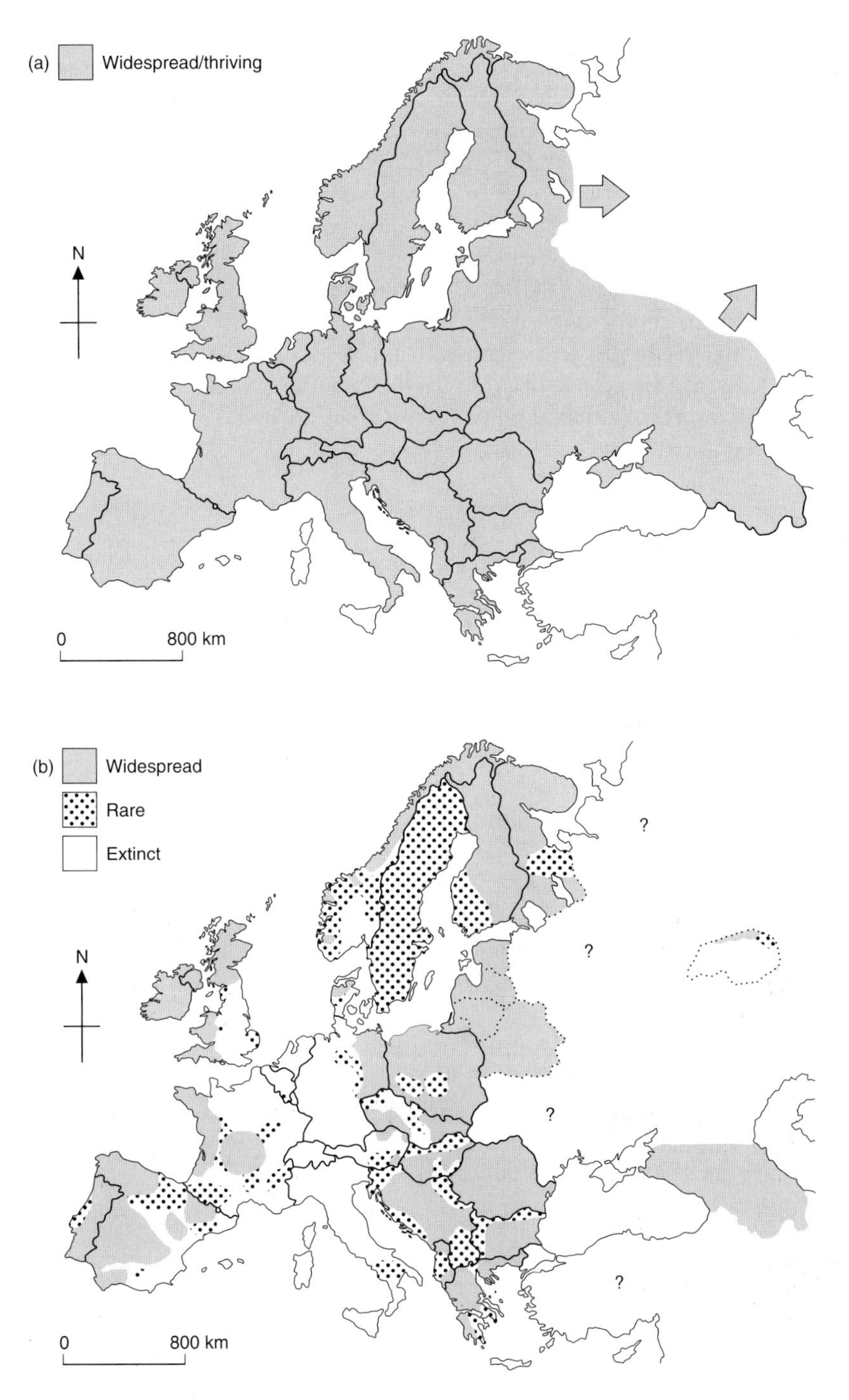

Fig. 1.4 Distribution of the otter in Europe (a) in 1950 and (b) in 1992 (adapted from Macdonald and Mason, 1994).

Otters are largely nocturnal and occur naturally at low densities, males patrolling home ranges of up to 40 km of waterway, so that the decline went largely unnoticed (Mason and Macdonald, 1986; Macdonald and Mason, 1994).

Several factors detrimental to the survival of otters, for example persecution and the destruction of waterside habitat, cannot explain the dramatic decline in range: the widespread introduction of a pollutant seems more likely (Mason and Macdonald, 1986). The 1950s saw the introduction of organochlorine pesticides, such as DDT and dieldrin, into agriculture, while during the same period the use of polychlorinated biphenyls (PCBs) in many industrial processes was increasing exponentially. Such compounds readily enter watercourses from agricultural run-off and industrial discharges to contaminate aquatic ecosystems. Organochlorines, especially PCBs, are also transported over wide areas by winds, to be washed into waterbodies with precipitation. Figure 1.5 provides an index of industrial output

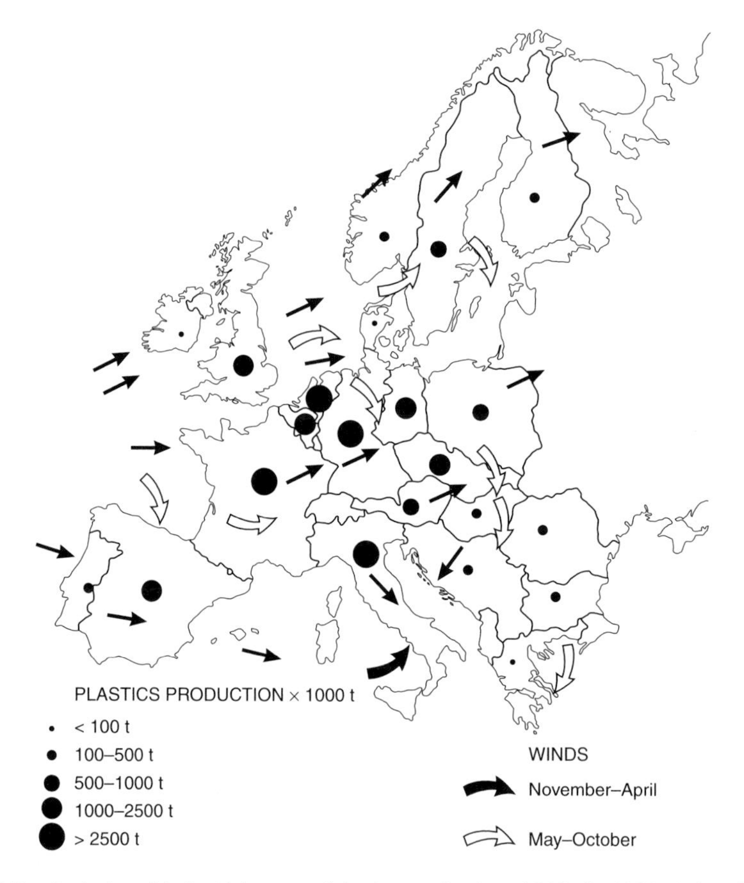

Fig. 1.5 An index of industrial output (plastics production, 1000 t in 1983–84) and prevailing wind patterns over Europe.

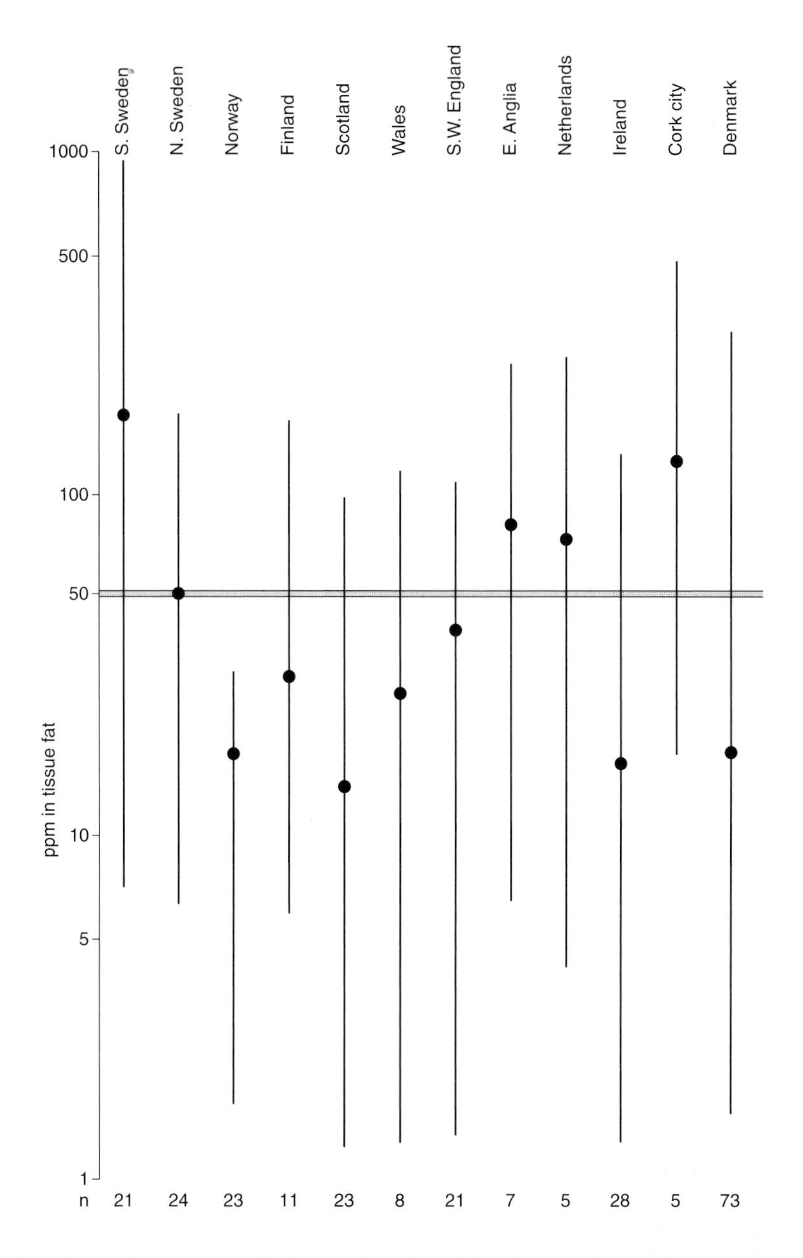

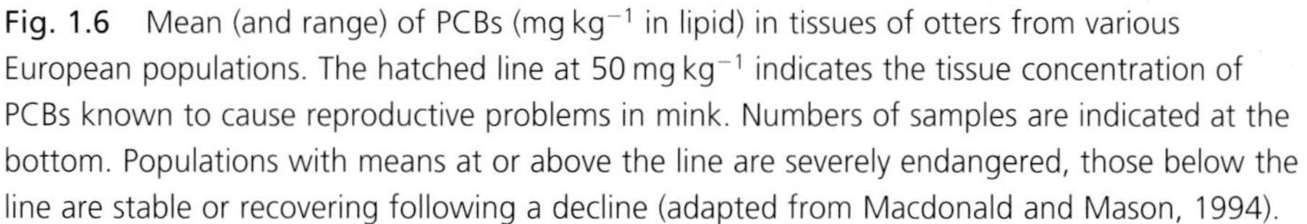

Fig. 1.6 Mean (and range) of PCBs (mg kg^{-1} in lipid) in tissues of otters from various European populations. The hatched line at 50 mg kg^{-1} indicates the tissue concentration of PCBs known to cause reproductive problems in mink. Numbers of samples are indicated at the bottom. Populations with means at or above the line are severely endangered, those below the line are stable or recovering following a decline (adapted from Macdonald and Mason, 1994).

for individual European countries, based on the production of plastics for 1983/84 and also shows the prevailing winds. If Fig. 1.5 is compared with Fig 1.4b it can be seen that otters are scarce or absent in those regions both with high industrial output *and* downwind of such industrial centres. The most intact populations of otters are upwind of the major industrial regions.

Organochlorines (pesticides and PCBs) are hydrophobic and lipid soluble, so they rapidly pass from water to living tissue and biomagnify in the food chain. Concentration factors of PCBs from water to predators may be as high as ten million times. Organochlorines are also highly persistent. In large concentrations they may directly kill otters under conditions of physiological stress, such as during periods of food shortage or pregnancy. In the long term, however, sublethal effects are probably more significant. In experiments with mammals it has been shown that PCBs cause disturbance to female reproductive function, affecting the reproductive tract, the neuro-endocrine system controlling puberty, oestrus and ovulation, and foetal and neonatal survival (Mason and Wren, 2001). Subsequent adult reproductive success may be affected by exposure in the uterus (p. 77). There are also effects on male reproduction.

Experiments have been conducted on the effects of PCBs on the American mink (*Mustela vison*), a species in the same family as the otter which also takes food from wetland habitats. Fewer young were born to females receiving 3.3 mg PCB kg^{-1} of food over 66 consecutive days and pups weighed 28 per cent less than controls. Only 1.7 per cent of young survived to five days, compared with 82 per cent of controls. Females receiving 11 mg PCB kg^{-1} food produced no young at all. At death females taking the lower dose of PCB had tissue concentrations of 50 mg kg^{-1} in muscle fat. Further experiments have shown that reproduction in mink is inhibited on a daily intake of PCB of only 25.2 μg and a lower rate of reproduction is recorded on a daily intake of only 2.5 μg PCB (see review in Mason and Wren, 2001).

Figure 1.6 shows the concentration of PCBs in tissues of otters found dead in various regions of Europe, compared with that concentration known to cause reproductive impairment in mink. It is clear that animals from those populations which are declining or are endangered have mean PCB concentrations greater than 50 mg kg^{-1} tissue weight, though there is a wide variation between individuals. One animal from southern Sweden had nearly 1000 mg PCB kg^{-1} in muscle fat.

In the British Isles there is an inverse relationship between the regional distribution of otters (the number of sites found positive for the species in standard field surveys) and the average concentration of PCBs in spraints (faeces); high concentrations of PCBs are found in areas where otters are scarce and low concentrations where otters are widespread (Mason, 1995). In situations of PCB pollution we might expect a slow decline and contraction in range of otters as animals cease to reproduce, while themselves surviving to old age. This seems to be the case.

Because experiments have not been done on otters, it cannot be proved that

PCBs have caused the observed decline. For ethical reasons, such experiments are unlikely ever to be conducted. However, recent studies using vitamin A as a biomarker (see p. 51) have confirmed that otters are as sensitive to PCBs as mink. In otters from Denmark, there was a strong negative correlation between concentrations of vitamin A and PCB concentrations expressed as dioxin equivalents (Murk *et al.*, 1998). The reduction in vitamin A coincided with a higher incidence of diseases, such as bacterial and viral infections, endoparasites, and pathological changes, otters with the highest PCB burdens often having multiple diseases. A similar relationship between low vitamin A levels and high concentrations of organochlorine compounds, including PCBs, has been found in otters from south-west England (Simpson *et al.*, 2000). Vitamin A plays an important role in resistance to infection. Current concentrations of PCBs were high enough to cause adverse effects on the health of otters.

Those otters with high levels of PCBs were found in rivers where PCBs were not recorded by regulatory authorities using routine analytical methods on water. As far as the water industry is concerned, there are no significant problems with PCBs, but it is quite clear that freshwaters are chronically polluted with PCBs over large areas of Europe. Nevertheless, because of their potential adverse effects on health and the environment, controls on manufacture and discharge of these compounds are in place and appear to be effective in the river environment. PCB concentrations have declined in otters in Britain over the last two decades (Mason, 1998), vitamin A levels have increased (Simpson *et al.*, 2000), and the species is recolonizing its former range in England. Similar declines in tissue concentrations of PCBs and extensions in range have also been reported in Denmark (Mason and Madsen, 1993) and Sweden (Roos *et al.*, 2001). Unfortunately, PCBs do not disappear, they merely transfer to the oceans. A relationship between PCB concentrations and the incidence of infectious diseases has recently been reported in harbour porpoises (*Phocoena phocoena*) from the coasts of Britain (Jepson *et al.*, 1999).

Contamination of food chains with PCBs and their biological effects provide an example of chronic pollution.

Pollution incident on the River Rhine

The River Rhine rises in the Swiss Alps and passes through Germany, for a time forming the border with France and the Netherlands, to discharge into the North Sea some 1320 km from its source (Fig. 1.7). The river is navigable for much of its length and passes through a number of major cities (Fig. 1.8). It receives several major tributaries, including the River Emscher which drains the industrial region of the Ruhr. The Rhine therefore receives many polluting discharges but much effort has been put into reducing levels of contamination. Friedrich and Müller (1984) concluded that the Rhine's ecosystem was on the way to recovery from very severe damage but the possibility of danger remains because of accidents associated with a river used so intensively.

Fig. 1.7 River Rhine at Köln (photograph by The International Commission for the Protection of the Rhine against Pollution).

Such an accident happened on 1 November 1986 when a fire broke out in a chemical warehouse near the Swiss city of Basel. Some 1300 t of chemicals were stored there, including 934 t of pesticide and 12 t of organic compounds containing mercury. Around 30 t of chemicals, including mercury and organophosphorus pesticides, were washed into the Rhine during the fire-fighting operation. The slick of pollution also contained rhodamine, a red dye, which enabled its passage down the river to be monitored. Water intakes to German towns were closed and water had to be brought in by truck. The Rhine also serves two important drinking water reservoirs for the Netherlands, which had to be protected, and the Dutch attempted quickly to channel as much as possible of the slick into the North Sea. In the Netherlands the slick was 200 km long, with a mercury content of $0.22\ \mu g\ l^{-1}$, three times the normal level in Rhine river water.

Daphnia, an important planktonic component of the food chain in the river, were destroyed as the slick passed. Half a million fish were estimated to have died, including almost all the eels (*Anguilla anguilla*) for 400 km downstream of Basel. The river for 300 km downstream of the city was described as biologically dead.

In the same month 1100 kg of the herbicide dichloro-phenoxyacetic acid leaked into the Rhine from a factory in Ludwigshafen (Fig. 1.8), requiring the shutting off of water intakes downstream, and up to 50 kg of the solvent chlorobenzol were spilled down a drain by another chemical works into the River

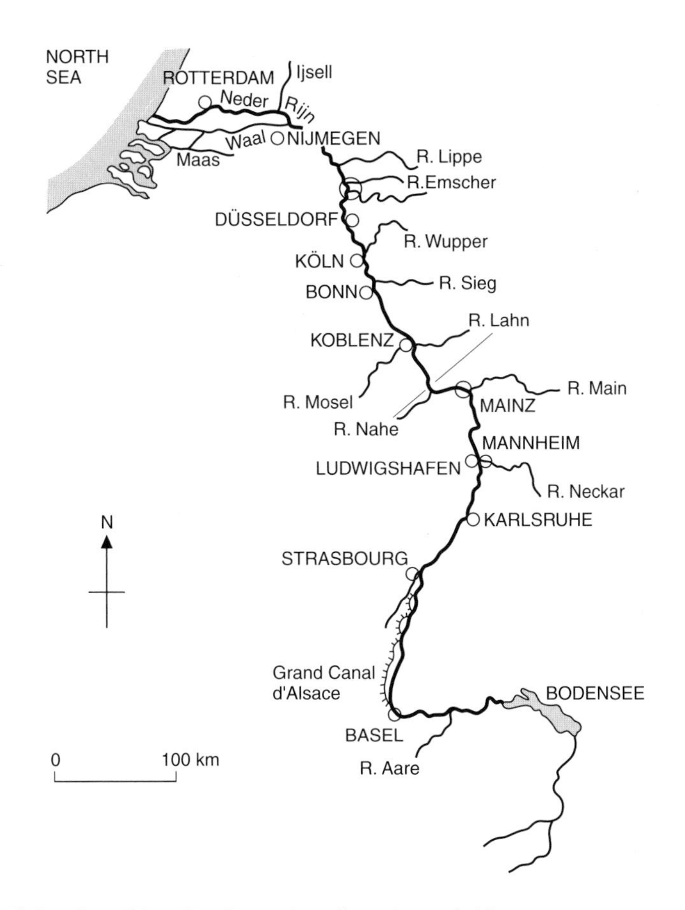

Fig. 1.8 Map of the River Rhine showing major tributaries and cities.

Main. It was thought by environmental groups that such accidents were only being announced because of the Basel fire, and that they are a problem of regular occurrence in the river.

It was estimated at the time that it would take between three and ten years for the Rhine to recover from the Basel accident, though within a year the International Commission for the Protection of the Rhine against Pollution (ICPR) announced that the river had practically recovered (Lelek and Kohler, 1990). Eel populations, however, were expected to take up to eight years to return to their former size. Pollutant residues in the sediments were also back to their former level, which were, though, described as alarmingly high. The rapid recovery of the Rhine was made possible by the many side channels and tributaries of the river, which were unaffected by the pollution and acted as sources of re-colonization.

The Rhine represents an example of a chronically polluted river which suffers periodic episodes of pollution, frustrating long-term attempts at clean-up. The

Basel accident resulted in the ICPR developing a Rhine Action Programme with a number of targets. Development of the entire Rhine ecosystem must be sustainable. This includes improvements in habitat and water quality to allow for the return of native fauna, including the migratory salmon (*Salmo salar*), once abundant but declining from the 1880s to extinction by 1940. The resident fish fauna is now quite diverse though numbers and biomass are often low. Some 155 macroinvertebrate species have been recorded since 1989, though a number of species with specialized requirements are still absent (Tittizer *et al.*, 1994). The first salmon and sea trout (*Salmo trutta*) have now been reported. A reliable supply of drinking water from the river must be guaranteed and there must be a substantial reduction in toxic chemicals, including those contained within the sediments. Eels from the Rhine, for example, contain some of the highest PCB levels ever reported in fish (de Boer and Hagel, 1994). There must be environmentally sound flood protection and, finally, the North Sea must be protected from pollution. A summary of the action programme, the cost of implementation of which is huge, is provided by Schulte-Wulwer-Leidig (1995).

WHY NEED WE BE CONCERNED ABOUT POLLUTION?

From the examples given above it is obvious that pollution is important because our resources are being damaged. When accidents occur the aquatic scientist can only assess the damage and suggest remedial action, such as restocking with fish, once the pollution has passed. The impact of accidents on the environment could often be minimized if hazard assessments were made of facilities dealing with dangerous chemicals and preventative measures taken. Often they are not.

Since human activities cause pollution we should be able to control much of it. Pollution control is extremely costly, however, and the benefit in resource terms may be far outweighed by the cost of control. Furthermore, while pollution may be caused by an individual or a company, the costs of remediation are very often borne by society, through increased taxes or increased costs of drinking water, so there is often little incentive not to pollute. Stringently applied laws, or a high degree of altruism, are required to control pollution.

Water supply and demand

In England and Wales the average domestic water consumption is about 160 litres per person per day (of which 32 per cent is used in toilet flushing, 17 per cent for baths/showers, 12 per cent for washing machines, the remaining 39 per cent being miscellaneous activities such as cooking, drinking, hand-washing and outside use in the garden). Water is also used by industry and agriculture. Most of the rainfall in England and Wales falls in the hills of the west and north, while most of the residual rainfall (the difference between precipitation and evapotranspiration), which is potentially available for use, falls in the winter. Moreover, regions of high population and industry, for example London, the Southeast and the Midlands, are in areas of low rainfall and demand for water is greatest during the summer. Furthermore, arable farming, with its requirements for irrigation, is also concentrated in the drier areas of the country.

This discrepancy between water availability and water use means that, during periods of drought, resources may not be able to meet demand, and considerable restrictions on water use can become necessary. Evapotranspiration exceeds precipitation in large areas of the world, including many developing countries, where rational water management is of paramount importance.

In England and Wales, the disparity between areas of high rainfall and areas where water is needed results in large quantities of water being abstracted for the public supply from lowland reaches of rivers. There is also considerable direct abstraction by industry and agriculture. Such water must be of an acceptable quality. Abstractions from the lower reaches of a river may be sustained by a controlled discharge from a reservoir, usually situated in the headwaters where rainfall is heavier, this being an economical way of transporting water to where it is needed. Water may also be transferred from one catchment to another that is short of water. Many rivers are being used as aqueducts in this way. Over-abstraction is an increasing problem and a growing number of rivers in England and Wales are suffering low flows, at least in part due to abstraction.

Effluent disposal

Domestic, industrial and agricultural users produce large quantities of waste products and waterways provide a cheap and effective way of disposing of many of these. During dry weather, the flow of some rivers consists almost entirely of treated effluents. The effluents of some towns become the water supplies of other towns downstream. There is the well-known saying that the water coming from the taps in London has already passed through five sets of kidneys! It is therefore essential that the effluent discharged into a watercourse is of high quality and the degree of pollution is such that the self-purifying capacity of the river (p. 109) is not overloaded.

In addition to providing a source of water and a sink for effluents, freshwaters have an important amenity role, including such activities as boating, angling and wildlife studies. Some of these pastimes require water of a very high quality. In England and Wales, angling is by far the largest participatory sport in the country. The service industries for these pastimes are locally very valuable. There are also lucrative commercial fisheries for salmon, migratory trout and eels. It is also very desirable that the natural communities of animals and plants in freshwaters be maintained.

Finally, the resources of the seas are vast and most of the pollutants travelling down rivers will eventually end up there. Animals such as Arctic seals and Antarctic penguins are already loaded with pollutants and we should not be complacent that the immense quantities of water in the oceans can absorb pollution indefinitely without effect.

SETTING PRIORITIES FOR POLLUTION CONTROL

There are, therefore, a number of reasons why the control of pollution is important. We do not, of course, have equal concern for all the components of our environment and this is especially so when the enormous costs of pollution abatement are taken into account. We place ourselves as the highest priority, where pollution may affect human health, followed by domestic livestock and crops. At the other extreme, if pollution kills organisms that are pests, then it might be considered positively beneficial. Some groups of organisms may be of fairly low priority in terms of human concern, but they may require water of an especially high quality. Nature reserves, for example, may need completely unpolluted water (though this is not necessarily so, it depends on what is being conserved), whereas grossly polluted water will suffice for shipping. Because of the efficiency of water treatment processes, water for potable supply need not be of the highest quality.

It would obviously not be economically feasible to clean all waters to such an extent that they would make pristine nature reserves and economic considerations may make it unrealistic to improve the quality of some waters for recreation and fisheries. Increasingly the concept is being adopted of maintaining water quality at a standard relating to the use to which that water is put. Thus a classification system for water uses is constructed and water quality criteria for these uses are formulated (see p. 298).

WHAT ARE POLLUTANTS?

In terms of the definition given at the beginning of this chapter almost anything produced by humans can be considered at some time to be a pollutant. Indeed, to

Table 1.1 Categories of pollutants found in freshwater

Acids and alkalis
Anions (e.g. sulphide, sulphite, cyanide)
Detergents
Domestic sewage and farm manures
Food processing wastes (including processes taking place on the farm)
Gases (e.g. chlorine, ammonia)
Heat
Metals (e.g. cadmium, lead, zinc)
Nutrients (especially phosphates, nitrates)
Oil and oil dispersants
Organic toxic wastes (e.g. formaldehyde, phenols)
Pathogens
Pesticides
Polychlorinated biphenyls
Radionuclides

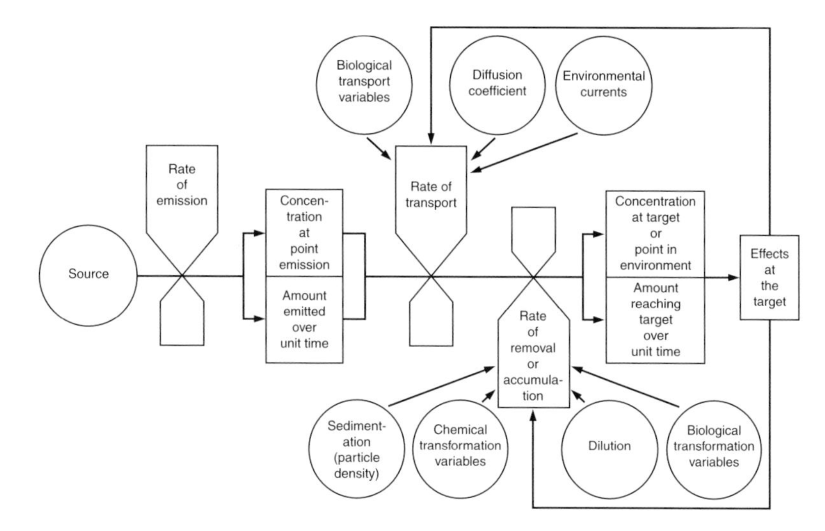

Fig. 1.9 A generalized pollutant pathway (from Holdgate, 1979).

the farmer whose land is about to be lost under a new reservoir scheme, pure water is itself a pollutant in almost every sense of the definition given. Substances that are essential to life (e.g. copper, zinc) can be highly toxic when present in large amounts.

Some 1500 substances have been listed as pollutants in freshwater ecosystems and a generalized list is given in Table 1.1. Some of the categories are not necessarily mutually exclusive. Domestic sewage, for example, may contain, in addition to oxidizable material, detergents, nutrients, metals, pathogens and a variety of other compounds.

Whether or not a compound will exert an effect on an organism or a community will depend on the concentration of that compound and the time of exposure to it (i.e. the dose). The effect of a pollutant on a target organism may be either acute or chronic. Acute effects occur rapidly, are clearly defined, often fatal and rarely reversible. Chronic effects develop after long exposure to low doses or long after exposure and may ultimately cause death. Sublethal doses result in the impairment of the physiological or behavioural processes of the organism (e.g. it may grow poorly, or fail to reproduce). Its overall fitness is reduced.

At the community or ecosystem level it is unlikely that pollution will cause irreversible effects, except possibly in the case of radioactive pollution. The effects of pollution are recorded in the loss of some species, with possibly a gain in others, generally a reduction in diversity but not necessarily numbers of individual species, and a change in the balance of such processes as predation, competition and materials cycling.

The generalized pathway of a pollutant from source to target is shown in Fig. 1.9. There are three important rate processes in the pathway: the rate of emission from the source of pollution; the rate of transport through the ecological system;

and the rate of removal or accumulation of the pollutant in the pathway. The rate of transport will depend on the diffusion rate of the pollutant and on a variety of environmental factors as well as properties of transport within organisms in the pathway. The rate of removal or accumulation will depend on rates of dilution or sedimentation and on chemical and biological transformations. These determine the dose reaching the target organisms. Processes within the target will either transport the pollutant to where it exerts an effect or will excrete the pollutant. We see how this basic pathway can be repeated along a food chain. Also, with slight terminological changes, Fig. 1.9 can illustrate the basic pathway from entry point to site of action within a target organism.

THE COMPLEXITY OF POLLUTION

The following chapters will show the general characteristics and effects of various types of pollutants. Only rarely, however, is a single pollutant present in a watercourse. Normally an effluent will consist of a variety of potentially harmful substances. Most watercourses will receive a number of effluent discharges. The effects of these will often be difficult or impossible to disentangle.

Pollutants occurring together may act completely independently on a target and the one exerting the greatest effect would then be the most important. One would not, for example, worry unduly about high levels of zinc in an effluent if the oxygen demand was so high that all life in the receiving stream was suffocated, but if the organic loading in the effluent was reduced such that the stream could support life, the concentration of zinc might then become important. The effects of pollutants might also be additive, antagonistic or synergistic (p. 32). These interactions will become apparent in later chapters.

It is worth remembering that the uses to which we put water (drinking water for ourselves and our livestock, irrigation, food processing etc.) often have biological implications. The effects that are perceived in natural communities might be considered as an early warning system for the potential effects of pollutants on ourselves.

THE PRECAUTIONARY PRINCIPLE

The complexity of pollution described above means that our knowledge of the ecological effects is often limited or unknown. We often do not have clear evidence of the linkages between chemicals in the environment and effects that we observe on species or ecosystems. PCBs and the otter discussed above provide an example. All the evidence points to an effect of PCBs on otter populations but we do not have conclusive scientific proof. To get this we would have to feed PCBs to otters at different doses and observe effects on health, reproduction and survival. Doses causing effects would then have to be compared with concentrations measured in foodstuffs in the wild to establish the linkage between effects observed in wild otters and those known to be induced by PCBs in the laboratory. The true sceptic may still not be satisfied because a wild otter taking a very varied diet, but exposed to the many rigours of the environment, may respond very differently to particular doses of PCBs from an otter pampered in the laboratory but poisoned very precisely. Most of us anyway would find such an experiment unethical.

When an environmental effect of a chemical seems likely, but the scientific proof is lacking, we invoke the precautionary principle. We can define it as the avoidance or reduction of risks to the environment before specific environmental hazards are encountered. We can use it to set stringent environmental standards for the release of a potentially damaging compound to the environment. Prevention is better than cure.

CHAPTER 2

AQUATIC TOXICOLOGY

On 6 July 1988 a lorry arrived at Lowermoor Water Treatment Works in Cornwall, southwest England, with a load of 20 t of aluminium sulphate, which is used in the treatment process to flocculate suspended matter from water. The works was unmanned. The aluminium sulphate was destined for a storage tank but tanks were unmarked. The driver, who was unfamiliar with the site, opened an aperture with his key and pumped his cargo out of the lorry. The tank receiving the chemical was the wrong one and the aluminium sulphate went straight into the water main serving 22 000 people in the vicinity of Camelford.

In less than five hours complaints were being received over the quality of water coming out of the taps (Craig and Craig, 1989). Milk was curdling in tea and the water stung lips and made fingers feel sticky. Over the next few days, before the cause of the problem was found, tap water was very acid (pH 4.2) and the concentration of aluminium was up to 4000 times the permitted level of 200 $\mu g\,l^{-1}$. Aluminium has been linked with diseases such as arthritis, brittle bone disease and Alzheimer's disease, a virulent form of senile dementia. The acid water leached out metals such as lead, zinc and copper from the pipes. A large proportion of the population of the Camelford area complained of symptoms such as sore throats, nausea and vomiting, muscle cramps and joint pains, skin rashes and even of hair turning green! The water authority consistently underplayed the seriousness of the accident, while at national level the relevant authorities remained inactive. The scientific periodical *New Scientist*, in an editorial on 21 January 1989, said of the Department of the Environment that 'its limp attitude to a major pollution incident was nothing short of a scandal'. It took almost six months for the Department of Health to set up a team of experts to investigate the long-term health effects of the incident, which were rightly causing great concern to those who had drunk the water. Their report (the Clayton Report) concluded that there would be no long-term health effects, though many who suffered symptoms remain sceptical. A more recent study has concluded that people who were exposed to contaminated water suffered considerable damage to cerebral function, which was not related to anxiety (Altmann *et al.*, 1999), though the

Fig. 2.1 An urban river landscape. Rivers in towns are now much cleaner because of the enforcement of strict pollution control legislation and the path along this river, running through the city of Leicester, is popular with walkers, cyclists and anglers, as well as providing a valuable corridor for wildlife. Derelict factories are converted into flats, restaurants and boutiques (photograph by the author).

methodology of this study has been criticized. If nothing else it shows the extreme difficulty of linking cause and effect in cases of environmental pollution, especially when effects occur over the long term.

In order to clear the public supply of polluted water, the authority flushed out the water distribution system overnight into the Rivers Camel and Allen. With immediate attention focused on the water supply to customers, the potential effects on watercourses and aquatic life were given no consideration. It was estimated that the average pH of the river water fell to 4.5 and the aluminium concentration averaged $100\,mg\,l^{-1}$, but was probably very variable. Many fish were found dead, covered with mucus and with bright red gills, characteristic of aluminium poisoning. Aluminium is toxic to fish at concentrations as low as $0.1\,mg\,l^{-1}$. The aluminium concentration in the gills of two fish sent for autopsy was 76 times higher than in control fish. The estimated total loss of Atlantic salmon (*Salmo salar*) and brown trout (*Salmo trutta*) was between 43 000 and 61 000 and substantial losses of other, economically less important species also occurred. It was concluded, however, that the incident had no measurable effect on the invertebrate fauna of the river. A detailed account of the development of this incident is provided by Rose (1990).

The Camelford example demonstrates how a single pollution episode can have a major impact on an otherwise clean river, as well as putting the health of the local population at risk. Unfortunately the River Camel suffered a further pollution episode in 1994 when farm slurry killed 100 000 trout, 500 in the river and the remainder in a fish farm fed by river water.

The majority of our watercourses receive a great variety of potential pollutants from industrial, agricultural and domestic effluents, and these complex situations become apparent when considering toxicity. Many industries discharge effluents to our urban waterways, where factories are concentrated because of their need for water for processing or discharging wastes (Fig. 2.1).

Agriculture and forestry also add many toxic pollutants to freshwaters. These additions may be indirect, such as run-off of herbicides and insecticides applied to land, while waste pesticides and their empty containers are carelessly dumped into ponds or streams with unfortunate effects. Landfill sites and toxic waste dumps, many of them in the past poorly constructed, managed and controlled, are a matter of great concern, especially with respect to groundwater contamination (Lisk, 1991). Clean groundwater is critical to many of our water resources.

Run-off from roads may be another source of chemicals to watercourses, particularly various hydrocarbons and metals (Maltby *et al.*, 1995). In many cases, however, the impact of chronic pollution is restricted to the immediate downstream stretch (Perdikaki and Mason, 1999).

Toxic chemicals are also used in the direct control of particular members of the freshwater community. The most widely used are herbicides to control water plants considered to be interfering with human use of watercourses. Other organisms may be directly poisoned by the herbicide or may be indirectly affected by the change in community structure resulting from the loss of plants. Insecticides are also

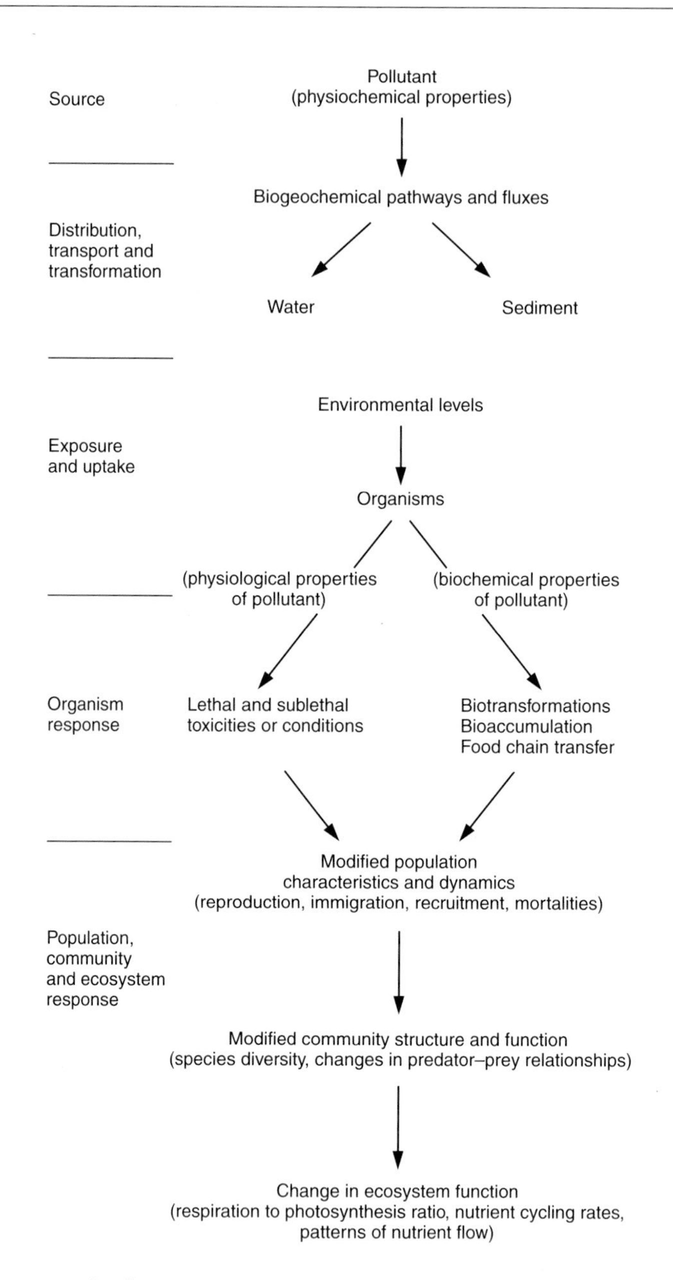

Fig. 2.2 Impact of pollutants on organisms and ecosystems (adapted from Connell and Miller, 1984).

applied directly to watercourses, for instance to destroy the larvae of mosquitoes, the vectors of malaria. In tropical Africa, DDT has been added in large quantities to rivers to kill larvae of the blackfly *Simulium damnosum*, the vector of the disease onchocerciasis (river blindness). Molluscicides are widely used in the tropics to control the snail vectors of schistosomiasis (p. 89). In some countries piscicides are applied to control fish. This may involve the entire fish community (e.g. with rotenone), the elimination of selective groups of fish (e.g. with antimycin) or the control of a particular species, such as larval sea lampreys (*Petromyzon marinus*) in the Great Lakes of North America, using trifluoro-methyl nitrophenol (TFM). Poachers often use poisons to illegally obtain fish, with severe consequences to the aquatic environment.

A generalized flow diagram of the effects of toxic pollutants on freshwater ecosystems is presented in Fig. 2.2.

TYPES OF TOXIC POLLUTANTS

The major types of toxic pollutants can be listed as follows:

- Metals, such as lead, nickel, cadmium, zinc, copper and mercury, arising from many industrial processes and some agricultural uses. The term *heavy metal* is somewhat imprecise, but includes most metals with an atomic number greater than 20, and excludes alkali metals, alkaline earths, lanthanides and actinides.
- Organic compounds, such as organochlorine pesticides, herbicides, polychlorinated biphenyls (PCBs), chlorinated aliphatic hydrocarbons, solvents, straight-chain surfactants, petroleum hydrocarbons, polynuclear aromatics, chlorinated dibenzodioxins, organometallic compounds, phenols, formaldehyde. They originate from a wide variety of industrial, agricultural and some domestic sources.
- Gases, such as chlorine, ammonia and methane.
- Anions, such as cyanides, fluorides, sulphides and sulphites.
- Acids and alkalis.

The most dangerous toxic compounds, such as heavy metals and organochlorines, have been placed by the European Union on a 'Black List', properties that influence their selection including toxicity, persistence and potential for bioaccumulation. The intention is to eliminate these from the environment. Less dangerous substances make up a 'Grey List'. The Environment Protection Agency in the United States has a list of 129 priority chemicals that pose potential serious risk to aquatic habitats.

Some potentially toxic compounds, such as heavy metals, are released into the aquatic environment from natural processes such as volcanic activity and the weathering of rocks. A number (e.g. copper, zinc) are essential, in small amounts, to life. Industrial processes have greatly increased the mobilization of many metals.

Human-induced releases of tin, lead and mercury, for example, are respectively 110 times, 13 times and 2.3 times greater than geological mineralization. The rate of manufacture of organic compounds, such as pesticides, has increased exponentially since the 1950s. Global pesticide usage is in excess of 2.2×10^9 kg per annum.

TOXICITY

Two general categories of toxic effects can be distinguished. *Acute* toxicity (a large dose of poison of short duration) is usually lethal, whereas *chronic* toxicity (a low dose of poison over a long time) may be either lethal or sublethal. There are a number of words in regular use in the study of toxic effects:

- **acute** – coming speedily to a crisis;
- **chronic** – continuing for a long time, lingering;
- **lethal** – causing death, or sufficient to cause it, by direct action;
- **sublethal** – below the level that directly causes death;
- **cumulative** – brought about, or increased in strength, by successive additions.

There are also a number of terms that are used to express quantitatively the results of toxicity studies:

- **Lethal concentration (LC)**, where death is the criterion of toxicity. The results are expressed with a number (LC50, LC75) that indicates the percentage of organisms killed at a particular concentration. The time of exposure is also important in studies of toxicity so this must also be stated. The 48 hour LC50 is the concentration of toxic material that kills 50 per cent of the test organisms in 48 hours.
- **Effective concentration (EC)** is the term used when an effect other than death is being studied, for example respiratory stress, developmental abnormalities, or behavioural changes. The results are expressed in a similar way to lethal concentration (e.g. 48-hour EC50).
- **Incipient lethal level** is the concentration at which acute toxicity ceases, usually taken as the concentration at which 50 per cent of the population of test organisms can live for an indefinite period of time.
- **Safe concentration** is the maximum concentration of a toxic substance that has no observable effect on a species after long-term exposure over one or more generations.
- **Maximum acceptable toxicant concentration (MATC)** is the concentration of a toxic material that may be present in a receiving water without causing harm to its productivity and its uses.

To develop standards to protect the aquatic environment it is necessary to determine what concentrations of particular substances cause toxic or sublethal effects to a range of organisms. There are, however, so many chemicals being produced and released to the environment that only a small proportion are tested on a range of organisms. There are also strong and increasing ethical concerns for the widespread use of animals in toxicity testing. A method of predicting their likely effect is to use the **quantitative structure–activity relationship (QSAR)**. On the basis of their physicochemical properties, the biological activities (bioaccumulation, toxicity, persistence, degradation) of molecules can be predicted. The most frequently used measure of chemical structure is the octanol–water partition coefficient, which is a measure of the equilibrium distribution of the chemical dissolved in a two-phase system of water and octanol:

$$\log_{10} P_{ow} = \log_{10} (P_o / P_w)$$

where P_o and P_w are the proportions of the total solute in the octanol and water respectively. Compounds with a high P_{ow} are fat-soluble, toxic and carcinogenic; they bioaccumulate and are persistent in the environment (e.g. DDT, $P_{ow} = 5.7$). Those with a low P_{ow} are water-soluble and relatively innocuous (e.g. alcohol, $P_{ow} = -0.3$). Toxicity tests will be most effectively directed at those compounds with a high P_{ow}. More details are found in Landis and Yu (1995) while Polloth and Mangelsdorf (1997) discuss the reliability of the QSAR approach in assessing toxicity.

Acute toxicity

Examples of acute toxicity curves for fishes are illustrated in Figs 2.3 and 2.4. Note that both axes are on a logarithmic scale. Figure 2.3 illustrates a curvilinear relationship, which has been observed for many toxic chemicals. The incipient LC50 can be obtained approximately as the asymptote and is about $25\,mg\,l^{-1}$ for ammonia. Figure 2.4 shows a linear relationship with metals and there is an abrupt asymptote, the incipient LC50 being estimated at concentrations of $50\,\mu g\,l^{-1}$ for copper and $600\,\mu g\,l^{-1}$ for zinc. The incipient LC50 can be obtained more precisely by log-probit methods (Abel, 1996). The incipient LC50 is a useful value in that it enables the toxicities of different pollutants to be easily compared and it forms a basic measuring unit for predicting the joint toxicity of two or more pollutants, as well as for describing sublethal effects.

A number of factors influence the responses of organisms in toxicity tests (Fig. 2.5). Different species may vary in their vulnerability to specific pollutants. Table 2.1 shows the toxicity of several pollutants to three common freshwater invertebrates. The amphipod *Gammarus* is generally the most sensitive organism to pollutants and the midge larva *Chironomus* the least, though *Chironomus* is most sensitive to ammonia. The most striking difference is the insensitivity of *Chironomus* to cadmium. The pesticide lindane was the most toxic compound of the four tested on *Chironomus*. Copper was found to be most toxic to *Gammarus* after 96 h but lindane was more toxic over 240 h. Two herbicides were found to be less toxic to either species than lindane or copper (Taylor *et al.*, 1991).

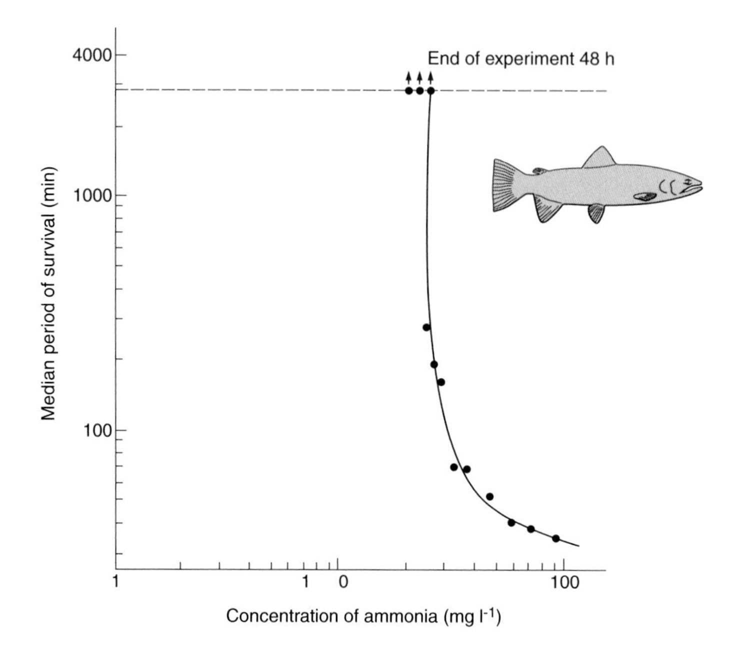

Fig. 2.3 A toxicity curve for trout in solutions of ammonia (NH_4Cl as mg l^{-1} N) at various concentrations (from Herbert, 1961).

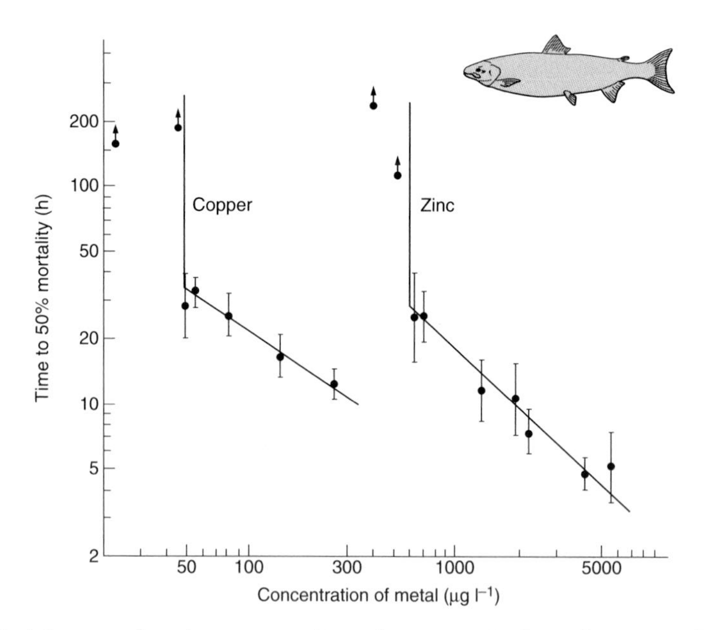

Fig. 2.4 Toxicity curves for salmon exposed to various concentrations of copper and zinc (from Sprague, 1964).

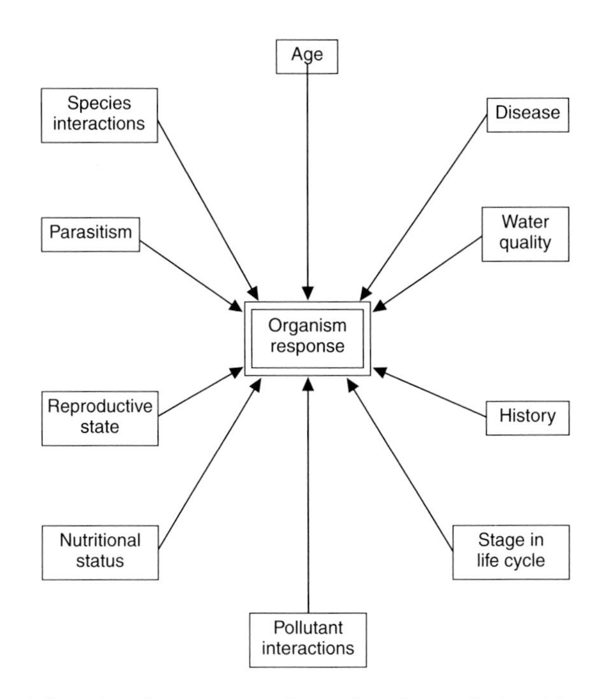

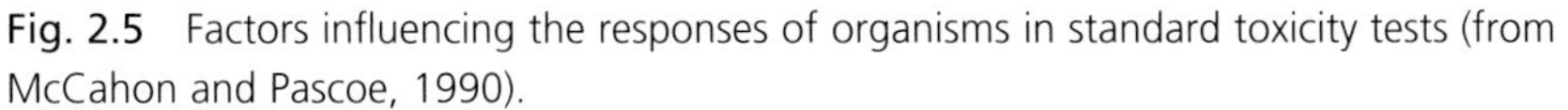
Fig. 2.5 Factors influencing the responses of organisms in standard toxicity tests (from McCahon and Pascoe, 1990).

The sensitivity of individuals of a particular species to a pollutant may be influenced by factors such as sex, age or size. Moulting *Gammarus* are more sensitive to cadmium than individuals between moults (McCahon and Pascoe, 1988b). Small *Asellus aquaticus* and *Gammarus pulex* were less tolerant of acidity than large individuals (Naylor *et al.*, 1990). First instar larvae of the caddis *Agapetus fuscipes* and the midge *Chironomus riparius* were much more sensitive to cadmium than later instars (McCahon *et al.*, 1989a; Pascoe *et al.*, 1989). However, the toxicity to the midge of the insecticide lindane is greatest at the pupation and emergence stages (Taylor *et al.*, 1993). In general the concentration of metals in invertebrates is inversely related to their body mass (van Hattum *et al.*, 1991). In fish, the embryo and larval stages are usually the most sensitive to pollutants (Kristensen, 1994).

Table 2.1 Median lethal concentrations (96-hour LC50) of four pollutants ($mg\,l^{-1}$) to three aquatic invertebrates (from data in McCahon and Pascoe, 1988a)

Species	Cadmium	Phenol	Ammonia	Lindane
Gammarus pulex	0.03	69	2.05	0.23
Asellus aquaticus	0.60	180	2.30	0.38
Chironomus riparius	>200	240	1.65	0.24

These differences in the tolerance to poisons within and between species make it dangerous to extrapolate from simple laboratory toxicity tests on a standardized organism to the field situation. The test organism may appear tolerant but a particular stage of its life cycle may be especially sensitive, and this is the crucial stage in relation to the success of a population exposed to an environmental pollutant. The developmental or larval stages of an animal are generally more sensitive to toxic pollutants than adults. In a review of toxicity studies on fish, Woltering (1984) reported that the survival of larvae was reduced in 57 per cent of studies, larval growth was inhibited in 36 per cent, reproduction of fish in 30 per cent and the hatchability of eggs in 19 per cent. By contrast, adult survival was reduced in 13 per cent of studies and adult growth in only 5 per cent. In general, therefore, early life stages are especially vulnerable.

Finally, the state of health may influence susceptibility to pollutants, diseased or parasitized individuals succumbing more quickly. As an example, experiments were conducted in a Welsh stream to simulate acid events by adding sulphuric acid and aluminium sulphate. An upstream stretch acted as a reference. Caged *Gammarus*, half of which were parasitized with the intermediate cystacanth stage of the acanthocephalan (spiny-headed worm) *Pomphorhynchus laevis*, were added to each section. There was almost no mortality in the reference section but animals quickly died in the experimental section, the time to death being significantly shorter in the parasitized animals (McCahon and Poulton, 1991). Parasitized *Gammarus* were also more sensitive to cadmium, but the parasite was not affected (Brown and Pascoe, 1989).

Environmental factors influencing toxicity

Environmental factors may modify the acute toxic effects of pollutants. Temperature is important because it not only influences the metabolic activity and behaviour of organisms, which may affect their exposure to a pollutant, but it also may alter the physical and chemical state of the pollutant. In general, toxicity increases with temperature but there are many exceptions. The accumulation of cadmium and copper, but not zinc, in *Asellus aquaticus* was dependent on temperature (van Hattum *et al.*, 1993). The time to death of rainbow trout (*Oncorhynchus mykiss*) exposed to phenol increased as temperature increased, but the LC50 decreased. Phenol causes paralysis and cardiovascular congestion, resulting in suffocation. The internal concentration of phenol is influenced by the relative rates of absorption and detoxification, both of which are directly proportional to temperature, but it is considered that temperature influences the rate of detoxification to a greater extent than the rate of absorption, at least at lower temperatures. Phenol therefore probably accumulates to higher levels at low temperatures, accounting for the greater toxicity in the cold (Brown *et al.*, 1967). Phenol was also more toxic at lower temperatures to *Asellus aquaticus* (Green *et al.*, 1988), because the rate of detoxification decreased more rapidly than the rate of absorption as the temperature was lowered. The caddis larva *Hydropsyche morosa* feeds largely on detritus during the spring and summer, which it collects from the water in nets that it spins. During the colder autumn and winter the animal spins few nets and grazes on attached algae. Mercury levels in the caddis are significantly higher in summer, the metal being accumulated from the contaminated detritus captured in the nets (Snyder and Hendricks, 1995).

The toxic effect of pollutants varies with the quality of water, pH and hardness being especially important. Hydrogen cyanide, for example, is especially toxic in the molecular form, so that any change in the pH that reduces the degree of dissociation will increase the toxicity of the solution without there being any change in the total concentration of cyanide. The toxicity of ammonia is also less at higher pH (Twitchen and Eddy, 1994).

The chemical speciation of some metals is markedly affected by pH. Metal 'species' can be grouped into three phases – an aqueous phase (free ions and dissolved complexes), a solid phase (particles and colloids) and a biological phase (incorporated into cells or adsorbed on to biological surfaces) (Gerhardt, 1993). Generally the ionic form of the metal is most toxic. Campbell and Stokes (1985) described two contrasting responses of an organism to a metal toxicity with declining pH:

1. If there is little change in speciation and metal binding is weak at the biological surface, a decrease in pH will decrease toxicity owing to competition for binding sites from hydrogen ions.
2. Where there is a marked effect on speciation and strong binding of the metal at the biological surface, the dominant effect of a decrease in pH will be to increase metal availability.

Zinc and copper show the first response, lead the second. Reviews of the influence of pH on the uptake and toxicity of metals are provided for fish by Spry and Wiener (1991) and for invertebrates by Wren and Stephenson (1991) and Gerhardt (1993). Tessier and Turner (1995) discuss the speciation and bioavailablity of metals in the environment.

Pollutants tend to be more toxic in soft waters (e.g. copper, lead, mercury, zinc). Figure 2.6 shows the relationship between the median survival time of rainbow trout and the concentration of zinc ions in waters of varying total hardness. There is a linear relationship between survival and concentration of zinc in soft water, whereas the relationship was curvilinear at intermediate and high levels of hardness. Low calcium concentrations in water enhance the toxicity of metals because the permeability of gill membranes is inversely related to the aqueous calcium concentration. Calcium competes with other metal cations for binding sites on the gill surface. The fish is thus protected in hard waters because the direct uptake of metal ions is reduced. Metals (e.g. lead) may be precipitated in hard water conditions and soluble complexes may be formed. The toxicity of zinc decreases with an increase in organic suspended solids because the metal is absorbed or adsorbed on to the suspended particles.

A number of poisons become more toxic at low oxygen concentrations because of an increase in respiratory rate, increasing the amount of poison the animal is

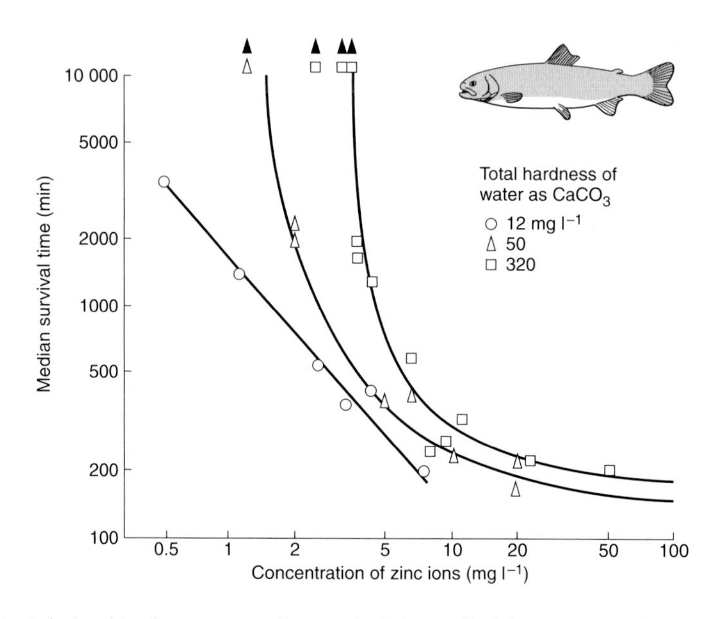

Fig. 2.6 Relationships between median survival times of rainbow trout and the concentrations of zinc ions at three levels of hardness (after Lloyd, 1960).

exposed to. The toxicity of several compounds to rainbow trout increased in direct proportion to the decrease in oxygen concentration in the water (Lloyd, 1992).

The time of exposure to a pollutant also influences toxicity, and this may be important in episodic pollution events. *Asellus aquaticus*, immobilized by brief exposures to phenol, was able to recover if placed in clean water. The rate of recovery was influenced by exposure time, exposure concentration and temperature (Green *et al.*, 1988).

Mayer and Ellersieck (1988) examined the effects of various factors on the toxicity of 410 compounds (75 per cent of them pesticides) in nearly 5000 toxicity tests with various species of invertebrates and fish. Only 20 per cent of chemicals showed a change in toxicity with pH, but this nevertheless caused the greatest average change in toxicity of any factor examined. Hardness had little effect on the toxicity of organic compounds. Temperature generally increased toxicity. Insects were usually the most sensitive group, followed by crustaceans, fish and amphibians.

Mixtures of poisons

Effluents are often complex mixtures of poisons. If two or more poisons are present together in an effluent they may exert a combined effect on an organism which is *additive*. Alternatively they may interfere with one another (*antagonism*), or their overall effect on an organism may be greater than when acting alone (*synergism*). A generalized scheme describing the combined effects of two toxic compounds is shown in Fig. 2.7. In this scheme the concentration of one unit of pollutant A produces the response in the absence of B, and one unit of B does the same in the absence of A. If, on combining the two pollutants, the response falls within the square, joint action is occurring, with the pollutants aiding one another. This joint action can be broken down into three special cases. If the response is produced by combinations represented by points on the diagonal (e.g. 0.5A + 0.5B), the effects are additive. If the response is produced by combinations falling in the lower triangle of the box (e.g. 0.5A + 0.2B), the effect is more than additive (synergistic), while if the response falls in the upper triangle (e.g. 0.8A + 0.7B), the effect is less than additive, though the pollutants are still working together in joint action. Antagonism occurs in the region outside the box when, for example, more than one unit of A is required to produce the effect in the presence of B. Antagonistic effects make the comparison between laboratory studies and field conditions difficult because animals in the field that should theoretically be dead very often are not.

An example of an additive interaction is the combined toxicity of zinc and cadmium to fish. Calcium is antagonistic to lead, zinc and aluminium. Copper is more than additive with chlorine, zinc, cadmium and mercury, while it decreases the toxicity of cyanide. The toxicity to the mayfly *Baetis rhodani* of phenol and ammonia at low concentrations is additive, but at higher concentrations the effect is more than additive (Khatami *et al.*, 1998). Figure 2.8 illustrates the toxicity to rainbow trout of copper and a detergent (sodium laurylbenzenesulphonate, LAS) acting

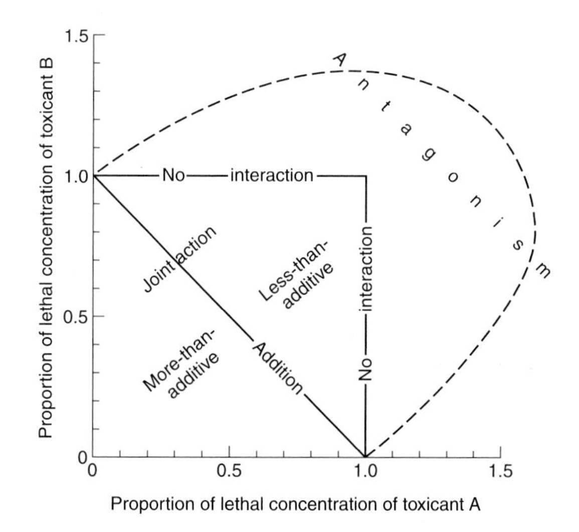

Fig. 2.7 Terms used to describe the combined effects of two pollutants (from Sprague, 1970).

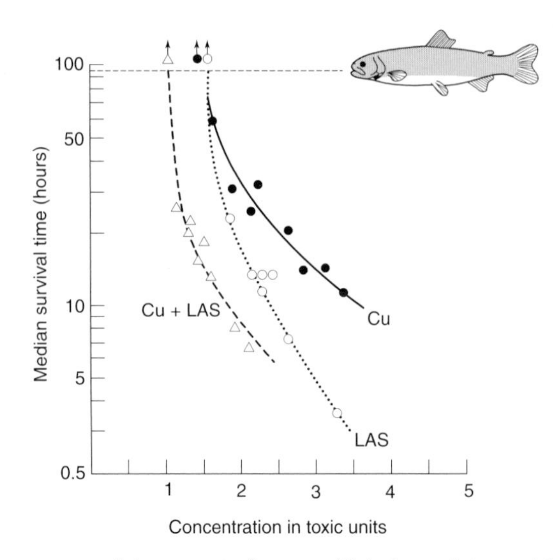

Fig. 2.8 Toxicity curves to rainbow trout of copper (Cu) alone, detergent (LAS) alone and in mixtures of various ratios of the two poisons (Cu + LAS) (from Calamari and Marchetti, 1973).

separately and in combination, and the toxicity can be seen to be markedly more than additive. It is thought that detergents reduce the surface tension on the gill membranes, thus increasing the permeability of the gill to the detergent and other poisons, but the physical effects are thought to be less than the chemical effects of the detergent.

Sublethal effects

Poisons are frequently present in freshwaters at concentrations too low to cause rapid death directly but they may impair the functioning of organisms. These sublethal effects may be observed at the biochemical, physiological, behavioural or life-cycle level. Many small changes in these parameters have been related to pollution but it is essential to show that they have ecological meaning, that they reduce the fitness of an organism in its environment and are not merely within its range of adaptation. The biochemical effects of pollution are basic, and these can then be related to the efficiency of tissues and organs, which can in turn be examined in relation to the performance of the organism and whether this has any adverse effect on the natural population (Sprague, 1971).

The early detection of specific molecular abnormalities in the tissues of organisms may provide an indication of exposure to pollutants long before any gross signs become apparent, and such biochemical indices may be valuable in signalling the development of sublethal abnormalities that could reduce the fitness of a population. Exposure of *Gammarus pulex* over ten days to low concentrations of copper and the insecticide lindane resulted in a significant increase in the concentration of protoporphyrin, which indicates a decline in the synthesis of haem, which in turn could lead to a decrease in the production of cytochromes and catalase, involved in various aspects of cell metabolism (Taylor *et al.*, 1998).

PCBs added to the diet of barbels (*Barbus barbus*) caused an increase in the enzyme cytochrome P450, while there was strong induction of the enzymes ethoxyresurufin *o*-deethylase (EROD) and ethoxycoumarin *o*-deethylase. The ultrastructure of the liver was examined by electron microscopy and changes to the rough endoplasmic reticulum, a reduction in glycogen and a dissolution of mitochondrial contents were observed (Hugla and Thomé, 1999). Damage to the hepatopancreas of *Gammarus pulex* has been recorded following exposure to copper, lindane and the herbicide 3,4-dichloroaniline (Blockwell *et al.*, 1996a).

Pollutants act at the biochemical level at a number of sites in the cell but an organism may be able to adapt by normal homeostatic mechanisms so that enzyme inhibition may not reduce overall fitness. Enzyme bioassay remains, however, a useful technique in looking for sublethal effects of toxic pollution.

It is suspected that the chronic exposure to pollutants increases the sensitivity of fish to infectious diseases by compromising the immune system. Carp (*Cyprinus carpio*) exposed in cages in a river of poor water quality showed a reduction in the

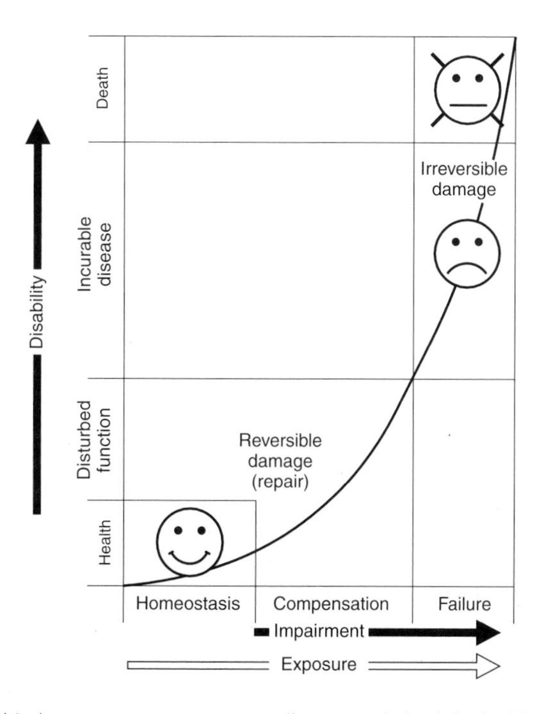

Fig. 2.9 Relationship between exposure to a pollutant and physiological impairment (after Fox, 1993).

proliferation ability of pronephric T lymphocytes and a decrease in spleen mass compared with fish from an unpolluted control site, i.e. there was a clear depression of immune function (Price *et al.*, 1997).

The effects of pollution on the genetic diversity of populations have been examined. Restriction fragment length polymorphisms were used to assess genetic variation in the genome of brown bullheads (*Aneiurus nebulosus*) from contaminated and clean sites in the Great Lakes region of North America. Genetic diversity estimates were always lower in bullhead populations from contaminated sites. Although these sites supported good bullhead populations, it was considered that severe pollution events in the past had reduced numbers and eliminated much of the genetic diversity (Murdoch and Hebert, 1994). This reduced genetic diversity may compromise the long-term survival of populations in contaminated sites.

Figure 2.9 illustrates a generalized relationship between physiological impairment following increasing exposure to pollutants and the consequent disablement of organisms. At low levels of pollution normal homeostatic mechanisms will maintain health. As levels rise, compensation occurs such that normal functions are maintained without significant metabolic costs, but at higher levels still, the organism becomes stressed and physiological breakdown occurs, an inability to repair the damage causing disablement. Physiological failure and death result from still higher loadings of pollutants.

The effects of pollutants on the respiration of fishes and invertebrates have received widespread attention. By cannulating the blood system of a fish it is possible to measure the concentrations of oxygen, metabolites and pollutants and hence understand more fully the mode of action of toxic pollutants. Using cannulation techniques it was found that zinc reduced the oxygen level of blood leaving the gills, but zinc injected into the blood system had no effect. Zinc, therefore, reduces the efficiency of oxygen transport across the gill membrane so that the fish dies of hypoxia (Skidmore, 1970). When the freshwater shrimp *Macrobrachium carcinus* was exposed to copper or zinc, the respiration and ammonia excretion rates were reduced, resulting in a reduction in the ratio of oxygen to nitrogen in the tissues and increasing the animal's dependence on carbohydrate and fat reserves (Correa, 1987). Respiration was found to be a sensitive indicator of copper stress in *Gammarus pulex*, effects being measurable at concentrations as low as $10.8\ \mu g\, l^{-1}$ (Kedwards *et al.*, 1996).

Growth can be examined as an integrated measure of the sublethal effects of toxic pollutants but, with homeostatic mechanisms operating, the effects may be minimal. Bleached kraft mill effluent from the paper industry, for example, has been variously described as reducing growth, stimulating growth and having no effect on growth in different fish species. Nevertheless the *scope for growth* has been shown to be a good indicator of environmental stress. The scope for growth (or production) is the difference between the energy intake and the energy lost in metabolism and excretion. A pollutant may reduce the scope for growth by increasing energy expenditure or reducing energy intake, but the latter is more common. For example, the exposure of *Gammarus pulex* to zinc, 3,4-dichloroaniline, oxygen, ammonia and chlorinated ethers all resulted in a significant reduction in the rate of feeding, and hence the scope for growth, but only ammonia significantly affected respiration (Maltby *et al.*, 1990a; Maltby, 1992). Copper and lindane have been found to impair growth in the same species (Maund *et al.*, 1992; Blockwell *et al.*, 1996b). Because growth is easily measured and is an important indicator of the success of natural populations it should be recorded in laboratory studies of the effects of pollution.

Pollutants may cause changes in swimming performance, orientation, coordination of movement, aggression and comfort behaviour of fish. There was, for example, an increase in aggression in dominant members of a population of bluegill sunfish (*Lepomis macrochirus*) when exposed to copper (Henry and Atchison, 1986). Atchison *et al.* (1987) and Weber and Spieler (1994) provide reviews of the effects of metals on the behaviour of fish.

The net-spinning behaviour of the caddis larva *Hydropsyche slossonae* has been found to be highly sensitive to concentrations of cadmium, 2,4-dichlorophenol and the insecticide malathion (Tessier *et al.*, 2000a,b,c). There are two types of anomalies in the net, a distortion in the midline meshes, and the production of irregular nets without any real structure (chaotic nets). Spotted frog (*Rana luteiventris*) tadpoles and snails *Physella columbiana*, but not caddis fly larvae, were found to show a reduction in anti-predator behaviour in the presence of heavy metals (Lefcort *et al.*, 1998, 2000), while the predation rate of the flatworm *Dendrocoelum lacteum* on *Asellus aquaticus* decreased on exposure to cadmium (Ham *et al.*, 1995).

Reproductive behaviour may also be affected. *Gammarus pulex* engages in precopulatory pairing for up to ten days before mating, the behaviour allowing the male to protect the female from the attention of other suitors. This pairing behaviour is disrupted by pollution (Poulton and Pascoe, 1990). The related American species *Hyalella azteca* shows a similar disruption of guarding behaviour in the presence of pollutants (Blockwell *et al.*, 1998). The courtship behaviour and parental care of some fish species is also influenced by pollutants (Jones and Reynolds, 1997).

Animals may also avoid polluted water. This has been shown, for example, with Atlantic salmon, where avoidance of copper has been demonstrated in laboratory studies and verified by field observations (Atchison *et al.*, 1987). *Gammarus pulex* exposed in the laboratory to sediments contaminated with copper spent more time in the water column (Taylor *et al.*, 1994), which in the field situation would be equated with downstream drift to avoid contaminated conditions.

Because the various life stages of an organism may be affected differently by a toxic chemical (p. 30), ideally we should study the species over its lifetime to find the weak link in its response to pollution. Such long-term experiments are essential in order to discover any carcinogenic or mutagenic effects of pollutants, or any teratogenic effects causing developmental abnormalities.

Tolerance

Populations may develop a tolerance to pollutants which enables them to survive in highly polluted environments. They may achieve this by functioning normally at high toxic loadings or by metabolizing and detoxifying pollutants. The mechanisms of tolerance to pollution are extremely complex, involving several metabolic systems, and species have solved the problem of tolerance to a particular pollutant in different ways.

Many rivers are polluted with heavy metals from old mine workings and some species of algae become very tolerant of contaminated conditions. A survey of 47 sites with different concentrations of zinc found the filamentous green alga *Hormidium rivulare* to be abundant everywhere, tolerating zinc concentrations as high as 30.2 mg Zn l^{-1}. The closely related *H. fluitans* only occurred at sites with zinc concentrations up to a mean of 5.59 mg Zn l^{-1}. *Hormidium* growing in streams high in zinc were more tolerant than populations from clean waters and this adaptation was genetically determined (Say *et al.*, 1977). The culturing of algae in increasing concentrations of metals leads to the rapid development of tolerance, resistance being due to the selection of spontaneous mutants within the cultures (Whitton and Shehata, 1982).

Tolerance to metals has also been recorded in invertebrates such as *Asellus aquaticus* and in fish. After exposure for 24 hours to a copper concentration of 0.55 mg l^{-1}, rainbow trout showed a 55 per cent inhibition of sodium uptake and a 49 per cent reduction in affinity for sodium, which resulted in an overall decrease in total sodium concentration of 12.5 per cent. Within one week, however, the concentration had returned to normal. The rate of sodium uptake was still inhibited but the rate of sodium loss was reduced. Copper accumulated in the liver and there was an increase in the concentration of sulphydryl-rich protein (Laurén and McDonald 1987a,b). The protein was considered to be a metallothionein. These low molecular weight proteins contain many sulphur-rich amino acids which bind and detoxify some metals. The pretreatment of an organism with low doses of a metal may stimulate metallothionein synthesis and provide tolerance during a subsequent exposure (Pascoe and Beattie, 1979).

Many bacteria quickly become tolerant to toxic pollutants, and the resistance gene is carried on plasmids. Plasmids are small, independent genetic elements, which may be transferred from cell to cell by bacterial conjugation or by transduction (which involves transfer by bacterial viruses). The transferred plasmid replicates rapidly and is passed on to the progeny, so that resistance can spread very quickly. Resistant bacteria might enhance the transport of pollutants, especially metals, in the environment by mechanisms such as solubilization, concentration and conversion to organometallic species and their respective elemental forms, thus constituting a potential environmental danger. For example, inorganic mercury is converted to organic methyl mercury in aquatic environments owing to the activities of bacteria and fungi and this process may occur under anaerobic (e.g. by *Clostridium*) or aerobic (e.g. by *Neurospora*, *Pseudomonas*) conditions. Methyl mercury is especially toxic to many animals.

Resistant bacteria can, however, also remove heavy metals from the environment. For example, in laboratory experiments a mercury-resistant strain of *Pseudomonas putida* was found to remove 90–98 per cent of mercury in wastewater from chloralkali factories (von Canstein *et al.*, 1999). Genetically engineered microbes (GEMs) could play a valuable role in detoxifying industrial wastes. Scragg (1999) and Timmis and Pieper (1999) discuss GEMs.

Accumulation

It is necessary to distinguish between the terms *biomagnification* and *bioconcentration* (or *bioaccumulation*). With biomagnification there are progressively greater amounts of contaminant along the food chain, carnivores containing larger concentrations than herbivores, which contain more than plants. Bioconcentration requires only uptake from water and is independent of trophic level. Organochlorine pesticides biomagnify along the food chain, but biomagnification is the exception for metals. Mercury provides one such exception. Bioconcentration occurs with many toxic pollutants, very high levels being accumulated in organisms from very low levels in water. Bioconcentration factors for some organic compounds in the fathead minnow (*Pimephales promelas*) are given in Table 2.2.

The rate of accumulation of pollutants will depend on factors both external and internal to the organism. The concentration of pollutant in the water is clearly important, and many species carry higher burdens of pollutants when living in contaminated waters. For example, metal concentrations in algae and bryophytes are significantly correlated with concentrations in water. There appear to be no consistent correlations, however, between environmental levels of metals, other than mercury, and concentrations in invertebrates and fish (Kelly, 1988; Barak and Mason, 1989; van Hattum *et al.*, 1991). Temperature influences the absorption, detoxification and excretion rates of pollutants but not necessarily to the same extent, so that the overall bioconcentration may vary with temperature. An increase in the bioconcentration of cadmium with temperature was found, for example, in stone loach (*Noemacheilus barbatulus*) (Douben, 1989).

Internal factors that influence bioconcentration include physiological condition. The concentration of lipophilic organochlorine compounds in fish is closely related to the fat content of different species. Fish with higher metabolic rates

Table 2.2 Bioconcentration factors of organic compounds in the fathead minnow (*Pimephales promelas*) after 32 days' exposure (from Veith *et al.*, 1979)

Compound	Bioconcentration factor
Lindane	180
Pentachlorophenol	770
Mirex	18 100
P,P'DDT	29 400
PCB (Aroclor 1260)	194 000

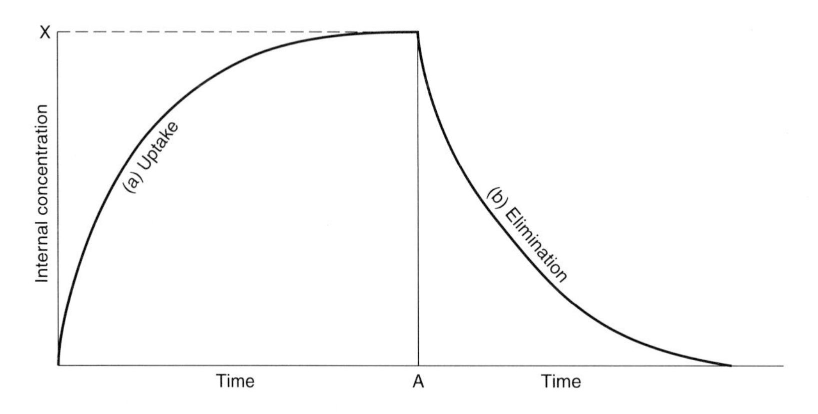

Fig. 2.10 A generalized curve for the uptake and elimination of a pollutant. The internal and external concentrations are equal at X, external concentrations are reduced to zero at time A (from Holdgate, 1979).

accumulate contaminants faster and, because feeding results in higher metabolism, a greater uptake of pollutants across the gills may occur in feeding as opposed to starved fish, as has been shown for cadmium uptake in the stone loach (Douben, 1989). Age, sex and the presence of competing pollutants in the water may also influence accumulation rates.

The accumulation of a pollutant is a function of both uptake and elimination, and a generalized curve is shown in Fig. 2.10. Xu and Pascoe (1993) provide a curve of a very similar shape to this based on the exposure of *Gammarus pulex* to zinc. Assuming uptake to be due solely to chemical diffusion, the process will continue until the internal concentration is equal to the concentration in the environment (point X in Fig. 2.10). Most pollutants, however, can be eliminated from the body, and this is an active biochemical and physiological process which cannot be described in simple diffusion terms. Pollutants may be oxidized in the liver by microsomal enzymes, the best known of which are the mixed function oxidases. These utilize NADPH (nicotinamide adenine dinucleotide phosphate) as a co-substrate. Two enzymes are involved in this type of oxidation, cytochrome P-450 and a flavoprotein, NADPH-cytochrome P-450 reductase, which are embedded in the phospholipid membranes of the endoplasmic reticulum. The flavoprotein transfers electrons from NADPH to cytochrome P-450. This then inserts one atom of molecular oxygen into the toxic compound and reduces the second oxygen atom to produce water, using the hydrogen atoms from the reduced NADP. Multiple cytochrome P-450 isoenzymes occur in animals and these have different substrate specifications. The metabolites produced may be more or less toxic, but they are generally water-soluble and can be eliminated from the body. More details of these processes can be found in Stegeman and Hahn (1994) and Roesijadi and Robinson (1994).

Conjugation or transformation may also occur to render compounds less toxic or more soluble. Both uptake and elimination of pollutants occur simultaneously and the plateau in Fig. 2.10 will depend on a balance of the factors determining the two processes.

Contaminants may be taken up through the gut from the food or directly from the water. With invertebrates, uptake from water is generally the most significant (Gerhardt, 1993). In *Asellus aquaticus*, for example, the majority of cadmium was taken up directly from the water (van Hattum *et al.*, 1989), as was zinc in *Gammarus pulex* (Xu and Pascoe, 1993). There are exceptions. The burrowing mayfly *Hexagenia rigida* obtained most of its cadmium and zinc from sediments ingested as food (Hare *et al.*, 1991). Copper appears to be taken up in invertebrates via the food, but cobalt from water (Beltman *et al.*, 1999).

EFFECTS OF TOXIC POLLUTION ON THE COMMUNITY

Toxic pollutants may exert differential mortality on populations within an aquatic community, exterminating some species, while having little direct effect on others. However, there may be marked indirect effects, the loss of one group of organisms having severe repercussions on other groups. Toxic pollutants in sewage effluents, for example, may destroy those bacteria responsible for the biodegradation of organic matter, with the result that the oxygen sag curve (p. 109) may extend considerably further downstream than would otherwise be the case. Animals tolerant of particular pollutants may be released from the constraints of competition and predation, building up large populations in a simplified community. Predators, alternatively, may find their prey eliminated. Because researchers generally look at only a small part of the ecosystem affected by a pollution event, the wider ramifications are poorly understood.

Experimental mesocosms have been used to determine the responses of a stream invertebrate community to mixtures of the metals copper, zinc, manganese and lead (Richardson and Kiffney, 2000). The most sensitive taxa in the stream were found to be the mayflies *Baetis*, *Ameletus* and *Paraleptophlebia*, while groups such as the stonefly family Nemouridae and oligochaete worms were only mildly affected at the highest concentrations. Chironomid larvae were unaffected by the metals and, making up more than 80 per cent of the benthos, resulted in no significant decline in overall benthic invertebrate densities as metal concentrations were increased.

A study was made of the fish community below a pulp and paper mill which was discharging effluent into the Mänttä watercourse in central Finland. The flow in the stream was small relative to the discharge of the effluent, resulting in a biological oxygen demand (BOD) (see p. 81) of $18\,mg\,l^{-1}$ in the receiving water in 1977. In 1986 an activated sludge plant (p. 99) was constructed and, by 1990, BOD averaged $8\,mg\,l^{-1}$. Effluents from pulp mills can contain up to 200 compounds, including chlorophenolics, fatty acids and resin acids, which may be acutely toxic and bioaccumulate in tissues. Dehydroabietic acid (DHAA) has an LC50 to fish of $1\,mg\,l^{-1}$.

The number of fish species recorded downstream of the discharge is shown in Fig. 2.11. In 1977 only roach (*Rutilus rutilus*) and ide (*Leuciscus idus*), two cyprinid

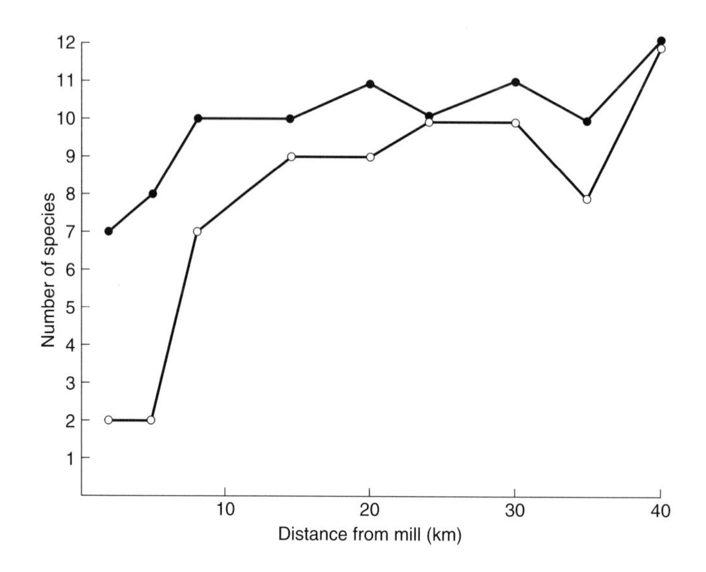

Fig. 2.11 Numbers of fish species at stations downstream of a pulp and paper mill before (o) and after (•) the installation of an effluent treatment plant (from data in Hakkari, 1992).

species, were present in the first 5 km, and the population was very small (Fig. 2.12), owing mainly to a lack of oxygen in the stream in the autumn. By 1990, following effluent treatment, the number of fish species in the first 5 km had increased to nine, including pike (*Esox lucius*) and perch (*Perca fluviatilis*). There had also been a substantial increase in the biomass of the community. Nevertheless there was no significant recruitment of young fish in the stream for 15 km below the mill, the population being maintained by immigration. Studies in Sweden of perch below pulp mill effluents have revealed a 10 per cent incidence of malformed embryos, with sharp bends in the posterior part of the spinal cord, and high mortality close to hatching, probably the result of acute toxicity (Karås *et al.*, 1991). In the Mänttä study area in both 1977 and 1990, the fish caught in the first 15 km below the discharge were tainted, i.e. they tasted 'off' when cooked and eaten (p. 225). Further downstream 60 per cent of fish were tainted in 1977, but none was in 1990. The installation of the treatment works at the mill had led to some improvement in the fishery, but recruitment remained poor, most likely because of a toxic chemical in the effluent which was not removed by treatment.

Pyrethroid insecticides, based on compounds found naturally in plants and which break down relatively rapidly after application, are generally considered to be less environmentally damaging than organochlorine and organophosphorus insecticides. They are, however, acutely toxic to aquatic life and there is particular concern about their use in sheep dips in upland areas, run-off causing considerable damage to otherwise pristine streams. The addition of a pyrethroid to mesocosms,

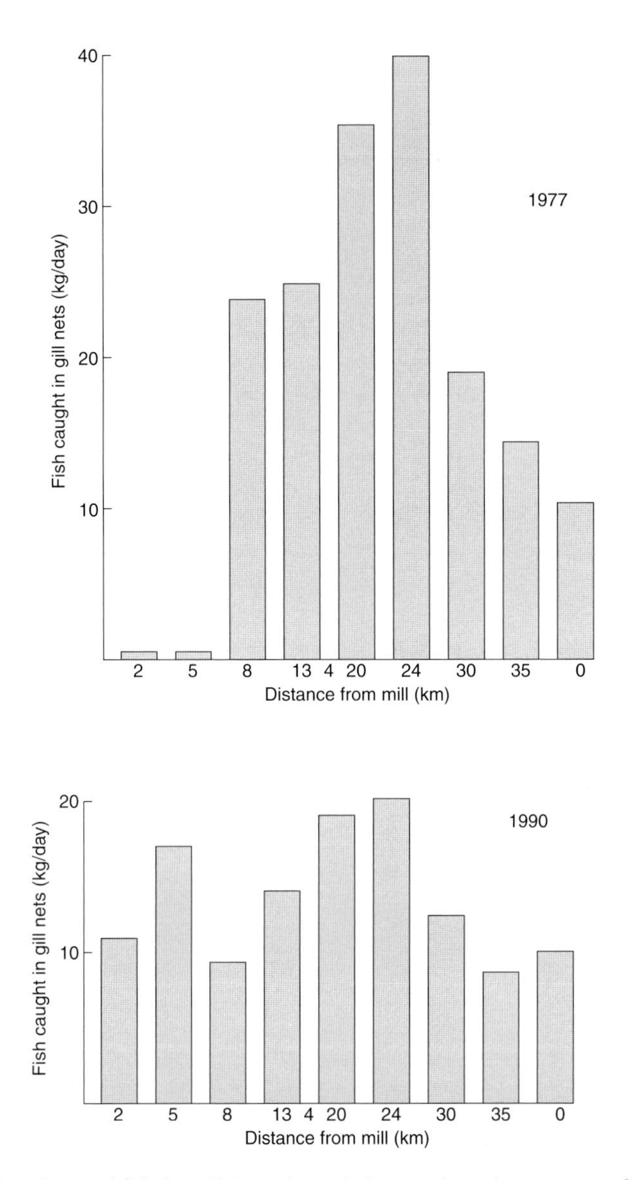

Fig. 2.12 Abundance of fish (kg gill-netted per day) at stations downstream of a pulp and paper mill before and after installation of an effluent treatment plant (from data in Hakkari, 1992).

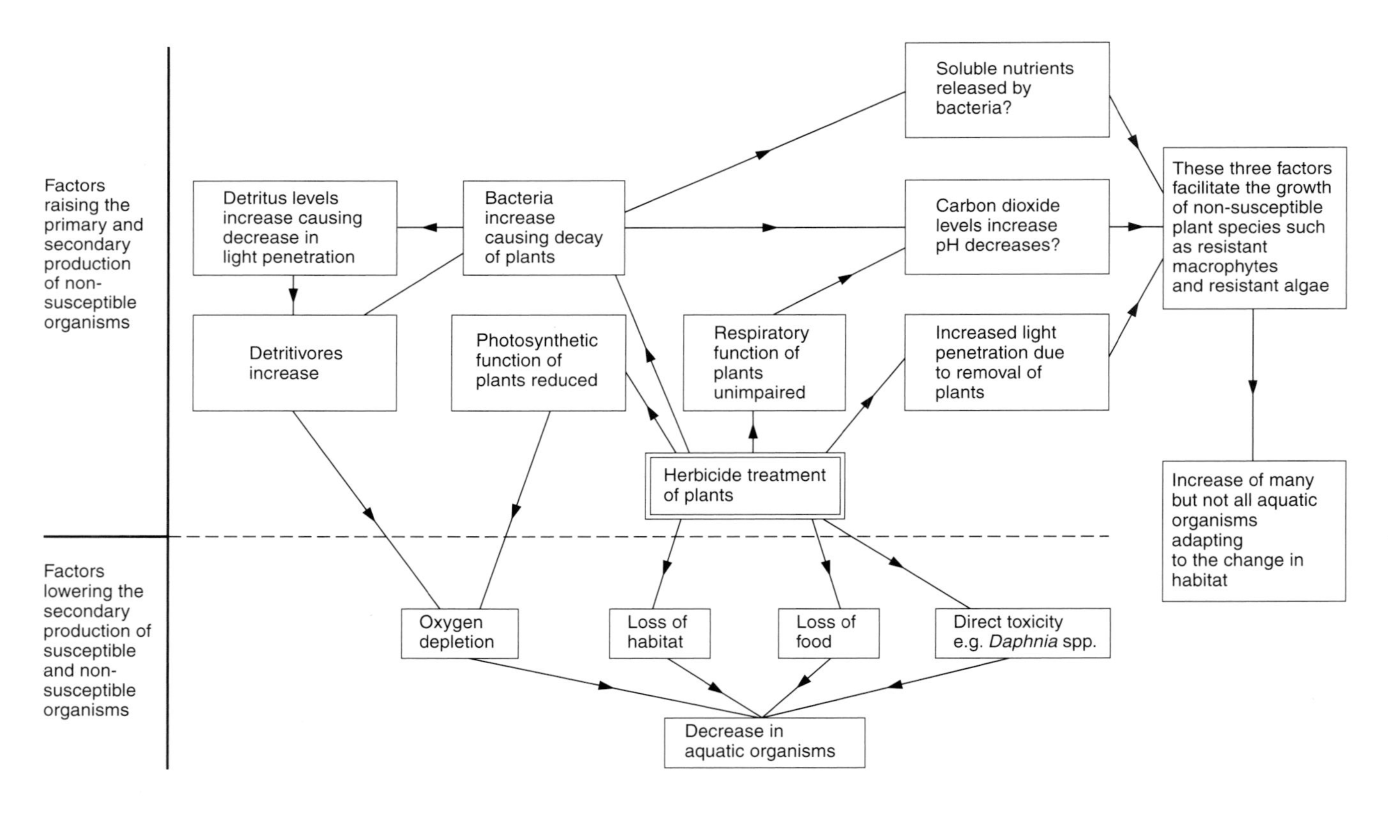

Fig. 2.13 Short-term ecological effects of herbicides in aquatic ecosystems (from Newbold, 1975).

which simulated fish-free eutrophic ponds, were compared with control mesocosms. There was a loss of many arthropods by direct mortality. However, there was a ten-fold increase in the oligochaete *Stylaria lacustris*, released from predation and competition. As the predatory insects gradually recolonized, the oligochaetes declined in numbers and were replaced by the ostracod *Herpetocypris reptans*, which is less susceptible to predation but uses similar food resources. Differences in exposed mesocosms were still marked more than two years after the exposure to the insecticide, showing that non-persistent chemicals can produce long-term changes at the ecosystem level (Woin, 1998). The run-off from agricultural land of pesticides, applied following established practice, has a considerable impact on the macroinvertebrates in the receiving streams, causing both increased mortality and increasing drift (Schultz and Liess, 1999a).

Herbicides are widely used to control water plants that may impede water flow during summer. The direct effect of herbicide addition is the loss of higher plants and non-target organisms, such as sensitive species of invertebrates and fish. Indirect effects include the death and decay of plants, with resultant changes in water chemistry and oxygen levels, a loss of habitat and a loss of food supply (Brooker and Edwards, 1975). Aquatic plants are often replaced by algae. The short-term ecological effects are shown in Fig. 2.13. Although herbicides may be acutely toxic to fish and invertebrates in laboratory studies, the concentrations in the field, when sensibly applied, are often too low to cause problems. The application of herbicides results in a rapid reduction in photosynthesis, with a consequent reduction in carbon dioxide uptake and oxygen output. As large quantities of plant material decompose, severe deoxygenation of the water may take place, resulting in fish kills. The release of nutrients stimulates algal growth. The available detritus may lead to substantial increases in benthic invertebrate populations, while those species associated with vegetation, such as molluscs and caddis flies, show sharp reductions in numbers.

Other examples of the effects of pollutants on communities are found elsewhere in the book.

TOXICITY TESTING

The preceding pages have described the principles of toxicology and the factors that influence toxicity. Because pollutants often occur in complex mixtures and interact, the chemical determination of pollutant levels alone frequently gives little indication of their potential biological effects. In a toxicity test, environmental conditions are carefully controlled so that the response of a test organism to particular pollutants can be defined, but the extrapolation from test to field situations requires caution. Toxicity tests must go hand in hand with field observations and field experiments in order to fully understand a pollution problem.

The selection of a suitable organism for routine toxicity testing will depend on a number of factors. The organism:

- must be sensitive to the material or environmental factors under consideration;
- must be widely distributed and readily available in good numbers throughout the year;
- should have economic, recreational or ecological importance both locally and nationally;
- should be easily cultured in the laboratory;
- should be in good condition, free from parasites or disease.

Small organisms with short generation times are generally preferred, though fish are also popular because of their physiology and their recreational and economic importance. The chief uses of toxicity tests are for preliminary screening of chemicals, for monitoring effluents to determine the risk to aquatic organisms and, for those effluents that are toxic, to determine which component is causing death so that it can receive special treatment.

Studies on toxicity can be conducted in the field, using caged organisms. In Slate River, Colorado, *in situ* tests were conducted to explain the field distributions of fish in relation to a metal-contaminated tributary, Coal Creek (Davies and Woodling, 1980). Coal Creek had zinc and copper concentrations as high as $9.9\,mg\,l^{-1}$ and $0.11\,mg\,l^{-1}$ respectively and concentrations were still high, compared to an upstream control, at the lowest site on Slate River, 15 km downstream of its confluence with Coal Creek. Upstream of Coal Creek, Slate River supported a thriving community of three salmonids – brown trout, rainbow trout and brook trout (*Salvelinus fontinalis*). Downstream, rainbow trout were recorded only occasionally, and only at the lowermost site, while populations of brown trout were reduced ten-fold. Brook trout were the most abundant throughout but did not occur in Coal Creek or at its confluence with Slate River. An over-abundance of gravid females 6.7 km below the confluence showed that brook trout were avoiding upstream migration to their spawning grounds because of elevated metal concentrations. Tests with caged fish of all three species showed that rainbow trout were the most sensitive, mortality occurring at all sites except the control, explaining the virtual absence of this species below the confluence. All species suffered total mortality in Slate River 50 m below the confluence when river levels were low and metal levels high. Mortality did not occur in brook or brown trout when water levels were higher and metals were diluted. Metals were therefore likely to be only periodically toxic to wild populations of these two species. An example of using caged invertebrates to assess toxicity is described on p. 197. Care in interpretation of the results in cage experiments is necessary because wild individuals, unlike caged animals, can take avoiding action from pollutants (Schultz and Liess, 1999b). The use of caged animals in the field makes heavy demands on labour, especially with fish, which require feeding, and vandalism is a recurring problem.

The simplest type of laboratory toxicity test is the static test, in which the organism is placed for 48–96 hours in a standard tank of the water under examination. There is normally a series of tanks with test water of different dilutions, usually in a logarithmic series. The organisms are removed at the end of the test period and mortality is recorded. In these tests, the poison may evaporate, degrade or adsorb on to the surface of the tank, so that toxicity may be underestimated. These tests are nevertheless useful when an effluent needs rapid evaluation. A more sophisticated method involves the periodic replacement of test water. Continuous flow systems are also used, but these require large volumes of clean water for dilution.

Traditionally, fish have been used in standard toxicity tests. In the United States, fathead minnow and bluegill sunfish, together with goldfish (*Carassius auritus*) and guppy (*Poecilia reticulata*), are the main test species, though there is a tendency now to use a range of fishes. Early work in Britain concentrated on brown and rainbow trout but the tropical harlequin fish (*Rasbora heteromorpha*) is also now used extensively because it is small and has a similar sensitivity to trout. Toxicity tests need to be replicated and this requires the maintenance of large numbers of fish, needing considerable volumes of clean water. The response time of fish to low concentrations of pollutant may also be slow. Furthermore there are growing ethical objections to using vertebrates in routine toxicological assessments and, in Britain, such experiments on fish must be licensed and are regularly inspected. There may therefore be considerable benefits in using other organisms for routine work. Eventually *in vitro* toxicity tests may be developed that replace vertebrates in routine toxicity tests. Freshly isolated rainbow trout hepatocytes (liver cells), for example, have been used, though they were found to be, in general, less sensitive to a range of chemicals than the cladoceran *Daphnia magna* (Lilius *et al.*, 1994). Solbé (1993) provides a review.

Algae and vascular plants, in particular duckweeds (*Lemna*), have also been used in toxicity tests. Duckweeds are small, have a simple structure of leaf and root, reproduce rapidly, are easy to culture and may be more effective than algae as test organisms (Wang, 1990). Of more complex plants, clones of *Vallisneria americana* have been used to identify point sources of pollution in the Detroit River (Lovett Doust *et al.*, 1994). Reviews of the use of primary producers in toxicity tests are provided by Lewis (1993) and Klaine and Lewis (1995).

Much recent work has considered *Daphnia*, especially the large *D. magna*, as a test organism. The genus is ubiquitous, has a pivotal role in many lake food webs and its ecology has been well studied. *Daphnia* is sensitive to pollutants, is easily cultured and has a high reproductive rate. *Daphnia magna* has been considered unrepresentative of other zooplankton species because of its large size and restricted natural habitats (Koivisto, 1995) but Guilhermino *et al.* (2000) consider it a valuable species for screening chemicals for toxicity, considerably reducing the number of vertebrates required for this purpose.

The amphipod *Gammarus pulex* has been proposed as a test organism because it is easy to culture and is sensitive to pollution (McCahon and Pascoe, 1988), while it has also been used in scope for growth studies (p. 36), both in the laboratory and in the field (Maltby *et al.*, 1990b; Crane *et al.*, 1996). A wide range of other invertebrates has been suggested for use in routine toxicity tests (Pascoe *et al.*, 1991).

For management purposes, the results of acute toxicity tests need to be related to field conditions in terms of safe concentrations of effluents and this has led to the introduction of application (or uncertainty) factors, many of which are purely arbitrary. For every poison it is assumed that there must be a concentration that is so low as to have a negligible toxic effect and this concentration has an approximate relation to the lethal threshold concentration. Multiplying the lethal threshold by an application factor should give an approximate indication of the acceptable concentration of the poison (Lloyd, 1992). Application factors vary from 0.01 for persistent or accumulative compounds to 0.3 for compounds of low toxicity that break down rapidly.

Despite their widespread use, toxicity tests using a single species are unlikely to yield reliable information on the potential effects of pollutants on the aquatic ecosystem, because there is no universally sensitive indicator species, while their relative sensitivities will vary according to the pollutant being tested for. Whitehouse *et al.* (1996) and Girling *et al.* (2000) evaluate methods and precision in toxicity testing. The loss of a sensitive key species from the aquatic ecosystem could markedly alter the processes occurring within it. The use of artificial channels or streams allows the effects of pollutants on entire communities to be studied (Hill *et al.*, 1994; Brooks *et al.*, 1996).

Automated biomonitors

As earlier pages in this chapter have shown, organisms show distinct physiological and behavioural responses to low levels of pollutants and these can be harnessed to devise automatic alarm systems. Automatic monitors should provide a rapid indication that water quality has deteriorated and they have potential use in monitoring river waters, especially those that are abstracted for potable supply, and for monitoring effluents from sewage treatment works and industrial plants. Such monitors must be on line to an operations control centre so that immediate action can be taken if an alarm is sounded.

A number of requirements must be met for an automatic biomonitor to be fully successful (Diamond *et al.*, 1988):

- the operation of the system should be continuous, being always alerted at such times as it is likely to be necessary;
- the alarm must be given soon enough so that any action required can be taken before toxic effects occur;
- the system should detect a wide range of toxic substances;
- the detector must respond unfailingly to the selected signal at the relevant level, always giving an alarm when the selected level is exceeded;
- the number of false alarms, owing to non-toxic variations in water quality, should be minimal;
- the monitoring systems should be easy to operate and the alarm signal should be clear, characteristic and easy to interpret;

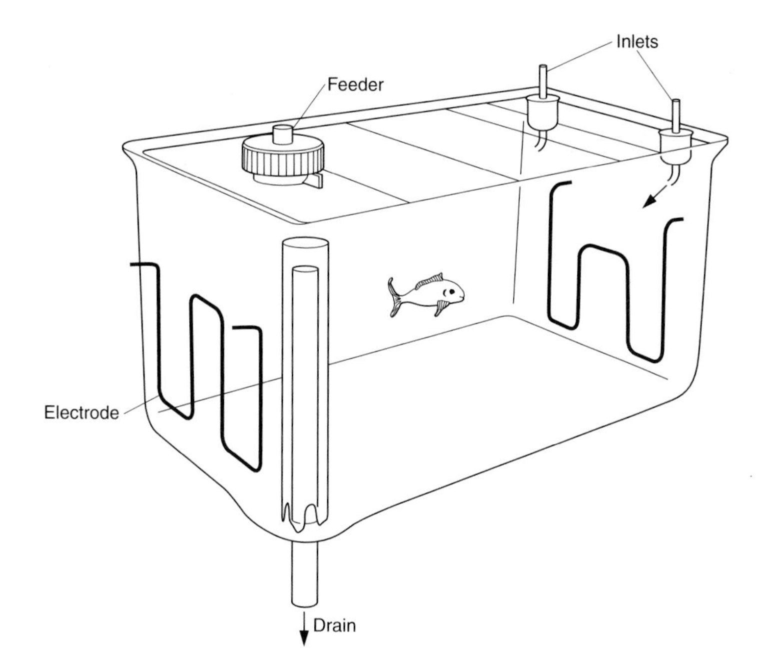

Fig. 2.14 An automatic fish monitor tank (from Westlake and Van der Schalie, 1977).

- the organism used as the sensor in the system should be fairly inexpensive and easy to acquire and culture;
- the apparatus should require as little maintenance as possible.

Much of the early work on biomonitors made use of fish and these have developed the furthest, commercial models being available. Fish alarm systems have monitored movement and respiratory activity in relation to levels of pollution. A typical tank, in which increasing opercular rhythms and movement of fish in relation to sublethal levels of pollutants are monitored, is illustrated in Fig. 2.14. To remove environmental effects, the experimental tanks are usually enclosed in a chamber, which reduces sound and vibrations and allows the photoperiod to be controlled, while water temperature is maintained constant. Food is given from an automatic feeder. Water is provided on a continuous flow basis. One fish is held per tank. The opercular movements of the fish result in a change of potential between electrodes, while the movement of a fish can be recorded as it swims across a light beam between two photocells. Infrared emitters can be used so that the natural rhythms of fish are not altered, while ultrasonic emitters are ideal for turbid waters (Morgan and Kuhn, 1988).

Figure 2.15 shows some typical results from a biomonitor, namely the ventilatory profiles of bluegill sunfish exposed to sublethal concentrations of zinc and trichlorethylene. Note the changes in amplitude of the signals. In the case of

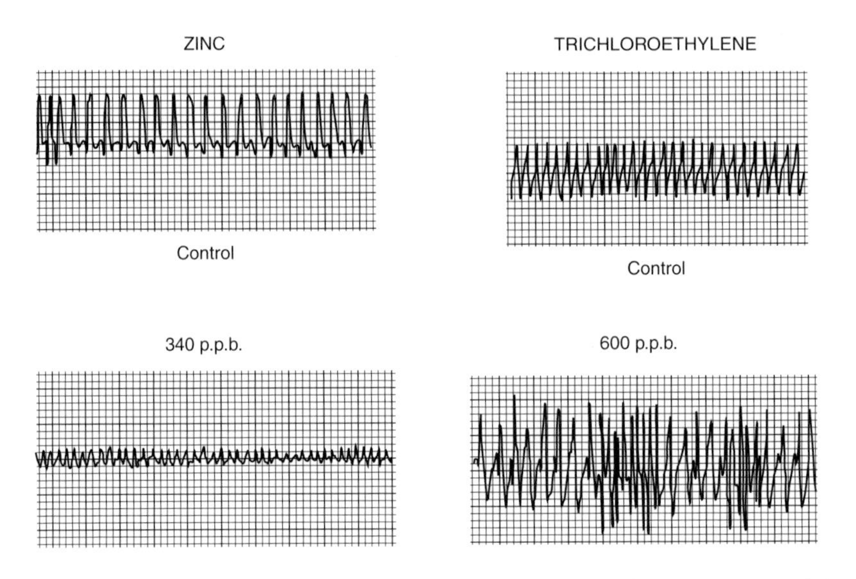

Fig. 2.15 Ventilatory profiles of bluegill sunfish exposed to sublethal concentrations of zinc and trichlorethylene (p.p.b. = parts per billion) (after Diamond *et al.*, 1988).

zinc the amplitude is very much decreased, whereas with trichlorethylene it is increased, while the signal is irregular. It may be possible eventually to identify the nature of the pollutant as well as the existence of chronically toxic conditions.

To be effective, a biomonitor should respond quickly to the presence of a pollutant. Table 2.3 shows the detection limits and detection times for several pollutants tested in the Water Research Centre Fish Monitor, which measures the ventilation frequency of rainbow trout. In this case one hour was considered a reasonable time for the detection of a pollutant. If a longer response time is acceptable, then limits of detection may be lower than in Table 2.3. The sensitivity of the biomonitor was found to be much greater than standard toxicity tests. For example the standard 96 hr LC50 for phenol is 9.3 $mg\,l^{-1}$ and for lindane 0.06 $mg\,l^{-1}$, showing that the monitor responds more quickly to lower concentrations. This biomonitor has been used in particular to monitor continuously the quality of water in lowland river intakes which, after treatment, is used for drinking water supply. It provides a high a degree of protection, although it responds rather poorly to compounds producing toxic symptoms that are likely to be chronic and long term in nature (Baldwin *et al.*, 1994).

The signals produced by a biomonitor are usually passed on to a computer which, from predetermined criteria, will assess whether the output indicates toxic conditions. If it does, an alarm system will be activated or corrective action, such as the closure of a water intake, will be taken. Remote biomonitors may be linked to a control station by telemetry or earth satellite.

Table 2.3 Limit of detection and detection time of a biomonitor to various pollutants, measuring ventilation rate of rainbow trout (after Evans *et al.*, 1986)

Pollutant	Detection limit (mg l^{-1})	Detection time (min)
Ammonia	0.11	17
Phenol	2.00	28
Paraquat	0.80	26
Lindane	0.0003	21
Pentachlorophenol	0.14	33
Formaldehyde	100	38

Ethical concerns about using fish were mentioned above. There have been a number of recent developments with automated biomonitors that use other organisms. Several attributes of bacterial enzyme systems have been exploited, for example to detect pesticides, heavy metals and phenols, and as a replacement for the standard biochemical oxygen demand test. Such *biosensors* are discussed by Scragg (1999). The most widely used bacterial test, however, is the commercially available Microtox test, which utilizes the bioluminescence of the marine *Photobacterium phosphoreum*. The reduction in light output is a measure of toxicity and the test compares well with toxicity tests using fish and other animals. It is sensitive, precise and reproducible. It has been used, for example, to examine the toxicity of organic compounds in the River Meuse and showed that, over much of the length of the river, the toxicity of the water was below environmentally acceptable standards (Polman and de Zwart, 1994). Cloning techniques are being applied to broaden the applications of bioluminescent bacteria in toxicity assessments (Layton *et al.*, 1999).

The movement of the flagellate *Euglena gracilis* has been developed as the ECOTOX test to act as an early warning system to detect deterioration in water quality (Tahedl and Häder, 1999). The locomotion and activity of *Daphnia*, and of tubificid worms (Leynen *et al.*, 1999) have also been investigated, while the valve movements of zebra mussels *Dreissena polymorpha* are also being exploited commercially, mussels rapidly closing their shells in response to environmental stress (Kraak *et al.*, 1994). Multiple species monitors, with invertebrates as well as fish, are also being developed, on the assumption that two or more species are more likely to respond to a wider range of pollutants than a single species (Cairns and Cherry, 1993; Landis and Yu, 1995). A review of automated biomonitors is provided by Gruber *et al.* (1994).

BIOMARKERS

We have seen how environmental contaminants may exert effects at various levels of biological organization, down to the biochemical and subcellular. These effects may be quite subtle and can be measured long before any outward toxic effects become apparent. They may be used as *biomarkers* to assess the health of an organism in its environment. A biomarker can be defined as 'a xenobiotically-induced variation in cellular or biochemical components or processes, structures, or functions that is measurable in a biological system or sample' (National Research Council, 1987). Many of the sublethal effects described earlier in this chapter can be categorized as biomarkers.

The measurement of biomarker responses can show that organisms have been exposed to pollutants at levels that exceed the normal detoxification and repair capabilities; this cannot be done by measuring only the body burden of a contaminant. Biomarkers may provide evidence that an organism has been exposed to a compound that does not bioaccumulate or that is rapidly metabolized (e.g. polynuclear aromatic hydrocarbons, PAHs). They provide a biologically relevant measure of toxicant interactions in target tissues, expressing the cumulative effects of these at the molecular or cellular level. Biomarkers may also provide more sensitive and precise indicators of ecological effects (such as population change) which are themselves difficult to measure because of the variability in monitoring in the field (McCarthy and Shugart, 1990).

Biomarkers have, however, a number of limitations. They may be specific to a species or phyletic group, there may be differences among species, they may show a slow response time or lack of consistent response, seasonal variation, low levels of precision or they may lack ecological relevance. There may also be differences between individuals from the same population, variability that is often ignored. It is difficult to relate biochemical analyses in a specific organ to the LC50 (Forbes and Forbes, 1994).

The biomarker must be part of a continuum of events that occurs between exposure to the contaminant and the resultant disease or irreversible effect. Biomarkers may be indicators of exposure effects (impairment) or adverse effects (disability) depending on the events they relate to (Fox, 1993). Biomarkers of impairment include:

- induction of the stress response;
- induction of detoxification systems;
- inhibition of specific enzymes;
- metabolic impairments that alter processes of synthesis or breakdown, or that deplete energy, vitamin or substrate stores;
- impaired growth;
- genetic damage or impaired repair;
- impairment of immune system;
- impaired or altered reproductive function;
- impaired tissue or organ function.

In using biomarkers it should be possible to detect effects of contaminants at an early stage, i.e. we can ask 'what levels of contaminants produce effects that are *likely* to lead to biological damage?' rather than 'what levels of contaminants *do* lead to biological damage?' (Depledge, 1989).

In Chapter 1 (p. 6) it was described how vitamin A concentrations in otter (*Lutra lutra)* livers were negatively correlated with disease levels and concentrations of PCBs, measured as dioxin equivalents (Murk *et al.*, 1998). The vitamin was being used as a biomarker of adverse effects. The toxic potencies (dioxin equivalents) were themselves determined by a novel bioassay – CALUX or Chemical-Activated LUciferase Expression assay (Garrison *et al.*, 1996). Simply, the luciferase gene from the firefly *Photinus pyralis* is transfected into a rat hepatoma cell line. These cells are then incubated with the test material (e.g. liver or blood) and the luciferase induction by dioxin, which is dose dependent, is measured as light. Very low levels of dioxin induce a response.

Porphyrin can be taken as a further example of the use of biomarkers. Porphyrins are involved in the synthesis of haem and a number of xenobiotics result in excessive production of porphyrins. Concentrations of porphyrins in the livers of herring gulls (*Larus argentatus*) from the Great Lakes have been determined and compared with samples of porphyrins in gull livers from the Atlantic coast of North America (Fox *et al.*, 1988). The highest porphyrin levels were found in samples from Hamilton Bay, Lake Ontario (38 times levels in Atlantic samples) and Green Bay, Lake Michigan (28 times) (see p. 75). It was not possible, however, to determine which compound was responsible for the increased porphyrin levels.

Zebra mussels from the polluted St. Lawrence River in Canada have also been used to assess the value of biomarkers. Five biomarkers were assessed in mussels from 13 sites along the river. Metallothionein-like proteins showed the greatest discriminating power between sites while the enzyme EROD (ethoxyresorufin *ortho*-deethylase) showed the greatest range of variation among sites. There was a positive relationship between metallothionein and the concentration of copper in mussel tissues but no other relationship between biomarkers and ten metals could be demonstrated (de Lafontaine *et al.*, 2000).

The sea is the final repository of many contaminants entering rivers and organisms living in estuaries may be especially vulnerable to their effects. A study of EROD activity has indicated that flounders (*Platichthys flesus*), and probably other species, in industrialized English estuaries are facing a significant threat to their long-term health and viability from anthropogenic contamination (CEFAS, 2000).

To be of use in biological monitoring, biomarkers must reflect ecological effects or forewarn us of likely effects, and ideally they should identify the specific pollutant causing the response, rather than record a general exposure to pollution, which anyway may be within the capabilities of the organism to withstand or adapt to. Detailed discussions of biomarkers can be found in McCarthy and Shugart (1990), Peakall (1992), Fox (1993), Walker *et al.* (1996), Peakall and Fairbrother (1998), Connell *et al.* (1999) and Handy and Depledge (1999).

Neoplasms and abnormalities

A number of chemicals are known from laboratory experiments to be carcinogenic to fish, inducing neoplasms or cancers. These include aflotoxins, azo-compounds, nitroso-compounds, PAHs, PCBs and a number of pesticides. Neoplasms, especially of the skin and liver, are found in fish in the wild, and in some areas incidence may be high. Bottom-living fish are most likely to be affected, especially if they come into contact with contaminated sediments. Proving a link between contaminants and neoplasms is, however, extraordinarily difficult. Factors such as age, life-history, feeding behaviour and interspecific responses act to confound relationships, as do the dose and duration of pollution, which will be unique to a particular geographical location. Any polluted site is likely to have a number of compounds present that could induce neoplasms.

The most comprehensive studies have been made in the coastal environment. For example, in Puget Sound, in the western United States, PAHs, PCBs and DDT have been linked to liver lesions in three species of benthic fish (Myers *et al.*, 1994). In the Great Lakes region, PAHs appear to be responsible for neoplasms in brown bullhead (*Ictalurus nebulosus*) and other fishes (Black and Baumann, 1991). Baumann

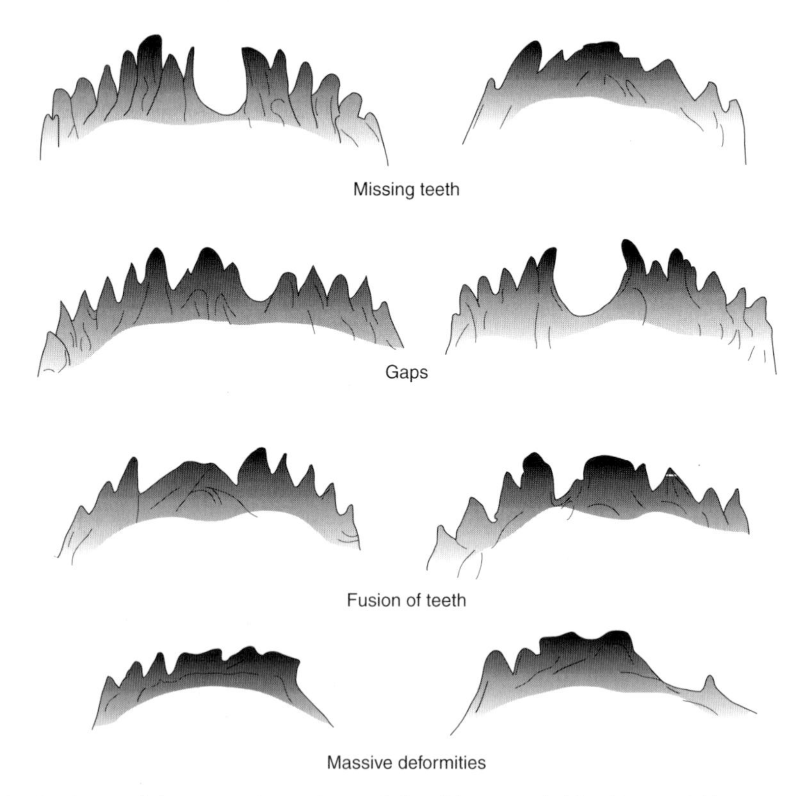

Fig. 2.16 Some of the types of mouthpart deformities recorded in chironomid larvae.

and Harshbarger (1995) report that, with the closure of a coke factory, PAHs in river sediments and brown bullheads declined sharply, while liver cancers in the fish declined by three-quarters. Damage to the immune system of fish is likely to be a major cause of their susceptibility to cancers. Reviews of neoplasms in fish are provided by Metcalfe (1994) and Moore and Myers (1994).

The larvae of chironomid midges live in sediments and are generally tolerant of pollution. Deformities in the antennae and mouthparts of chironomids have been described that have been related to pollution. The types of mouthpart deformities recorded are illustrated in Fig. 2.16, though care is required in distinguishing deformities from normal wear and tear. In the Buffalo River, New York, 29 per cent of all specimens of *Chironomus* exhibited deformities, while individual sites showed up to 67 per cent deformities (Diggins and Stewart, 1993). Some 26 per cent of chironomids collected from a sediment of the Niagara River contaminated with metals and oily wastes were deformed. In the laboratory, 10.3 per cent of chironomids cultured in the contaminated sediment, from which oil had been removed, showed deformities, compared with 2.2 per cent in those reared in clean sediments (Dickman and Rygiel, 1996), demonstrating the importance of metals in causing effects. Copper, lead, xylene and nonylphenol have also caused abnormalities to develop following chronic exposure in the laboratory and, with xylene, in the field (de Bisthoven *et al.*, 1997, 1998; Meregalli *et al.*, 2001).

Tadpoles of the bullfrog (*Rana catesbiana*) collected from a site contaminated with heavy metals, also showed a reduction in teeth, with tooth deformities. When fed on periphyton in the laboratory, the deformed tadpoles had a lower growth rate than normal tadpoles, though they did equally well when particulate food was also supplied (Rowe *et al.*, 1996).

ENDOCRINE DISRUPTORS

In Chapter 1 we saw how contaminants affected the reproduction and health of otters, leading to a rapid population decline. Earlier in this chapter we have examined the effects of toxic chemicals on various measures of the health of organisms, while the concluding pages of the chapter will consider further examples. Many of these effects may be caused by compounds that influence the normal functioning of hormones. Some of the biological impacts of these compounds have been known individually for decades, but they have recently been grouped together under the heading of endocrine disruptors. An endocrine disruptor can be defined as *an exogenous substance that causes adverse health effects in an intact organism, or its progeny, consequent to endocrine function.* These compounds may interfere with embryonic and early postnatal development, and the reproductive, endocrine, immune and nervous systems of animals (Colborn and Clement, 1992). This may lead to early death but often the effects are not seen until adulthood, with a loss of fertility. Most of the research so far has been concerned with the disruption of sex hormones but it is likely that other hormone systems can also be affected.

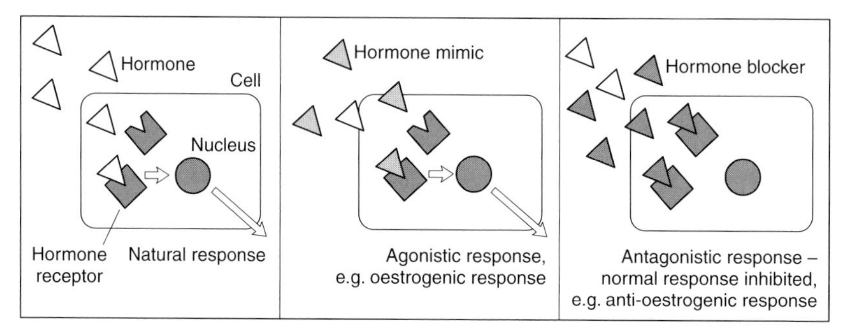

Fig. 2.17 How endocrine disruptors interfere with the normal functioning of the endocrine system (from Environment Agency, 1998).

The basic processes of endocrine disruption are illustrated in Fig. 2.17. A hormone is a regulatory chemical produced by an endocrine gland and released into the bloodstream. Although it is distributed to all cells in the body, only target cells for a particular hormone can respond. The hormone binds to a protein receptor, which is then able to bind to specific areas of the DNA in a cell, known as hormone response elements. Bound to the receptor–hormone complex the response elements are activated to produce messenger RNA, which codes for the production of specific enzymes. These alter the metabolism of the target cell in a specific fashion. A pollutant may act as a hormone mimic, binding to the receptor to produce a false response (an agonistic effect), or it may bind to the receptor and prevent the natural hormone from producing a response (antagonistic effect). So mimics may be described, for example, as oestrogenic or androgenic and blockers as anti-oestrogenic or anti-androgenic, if they influence, respectively, female or male hormones.

Many substances can potentially disrupt the endocrine system. Some occur naturally in plants (phytoestrogens) or are naturally occurring sex hormones, which may become concentrated in the environment via discharges of sewage effluent. Synthetic steroids used as contraceptives may be similarly concentrated. PCBs, dioxins, some insecticides and herbicides, and some heavy metals are endocrine disruptors, as are a group of compounds, or their precursors, used as detergents, plasticizers and resins (e.g. alkylphenols, alkylphenol ethoxylates, phthalates, bi-phenolic compounds). Many more compounds are endocrine disruptors. They may enter the environment in sewage or industrial effluents, or from diffuse sources, while some articles we use every day, such as plastic drinks bottles and sandwich wrappers, may also contain them.

Male rainbow trout were caged at various distances from sewage treatment works in the River Lea north of London. Vitellogenin levels in the fish were elevated immediately below the sewage works and an increase could still be detected up to 15 km below the discharge (Purdom *et al.*, 1994). Vitellogenin is normally produced in the liver of female fish in response to the hormone oestradiol and is incorporated into the yolk of developing eggs, providing food for larval fish before they feed freely. Clearly some chemical in the sewage effluent was behaving like an oestrogen and vitellogenin can be used as a biomarker for oestrogenic effects.

The feminization of wild fish has since been shown to be widespread (Jobling *et al.*, 1998). Roach were collected from sites on rivers both upstream and downstream of sewage treatment works. Fish were also collected from control sites that received no sewage effluent, though there remained the possibility that they could have been contaminated from diffuse sources, such as road or agricultural run-off, as well as from the air. The results are summarized in Fig. 2.18. A surprisingly large number of male roach showed intersexuality, with gonads showing both male and female characteristics. Intersex was recorded at all sites, including the controls, though it was significantly higher below sewage works. At two of these sites all the male roach examined showed intersexuality. Vitellogenin concentrations in male fish downstream of sewage works was significantly higher than in fish from upstream sites, indicating that the fish had been exposed to oestrogens. The gonads of males with high levels of vitellogenin were generally smaller. Overall there was a direct relationship between the degree of intersex in the roach population of the river and the concentration of sewage in the river, as measured by the population served by the sewage treatment works. Taken together these facts provide compelling evidence that some chemical (or chemicals) in sewage effluent is influencing the sexuality of male fish.

The oestrogenic components of sewage effluent have been shown to vary some ten-fold over time (Rodgers-Gray *et al.*, 2000). A significant induction of vitellogenin in male roach occurred after one month's exposure to a sewage effluent concentration of 38 per cent, while a concentration of 9 per cent induced vitellogenin production after four months' exposure. The vitellogenic response was thus both time and dose dependent. Studies with fathead minnow have confirmed that low concentrations of oestrogenic compounds have profound effects on the

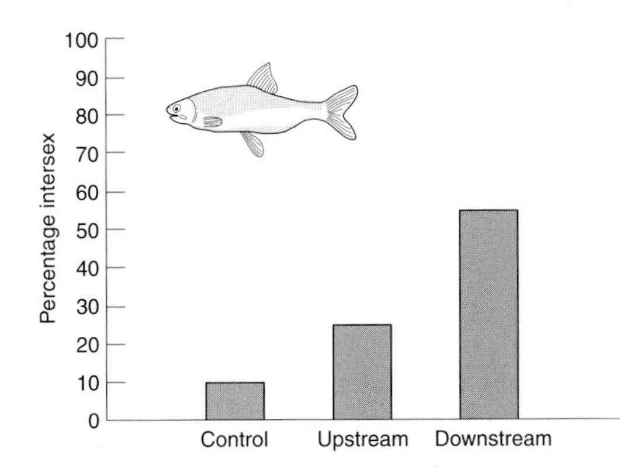

Fig. 2.18 The incidence of intersexuality in male roach collected from sites in rivers above and below discharges of sewage effluent, and from control sites not receiving effluent (adapted from Jobling *et al.*, 1998).

reproductive development of male fish (Panter *et al.*, 1998). Kime (1995, 1999) reviews the effects of pollutants on the reproduction of fish and methods for their study.

Oestrogenic compounds may not only affect the ability of male fish to produce sperm but they may also influence their ability to attract females in sexual displays. Low levels of oestrogenic compounds caused a dramatic decrease in the rate and intensity of sexual displays in male guppies (Bayley *et al.*, 1999). The sexual display of guppies, a species easy to keep in the laboratory, could be used as a biomarker for oestrogenic effects.

Oestrogenic effects are not confined to fish. Developmental abnormalities in embryos and hatchlings of snapping turtles (*Chelydra serpentina*) from the Great Lakes area of North America were linked especially to concentrations of dibenzodioxins and dibenzofurans (Bishop *et al.*, 1998). In Lake Apopka, Florida, abnormalities in hatchling and juvenile alligators (*Alligator mississippiensis*) have included reduced hatching success, modifications to the structure of gonads, altered steroid hormone concentrations in the blood of both males and females, and a reduction in penis size. The evidence suggests that the observed effects are caused by exposure to endocrine disruptors during embryo development rather than those chemicals measured in the alligators themselves (Guillette *et al.*, 1999). A range of endocrine disruptors, including pesticides and PCBs, have been measured in the blood of alligators but it is not known which compound or compounds are having the adverse influence. Endocrine disruptive effects on birds and mammals will be described below (p. 69). Sumpter (1998) and Tyler *et al.* (1998) provide critical reviews of the environmental impact of endocrine disruptors.

There is evidence of falling sperm counts (a 50 per cent drop over the last 50 years) and of increases in testicular cancer in men (Sharpe and Skakkebaek, 1993). These may be related to oestrogen mimics released into the environment though the data have been disputed by other workers. In support of the hypothesis, female laboratory rats, fed nonylphenols at concentrations similar to those experienced by humans, produced male offspring that, at maturity, had significantly smaller testes and lower sperm counts (Sharpe *et al.*, 1998).

The problem with managing the effects of endocrine disruptors is that we know so little about which chemicals cause what effects. The majority have not been routinely monitored for in the environment and they are used in a whole range of products and enter the aquatic environment from a diversity of sources. How widespread their impact is on the environment is unknown (Matthiessen, 2000). Most research has concentrated on known hotspots. A great deal more research is urgently required before quality standards can be sensibly applied to limit the discharge of the most damaging chemicals.

HEAVY METALS

Heavy metals are conservative pollutants in that they are not broken down, so that they effectively become permanent additions to the aquatic environment. They accumulate in organisms and some may biomagnify in food chains (p. 39), though biomagnification is the exception rather than the rule (Goodyear and McNeill, 1999; Mason *et al.*, 2000). The major uptake route for many aquatic organisms is directly from the water so that, to a certain extent, tissue concentrations reflect concentrations in water. Carnivores at the top of the food chain, however, such as many aquatic bird and mammal species, including humans, obtain most of their pollutant burden from aquatic ecosystems by ingestion, especially of fish, so there exists the potential for considerable biomagnification.

Some heavy metals are harmful to health at levels recorded in the environment. They may suppress the immune system, leading to increased susceptibility to disease, while some may be carcinogenic (Peakall, 1992). The heavy metals of most widespread concern to human health are mercury, cadmium and lead. It has been suggested that over one billion (10^9) human beings are exposed to elevated concentrations of toxic metals and metalloids in the environment and several million people may be suffering from subclinical metal poisoning (Nriagu, 1988).

Arsenic has recently been found to be a major health problem in Bangladesh and neighbouring parts of India, because the tubewells from which some 97 per cent of the population obtain their drinking water are contaminated. The wells, of which there are now more than eight million, were constructed by aid agencies to provide a drinking supply free from the risk of disease, which was killing hundreds of thousands of children each year. The precise cause of this unexpected contamination is as yet unknown but it may be due to the oxidation of arsenic-rich pyrite in the aquifer which has been exposed by over-exploitation of groundwater for irrigation. Some 28 per cent of samples from a total of 30 000 wells exceeded the public health standard of $0.05\,mg\,l^{-1}$. The vast majority of samples of hair, nail, skin and urine contained very high levels of arsenic. As many as 85 million people are at risk from arsenic and it is estimated that one in ten people who drink the water will ultimately die of some form of cancer, especially of the skin, lung or bladder. Serious poisoning has so far been reported from more than 7000 people, with symptoms such as conjunctivitis, skin lesions and melanomas. It is likely to develop into the biggest poisoning calamity the world has known (Karim, 2000).

Tin, in its organic form of tributyl tin, has been widely used as an anti-fouling agent on the hulls of boats. It is extremely toxic to life and has caused heavy mortality to aquatic life in coastal waters, especially in the neighbourhood of marinas. It causes imposex in shellfish, that is, the expression of both male and female characters, i.e. it is an endocrine disruptor. It has had a severe impact on some shellfisheries. However, there is little evidence that freshwater life has been affected by tin. Reviews of the environmental impact of organotin can be found in Champ and Seligman (1996).

Mercury

Environmental pollution by heavy metals became widely recognized with the Minamata disaster in Japan in the early 1950s. In 1932 a factory producing acetaldehyde and vinyl chloride began using mercuric oxide as a catalyst and effluents discharged into Minamata Bay contained mercury. Seafood was a major part of the diet of the local population. In the early 1950s a nervous disease began to affect dogs, cats and pigs and many died. In 1956 there was the first recorded human case, a young girl suffering from speech disturbances, delirium and difficulties in walking. A number of similar cases were detected over the ensuing months and an investigation team was set up, which established in 1956 that the cause of the illness was the consumption of seafood from Minamata Bay. In 1958, when the number of victims had exceeded 50, 21 of whom had died, a ban was placed on the sale of fish from Minamata Bay, though there was no restriction on catching fish. No controls were placed on the factory discharge. Concentrations of mercury in mud in the drainage canal from the factory were as high as 2000 mg kg^{-1} wet weight and in sediments in Minamata Bay of the order of 10–100 mg kg^{-1} wet weight. Mercury levels in fish and shellfish ranged from 5 to 40 mg kg^{-1} wet weight. It appears that by the end of 1959 the company itself had good reason to believe that its wastewater was the cause of Minamata Disease, but that information was withheld. The discharge of mercury was stopped only in 1968, when the production unit became uneconomic. By 1988 there were 2209 confirmed cases of Minamata Disease and 730 people had died. There is still serious mercury pollution in Minamata Bay but no controls on fishing have been enacted by the government. A detailed summary is provided by Laws (1993).

There are two natural sources of mercury in the environment. The weathering of mercury-bearing rocks releases about 3500 t yr^{-1} and 25 000–150 000 t yr^{-1} are released as gases from volcanic areas. The burning of fossil fuels releases a further 3000 t yr^{-1}. The world production of mercury is around 10 000 t yr^{-1} and it has a wide range of uses. Much goes into electrical apparatus and into the chlor-alkali industry, which produces chlorine and sodium hydroxide electrolytically, using mercury as a cathode. This industry is responsible for much contamination of lakes and rivers. The paper and pulp industry previously used mercury as a slimicide,

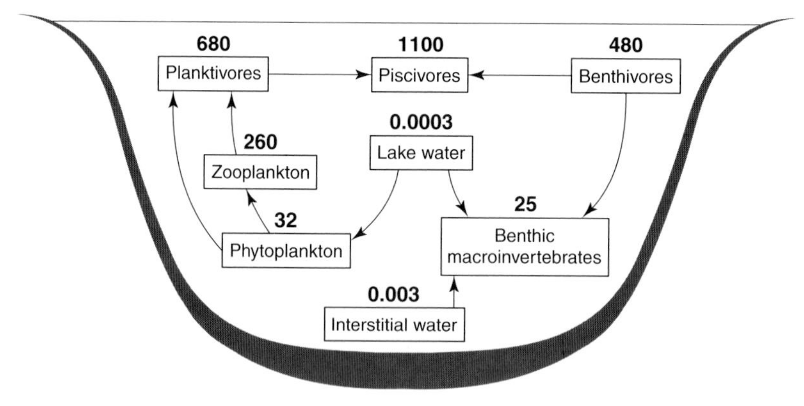

Fig. 2.19 Mercury concentrations (µg kg^{-1} wet weight) in the Lake Onondaga food web (from Becker and Bigham, 1995).

again resulting in much environmental contamination, but this has now largely ceased. Paints, pharmaceuticals, dental applications and precision instruments also use mercury. The use of mercury as a fungicide in agriculture and horticulture may also lead to contamination of the environment.

In the aquatic environment, inorganic mercury (Hg^{2+}) is converted by micro-organisms to the highly toxic methyl mercury ($Hg(CH_3)_2$), which is more readily taken up by tissues and dissolves in fat. Some 95 per cent of methyl mercury is absorbed by the gut and most of it is retained in the body, less than 1 per cent being excreted. Over 90 per cent of the body burden in fish is in the form of methyl mercury. Mercury biomagnifies along the food chain, an example from Lake Onondaga, New York, which received mercury inputs in the past, being illustrated in Fig. 2.19. There is a marked increase in concentration from phytoplankton and invertebrates to fish, with carnivorous species, such as walleye (*Stizistedion vitreum*) having greater loads than planktivorous or benthic feeding fish. The bioaccumulation factor increased with higher trophic level in both pelagic and benthic components of the food web, ranging from 8.3×10^4 for benthic macroinvertebrates to 3.7×10^6 for piscivorous fish.

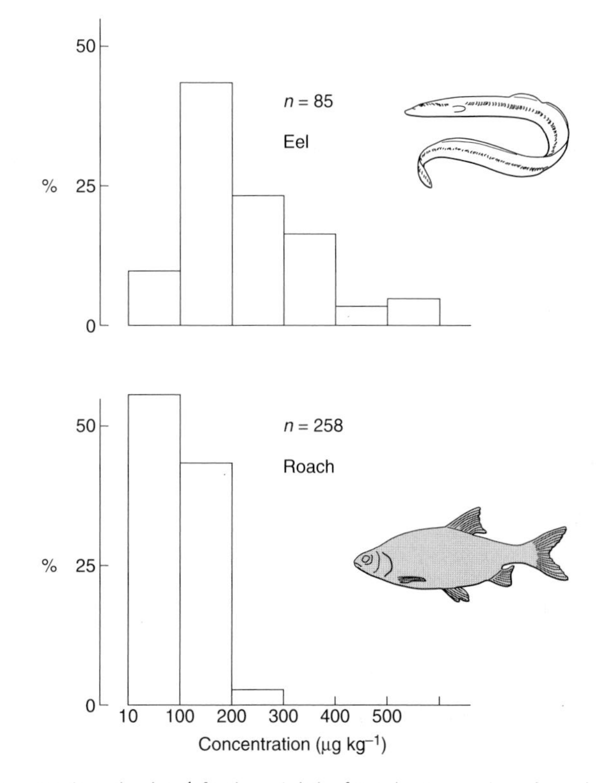

Fig. 2.20 Concentrations (μg kg^{-1} fresh weight) of total mercury in eels and roach from freshwaters in Great Britain (from data in Mason, 1987).

There have been many measurements of the concentrations of mercury in fish and bioconcentration factors from water in excess of 10^5 for total mercury and 10^6 for methyl mercury have been recorded (Zillioux *et al.*, 1993). Different studies are often difficult to compare, however, because factors such as age, size and condition of the fish may have a great influence on the amount of mercury which has accumulated. Mercury, for example, is directly correlated with fish length (which is closely related to age) in both eels (*Anguilla anguilla*) and roach, although eels, which spend much of their lives in intimate contact with sediments, generally have higher concentrations (Barak and Mason, 1990a). A comparison of mercury concentrations in eel and roach is shown in Fig. 2.20. Apart from eels, concentrations of mercury in fish species are related more to size than trophic position and open water species generally have lower amounts than eels (Barak and Mason, 1990b), emphasizing the importance of sediments as a source of contamination. After adjusting for length, however, concentrations of mercury in pike may be considerably greater than in other species living in open water (e.g. Wren and MacCrimmon, 1986). The mercury concentration in fish is inversely related to pH or alkalinity of lake water (Spry and Wiener, 1991). This is due, at least in part, to the microbial production of methyl mercury, itself inversely correlated with pH at the sediment–water interface.

Fish can take up heavy metals either in their diet or through their gills, though experimental studies have shown that the former is the predominant route, at around 60 per cent for pike and yellow perch (*Perca flavescens*) (Dallinger *et al.*, 1987). Although mercury is toxic to fish and accumulates to significant concentrations in wild populations, there is little evidence that fish themselves are disadvantaged by their burdens. However, a recent study has shown in the walleye that both hatching success and larval heart rate decline with increasing methyl mercury

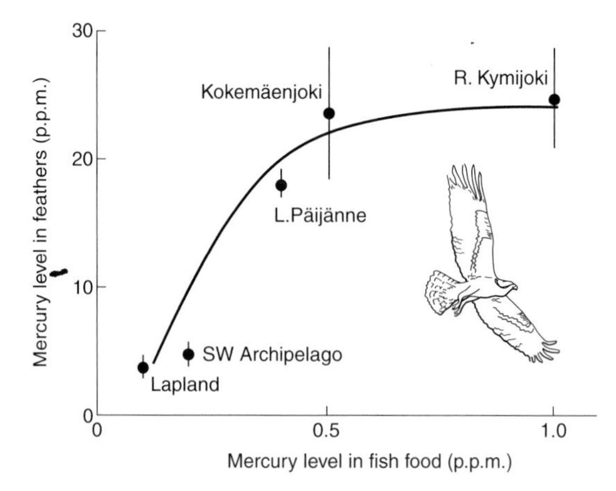

Fig. 2.21 Accumulation of mercury from fish eaten by osprey nestlings into feathers in five areas of Finland (adapted from Häkkinen and Häsänen, 1980).

concentrations in the water (Latif *et al.*, 2001) Piscivorous vertebrates, feeding on fish from contaminated areas, are certainly at risk. For example great northern divers (common loons, *Gavia immer*) laid fewer eggs and more readily deserted nests when mercury in their fish prey averaged 0.3–0.4 mg kg^{-1} wet weight. If prey concentrations exceeded 0.4 mg kg^{-1} territory occupancy was severely reduced and almost no eggs were laid (Barr, 1986).

The osprey (*Pandion haliaetus*) is a fish-eating bird of prey, taking its food by plunging, talons first, into the water. Figure 2.21 compares the concentrations of mercury in the feathers of nestling ospreys with mercury in their fish prey. Low concentrations of mercury in fish from two uncontaminated areas were reflected in low concentrations in feathers. At the three contaminated sites higher concentrations of mercury in feathers were found, but not in proportion to the increased amounts in the fish prey. There is therefore a limit to the amount of mercury that can be acquired by feathers, though transferring the metal to the plumage can be seen as a method of removing it from the body.

Ospreys declined in Finland and Sweden during the twentieth century, but there was no difference in breeding success between contaminated and uncontaminated areas, and a mercury ban in Sweden, which was followed by a decrease in the contamination of nestlings, did not lead to an increase in breeding success. This indicates that mercury was not responsible for the decline and other pollutants, such as organochlorines, may have been more significant.

Mercury is often recorded in aquatic birds at high levels and is cited as a possible cause for their decline (e.g. Spalding *et al.*, 1994). Birds such as the endangered wood stork (*Mycteria americana*) carry levels that, in theory, should cause sublethal effects (Gariboldi *et al.*, 1998). However, there is no evidence of increased mortality with increased body burdens and field dosing experiments on chicks of great egrets (*Ardea albus*) failed to increase mortality over their first eight months of life (Sepúlveda *et al.*, 1999). Other contaminants, such as organochlorines, are almost certainly present in these populations and are more likely to exert population effects (see below).

Even low concentrations of mercury may be considered potentially harmful to humans. Methyl mercury is believed to inhibit enzyme activity in the cerebellum, which is responsible for neuron growth in early developmental stages. Chronic exposure to mercury leads to mental retardation. Clinically observable effects on adults occur at blood levels of 0.2–0.5 $\mu g\, ml^{-1}$ and body concentrations of 0.5–0.8 mg kg^{-1}, or at concentrations in hair as low as 15–20 mg kg^{-1}. The toxic effects of methyl mercury are dominated by neurological disturbances. A tolerable weekly intake for mercury is set at 0.005 mg kg^{-1} body weight but people eating fish from contaminated lakes are liable to exceed this level significantly. Increased concentrations of mercury were recorded in hair from people eating fish from some Finnish lakes, with the highest concentrations (30 mg kg^{-1}) in middle-aged people who ate substantial amounts of fish (Lodenius *et al.*, 1983).

The European Union has suggested that a maximum average mercury content for fish taken for consumption should be 300 $\mu g\, kg^{-1}$. A reference back to Fig. 2.20

will show that 25 per cent of eels sampled in Britain exceeded this level. Nriagu (1988) suggested that, worldwide, some 40 000–80 000 individuals may be suffering from mercury intoxication.

Zillioux *et al.* (1993), Wren *et al.* (1995) and Boening (2000) provide reviews of the impact of mercury on freshwater ecosystems.

Cadmium

Japan was also host to the first reported case of environmental cadmium poisoning. In 1955 doctors recorded a disease, which they later called *itai-itai* (ouch-ouch), characterized by severe back and joint pains, a duck-like gait, kidney lesions, protein and sugar in the urine and a decalcification of the bones, leading sometimes to multiple fractures. It was especially prevalent in women over 40. A mining company was releasing cadmium-laden effluent into the River Jintsu, which was used to irrigate paddy fields downstream. Victims were confined to this region. There has been some controversy as to whether other factors (such as zinc or dietary deficiency) were involved, but a rapid decline in the incidence of itai-itai occurred when effluent controls were introduced. Some 200 people were afflicted, half of whom died.

World production of cadmium is about 21 000 $t\,yr^{-1}$, of which the main uses are in electroplating, as pigments and as stabilizers for plastics. Mine drainage, sewage sludge applied to land and phosphate fertilizers are also significant sources.

Cadmium is highly toxic to some forms of life. The 48 hr LC50 values for invertebrates range from 7.0 to 34 000 $\mu g\,l^{-1}$, with cladocerans being especially sensitive. Sublethal effects on invertebrates have been recorded at concentrations in the range 0.2–3.0 $\mu g\,l^{-1}$ in laboratory and field studies (Wren *et al.*, 1995). In a study of the cadmium concentrations in the benthic amphipod *Hyalella azteca*, 81 per cent of the variability could be accounted for by three parameters of lake water chemistry: concentration of calcium ions, total cadmium in water and dissolved organic carbon (Stephenson and Mackie, 1988). Experiments with the snail *Lymnaea stagnalis* showed a delay in egg development and hatching at cadmium concentrations of 25–100 $\mu g\,l^{-1}$, while increasing levels above this halted development at various stages of embryogenesis, a concentration of 400 $\mu g\,l^{-1}$ preventing the first cleavage of the egg (Gomot, 1998). The worm *Tubifex tubifex* can accumulate large amounts of cadmium, and because these worms occur in dense populations, they could potentially pose a threat to predators such as fish (Bouché *et al.*, 2000).

Sublethal effects of cadmium toxicity have been reported from wild populations of fish. Perch from a Swedish river contaminated with cadmium had a markedly enhanced lymphocyte count, slight anaemia and changes in the concentrations of potassium and magnesium in the blood (Larsson *et al.*, 1985). Residues of cadmium in the livers of exposed perch were six to eight times those of controls. Cadmium levels in fish from the polluted River Lot in France reflected concentrations in the water and were influenced by feeding habits (Andres *et al.*, 2000). However, although cadmium bioaccumulates in tissues, it does not appear to biomagnify along food chains (Spry and Wiener, 1991). Concentrations in fish flesh are generally less than 0.5 mg kg^{-1} wet weight, though levels are higher in liver and kidney.

Chicks of little blue herons (*Egretta caerulea*) exposed to cadmium in Louisiana wetlands showed slower growth rates than non-exposed birds (Spahn and Sherry, 1999). More than 80 per cent of the body burden of cadmium in vertebrates is localized in the liver and kidney, where it binds to low molecular weight metallothionein. Cadmium is the only metal that clearly accumulates with age and, as cadmium concentrations in the kidney approach 200 $\mu g\, g^{-1}$, the ability of the metallothionein to protect cells begins to diminish. This critical threshold in humans can be achieved on a rather low daily intake over an extended time and results in proteins being excreted into the urine, and in kidney damage. Nriagu (1988) estimated that more than 500 000 people may be at risk from cadmium-induced kidney damage worldwide, though freshwaters are only one of a number of routes of exposure, a major source being smoking. The measurement of cadmium in urine from a sample of more than 22 000 people in the United States revealed that some 2.3 per cent of the population had concentrations higher than 2 $\mu g\, g^{-1}$ and 0.2 per cent had concentrations greater than 5 $\mu g\, g^{-1}$, the current health-based exposure limit adopted by the World Health Organization (Paschal *et al.*, 2000). The amount of cadmium ingested by humans is very close to the tolerable weekly intake, so that releases of cadmium to the environment need to be more tightly controlled.

Lead

Some 4 million tonnes of lead are mined annually worldwide. The metal has a wide range of uses in pipes, battery cases, paints etc. and tetraethyl lead is used as a petrol additive. The latter has been the most important source of organic lead to the environment and has been widely dispersed, with contamination increasing sharply from the 1950s, even in remote areas, with the growth in the number of automobiles.

Lead concentrates in organisms, bioconcentration factors in mosses, for example, being in the order of 3000–5000. There appears, however, to be no biomagnification along food chains (Spry and Wiener, 1991). A detailed study of the distribution of lead in macrophytes from Shoal Lake, Manitoba, has shown that different species vary considerably in their accumulation of the metal. The pondweeds *Potamogeton gramineus* and *P. richardsonii* contained concentrations of 45 mg kg^{-1} dry mass in shoot tissue, whereas four other species of *Potamogeton* contained less than 20 mg kg^{-1}, other genera of macrophytes having lower concentrations still. Lead concentrations varied with depth but not season (Reimer, 1989). Invertebrates from lowland rivers, with generally low levels of contamination, had bioconcentration factors of between 32 and 360, accumulation being unrelated to trophic position (Barak and Mason, 1989).

The state of Missouri has been a major producer of lead since the early 1800s and the residues (tailings) are rich in heavy metals, resulting in considerable contamination of freshwater ecosystems. In 1977 the dam of a tailings pond burst,

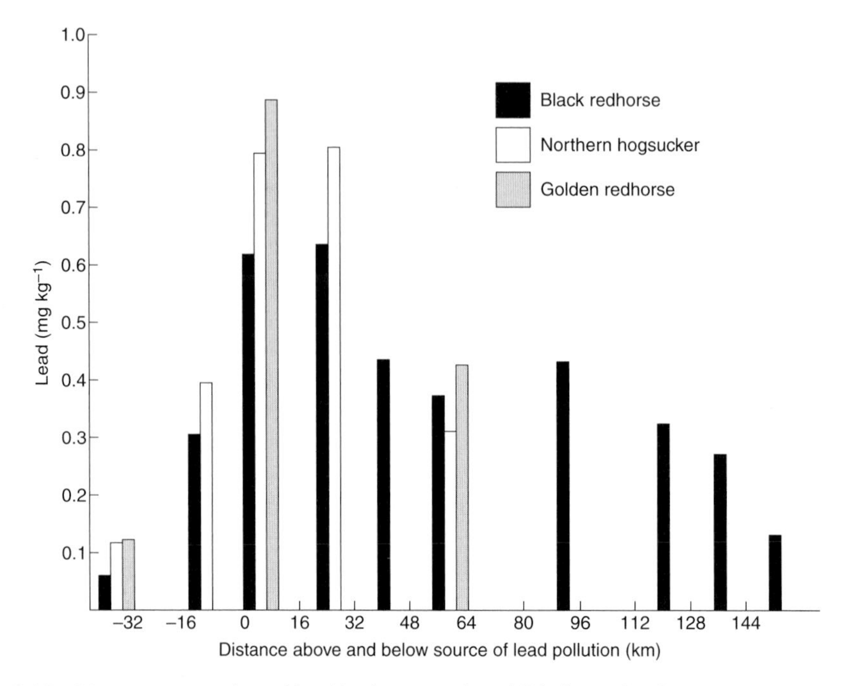

Fig. 2.22 Mean concentration of lead in three species of fish from Big River, Missouri (adapted from Czarnezki, 1985).

releasing contaminated material into Big River. Fish were sampled from a number of localities upstream and downstream of the source of pollution for a period of 18 months from December 1979 to see if lead had accumulated in tissues. The results for three species, black redhorse (*Moxostoma duquesnei*), golden redhorse (*M. erythrurum*) and northern hogsucker (*Hypentelium nigricans*), are summarized in Fig. 2.22. Lead concentrations in fish flesh increased sharply at the source of pollution and remained above the recommended level for human consumption (0.3 mg kg^{-1}) for 125 km downstream. There was no correlation between lead concentrations in flesh and the length of fish in this study, as has also been found with other species, such as eel and roach (Barak and Mason, 1990a). Lead concentrations in the calcified otoliths and opercula of Arctic char (*Salvelinus alpina*), however, have been shown to increase with age (Köck *et al.*, 1996).

Waterfowl are the group most affected by the chronic contamination of aquatic ecosystems with lead. Those wildfowl that manage to avoid being killed or maimed by hunters may nevertheless ingest large quantities of spent shot while feeding. In the United States, where the problem has been most investigated, some 6000 t of spent shot are deposited each year and a minimum of 2.4 million waterfowl, out of a total North American population of 100 million, die annually of lead poisoning due to ingestion of shot, though the introduction of steel shot is reducing this mortality. Many other waterfowl suffer sublethal effects, making them vulnerable to

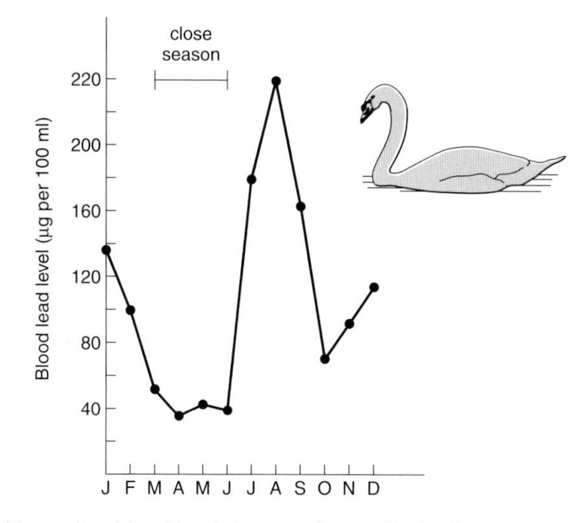

Fig. 2.23 Monthly median blood levels in 1981 from a flock of mute swans on the River Thames (adapted from Birkhead and Perrins, 1986).

disease and predation. The adverse physiological effects include anaemia, muscular paralysis and damage to the nervous system, liver and kidney. Chicks of little blue herons exposed to lead in their food showed a reduced survival rate (Spahn and Sherry, 1999).

The mute swan (*Cygnus olor*) is a familiar bird in Britain and many non-breeding flocks frequent urban stretches of river, where they are fed by the public. Lowland rivers are also used extensively by recreational anglers, who attach lead weights and shot to nylon lines to hold the bait beneath the water. Much of this is lost or discarded and swans ingest it while collecting food from the river bottom. The ingested lead is quickly ground down by the grit in the gizzard and is absorbed into the bloodstream. Of more than 1500 swans post-mortemed over the period 1981–84, 60 per cent had died of lead poisoning.

In the 1970s, up to 90 per cent of swans on some rivers were killed by lead poisoning and the annual death rate in England alone was estimated at 3370–4190 swans (Birkhead and Perrins, 1986). Lead levels in the blood of living swans on the River Thames are at their lowest during the close season for fishing and increase rapidly in June when angling recommences (Fig. 2.23). A blood lead level of 40 μg/100 ml indicates lead poisoning. The swans on the River Thames were below this level only during the angling close season. Swans suffering lead poisoning develop problems with the neuromuscular system. In particular they are unable to maintain the upright posture of their necks (they develop 'kinky necks') and cannot propel food through the gullet, eventually dying of starvation (Fig. 2.24).

Fig. 2.24 Young mute swan with a kinky neck caused by lead poisoning (photograph by Chris Perrins).

Many swans also die through collisions with power lines (Brown *et al.*, 1992). In Ireland such birds have been shown to have elevated concentrations of lead in their tissues, though lower than those dying of acute lead poisoning (O'Halloran *et al.*, 1989). Sublethal levels of lead may affect the coordination of swans, increasing their chances of collisions during flight. This is likely to occur, however, only in the very early stages of lead poisoning for swans quickly lose condition, with wasting of the muscles, and are unable to take flight (Perrins and Sears, 1991).

A voluntary ban on the use of lead for angling was introduced in 1985 and the ban became law in 1987. In the early 1980s 84–89 per cent of swans on parts on the River Thames had abnormally high levels of blood lead. This percentage had fallen to 44 per cent by 1987 and 24 per cent by 1988 (Sears, 1989). The legislation banning lead weights has had a dramatic effect in reducing the incidence of lead poisoning in swans and the population has since steadily increased. Nevertheless lead still persists in these rivers and, in 2000, some swans on the River Thames still had elevated lead levels (C.M. Perrins, personal communication).

Humans have used lead for many purposes since antiquity and an increase in lead in the Greenland ice cap reveals evidence of extensive smelting activity between 500 BC and AD 300: lead ore was smelted in open furnaces to extract silver. Its widespread use in Roman times has led to the suggestion that the fall of the Roman Empire was the result of lead poisoning, especially as wine was stored in lead containers.

Variable amounts of lead are found in food and a major aquatic source is lead pipes used to provide water to homes; lead is especially soluble in soft and slightly acid waters. Air pollution, particularly from vehicles, is a further major source. While the introduction of lead-free petrol has reduced pollution from exhausts in developed countries, leaded petrol remains the norm in many economically developing countries. The adult gut absorbs 10–20 per cent of ingested lead, though children have a greater efficiency of uptake, and this provides the principal route into the body. Although less lead enters the body via the lungs, the percentage absorption is greater, around 40 per cent (Moore, 1986).

Exposure to relatively low levels of lead has been associated with metabolic and neuropsychological disorders, which include anaemia and lowered IQ. Children are especially at risk. The threshold for possible medical intervention in the United States has been set at a concentration of lead in blood of 250–300 μg/100 ml; it has been estimated that 590 000 young children in the country exceed that level and many may be suffering from lead poisoning. Worldwide as many as 130–200 million people may exceed this threshold and be at risk (Nriagu, 1988), including up to 90 per cent of children in many African cities (Pearce, 1996). Recent work, however, has suggested that levels of lead in blood lower than 25 μg/100 ml can affect children (Joyce, 1990). Furthermore, behavioural impairment in children, associated with raised lead levels, has been shown to be permanent (Needleman *et al.*, 1990), affecting the lifetime success of individuals. A detailed review of lead and health is provided by Smith *et al.* (1989).

Tap water is a significant source of lead and in the United States it has been suggested that a reduction in the allowable concentration of lead in potable water from $50\,\mu g\,l^{-1}$ to $20\,\mu g\,l^{-1}$ would benefit 42 million people (Levin, 1987). The European Union currently also has a limit of $50\,\mu g\,l^{-1}$ for potable water, though the World Health Organization has recommended a health based guideline of 10 $\mu g\,l^{-1}$. Nriagu (1988) considers that lead poisoning must be regarded as the most prevalent public health problem in many parts of the world.

ORGANOCHLORINES

Organochlorine insecticides and polychlorinated biphenyls (PCBs) are hydrophobic, fat soluble and biologically stable so that they accumulate in body fats. They also biomagnify along food chains and concentration factors from water to top predators may be as high as ten million times. Organochlorines have led to widespread declines in some of these top carnivores, the decline of the otter having been already discussed (p. 6). Dispersal by air and water leads to contamination of regions remote from sources of production or use (e.g. Lockhart *et al.*, 1992; Pearce, 1997a). A vivid description of the global movement and biomagnification of PCBs is found in Chapter 6 of *Our Stolen Future* (Colborn *et al.*, 1996). Many of these compounds are potent endocrine disruptors. For example, DDE, a metabolite of DDT and widely distributed in the environment, inhibits androgen binding to the androgen receptor (i.e. it is anti-androgenic), causing abnormalities in male sexual development (Kelce *et al.*, 1995). These persistent chemicals eventually end up in the oceans, where they biomagnify in food chains, and where the full impact on marine life is not expected to peak for several decades, despite controls on their use (Motluk, 1995).

Because of their persistence, restrictions have been placed on the use of organochlorines and PCBs in the developed world over the past three decades, but concentrations from historical and current uses are still sufficiently high to pose problems to sensitive species. For example, DDT, dieldrin and lindane are still widely used in control programmes for disease vectors such as mosquitoes, the majority being applied to freshwater ecosystems. Some of the replacement pesticides, while not persistent, can be very damaging to the aquatic environment. Synthetic pyrethroids have largely replaced organophosphorus (OP) pesticides (which themselves replaced dieldrin) for dipping sheep, because of considerable health concerns over OPs. However, pyrethroids are some one hundred times more toxic to aquatic invertebrates than OPs and run-off from sheep dips may kill all the invertebrates in long stretches of river, removing the food supply for fishes (Pearce, 1997b). Even the drips from dipped sheep grazing by river banks can provide sufficient pesticide to kill aquatic invertebrates.

Pesticides

As well as being applied directly to watercourses, as described above, pesticides enter aquatic ecosystems in run-off from farmland. Sewage and industrial effluents are a further source of pesticides, while atmospheric transport, followed by precipitation in rainfall, is a major route of entry. Aerosols produced during crop-spraying can be dispersed over vast distances by winds.

The toxicity of organochlorines to fish can be quite high. For example, the 96 hr LC50 of DDT to various species ranges from 1 to 30 $\mu g\, l^{-1}$; values for invertebrates are similar. At sublethal concentrations, organochlorine pesticides result in impaired learning behaviour, slowed reflexes and a reduction in reproductive success.

The earliest example of biomagnification came from Clear Lake, California. Between 1949 and 1957 the lake was repeatedly sprayed with DDD, an insecticide related to DDT, to control populations of *Chaoborus astictopus*, a non-biting phantom midge whose swarms irritated bathers and anglers. An application of DDD to the lake water of 14 $\mu g\, l^{-1}$ in 1949 resulted in a 99 per cent kill of midge larvae, but they had recovered by 1951. A second treatment in 1954 was at a higher concentration of 20 $\mu g\, l^{-1}$, as was a third in 1957. Following the 1954 application 100 western grebes (*Aechmorphus occidentalis*) were found dead in the lake. The transfer of DDD from water to this top predator is illustrated in Fig. 2.25. The amount of DDD in the fat of the birds averaged 1600 $mg\, kg^{-1}$, a bioconcentration factor of 80 000. The population of grebes fell from 3000 to only 30 pairs by the end of the 1950s, and most of these were sterile. An organophosphorus insecticide, which does not biomagnify and is non-persistent, was substituted for DDD and the grebe

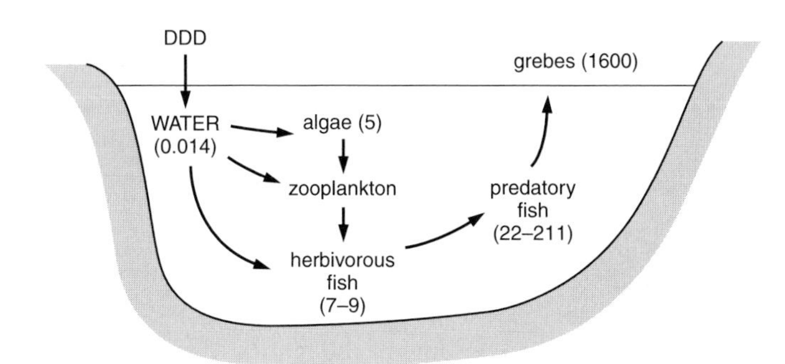

Fig. 2.25 Transfer of an organochlorine pesticide, DDD, through the food chain of Clear Lake. Numbers are concentrations in mg kg^{-1} (from data in Hunt and Bischoff, 1960).

population gradually recovered during the 1960s. This example was one of the cornerstones of evidence of the environmental damage wrought by persistent chemicals in the book *Silent Spring* (Carson, 1962), *the* seminal work that led to the formation of the environmental movement.

Organochlorine pesticides are widely dispersed in the environment and have had a significant impact on populations of aquatic wildlife, especially of birds. DDT affects the calcium metabolism of birds, resulting in the thinning of egg shells, which then break more easily. In studies of the night heron (*Nycticorax nycticorax*), concentrations of DDE exceeding 8 mg kg^{-1} in eggs were associated with decreased clutch size, increased incidence of cracked eggs and a lower number of fledged young. In eight colonies in the western United States, from 10 per cent to 59 per cent of nests had some eggs with DDE concentrations in excess of 8 mg kg^{-1} and in one colony, at Ruby Lake, the mean concentration averaged 8 mg kg^{-1}. Productivity in this colony was below the level necessary for population maintenance (Fleming *et al.*, 1983).

Eggs of red-necked grebes (*Podiceps grisegena*) from Manitoba, Canada, contained, on average, three times as much DDE as eggs from Saskatchewan and seven times as much as eggs from Alberta (PCBs were 7 times higher and 16 times higher as those respectively from Saskatchewan and Alberta). Thickness in egg shells from Manitoba declined significantly in the main period of DDE usage (1948–71) compared with earlier years and egg shells were still thinner in 1972–82. Declines in thickness in the other two areas were less and recovery was complete. The study sites were all remote from sources of industrial pollution and it was considered that contaminants were accumulated on the wintering grounds. Manitoba grebes winter on the Atlantic coast and in the Great Lakes, and Saskatchewan birds probably do. Alberta grebes winter on the Pacific coast (Forsyth *et al.*, 1994). This shows that organochlorines can be widely dispersed by living organisms as well as by wind and water. In Europe internationally important wetlands for wildlife conservation, such as the Coto Doñana in Spain, the Danube delta and Lake Mikri Prespa in Greece are contaminated with organochlorine pesticides at levels of considerable concern.

A reduction in organochlorine usage has led to a reduced contamination of the aquatic environment by some compounds. In spottail shiners in the Great Lakes of North America DDT and PCB residues have declined steadily at most sites over the period 1975–94 (Suns *et al.*, 1993; Scheider *et al.*, 1998). Similar declines in pesticides have been recorded in fish in Europe (e.g. Atuma *et al.*, 1996; Brevik *et al.*, 1996).

The human population, of course, is also widely contaminated with organochlorine pesticides. In samples of breast milk from Hong Kong, DDT and DDE concentrations had the ranges 0.67–4.04 and 4.07–22.96 $mg\,kg^{-1}$ fat mass respectively (Ip and Phillips, 1989). These are some of the highest values reported in the literature and reflect the importance of seafood in the diet of ethnic Chinese. In an earlier summary of data for 20 countries, mean concentrations of DDT in breast milk ($mg\,kg^{-1}$ fat) ranged from 1.7 (Australia) to 19.5 (India), with means from seven countries exceeding 10 $mg\,kg^{-1}$ (Schüpbach, 1981). The maximum allowable concentration of DDT in human foodstuffs is 0.74 $mg\,kg^{-1}$, so even the lowest mean in breast milk was more than twice this standard. It may be that, in cases where organochlorine concentrations in breast milk are high, the benefits to infants of breast-feeding are outweighed by the transfer of substantial amounts of contaminants. Fortunately there is evidence that concentrations of DDE and PCBs in human adipose tissue are declining, for example in Japan and Canada (Loganathan *et al.*, 1993; Mes, 1990).

PCBs

PCBs have great chemical stability, a low flammability and good heat conducting properties, but a low electrical conductivity and a high dielectric constant. These properties have ensured that they have been widely used in transformers and capacitors. They have also been used in heat exchangers, hydraulic systems, vacuum pumps and in lubricating oils and as plasticizers in paints and inks. They enter the general environment by leakage from supposedly closed systems, from landfill sites, from incineration of waste and in sewage effluents. Small localized sources can contaminate entire rivers; for example a small breaker's yard was a significant point source of PCBs on the River Mule in Wales (Ormerod *et al.*, 2000). PCBs are dispersed widely in the atmosphere. An analysis of sediment cores from Esthwaite Water in the English Lake District showed that PCB fluxes increased slowly from the late 1920s, then increased rapidly from the late 1940s, peaking during the late 1950s and early 1960s. Following restrictions on their manufacture and use, fluxes declined sharply (Sanders *et al.*, 1992). Some of the compounds that have replaced them, such as polybrominated biphenyls used as flame retardants, may prove just as damaging.

PCBs were commercially produced as mixtures and they also contained highly toxic impurities, namely dioxins and furans. Early studies were conducted on total PCBs, comparing amounts in tissues with commercial mixtures (e.g. Aroclor 1260). Advances in the analysis of PCBs over the last two decades have enabled the differentiation of individual isomers (congeners) in environmental samples. There are 209 PCB congeners which differ in the number of chlorine atoms (1–10) and their position in the biphenyl ring. Toxicity varies largely with the number of chlorine atoms and their positions in the ring, planar (flat) PCBs being especially toxic. The less toxic congeners may be slowly metabolized but this may result in the PCB mixture eaten in food by top carnivores being more toxic than technical mixtures, because fish (their prey) are less able to metabolize the more toxic congeners.

The mechanism of action of the most toxic congeners is similar to that of dioxin (2,3,7,8-tetrachlorodibenzo-*p*-dioxin, or 2,3,7,8-TCDD), generally considered to be the most potent synthetic environmental contaminant. Congeners bind to a receptor protein (arylhydrocarbon, Ah) in the cell, which subsequently translocates to the nucleus. This PCB–receptor complex increases the production of mRNA which controls the production of mixed-function oxidase enzymes involved in the metabolism of fat-soluble xenobiotics. The metabolites produced may be harmless compounds, which can be excreted, or they may be active intermediates, which may react with critical receptors, resulting in direct toxic effects or the initiation of cancer. The toxic effects observed in laboratory animals include liver damage, skin disorders, reproductive failure, growth of the thymus gland, loss of body mass, damage to the immune system and the production of deformities in young (teratogenesis).

Because PCBs occur as complex mixtures within the biota we need a method for assessing overall toxicity. One approach is to calculate toxic equivalency factors (TEFs) (Safe, 1987, 1990). The ability of each compound to induce particular liver enzymes, which is generally correlated with other toxic effects, is determined and compared with the induction caused by dioxin (TCDD). The TEF is:

$$\text{TEF} = \frac{\text{EC50(TCDD)}}{\text{EC50(compound)}}$$

EC50 is defined on p. 26. The TEF is then multiplied by the concentration in a tissue to determine the toxic equivalent:

$$\text{TEQ} = \text{TEF} \times \text{concentration}$$

The TEQs can then be added together to find the total toxic equivalence of the congeners within a tissue and determine which individual congeners are likely to be toxicologically of most significance. Because they occur in much higher concentrations than the highly toxic dioxins and furans, the non-*ortho* and mono-*ortho* PCB congeners are the main contributors to dioxin-like toxicity in environmental samples.

The Hudson River rises in the Adirondack Mountains of New York State and discharges into the Atlantic Ocean via New York City. Two large generating stations, at Hudson Falls and Fort Edward, used PCBs in their capacitors and a substantial amount found its way into the environment. Much of the PCB was retained by a dam, but this was removed in 1973, releasing large amounts of contaminated sediment into the river. Other sources of PCB were municipal waste tips

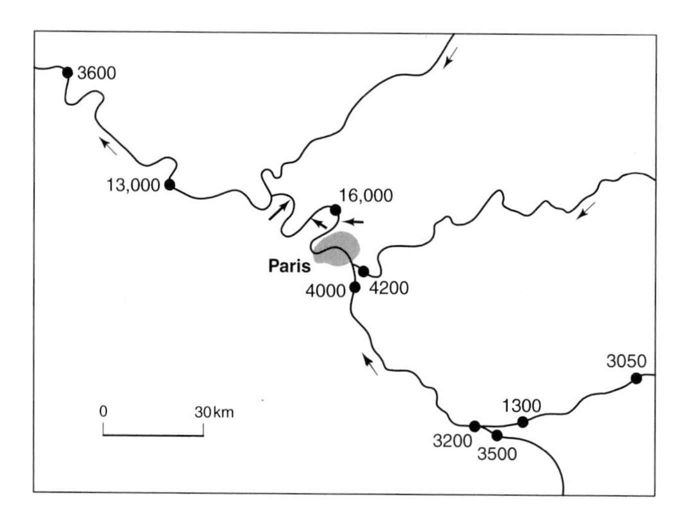

Fig. 2.26 Total PCBs (μg kg^{-1} dry weight) in tissues of roach at stations in the River Seine catchment above and below Paris • Bold arrows indicate points of effluent discharge. (adapted from data in Chevreuil *et al.*, 1995).

and the burning of refuse. It was estimated that the Hudson River basin contained 1351 t PCB, of which less than 0.1 per cent was held within the biota. Nevertheless all but three species of fish had a PCB concentration exceeding 5 mg kg^{-1}, the tolerance limit for edible fish, and angling was banned along long sections of the river. The shellfish industry, worth many millions of dollars annually, was completely closed (Sanders, 1989).

Sediments are the greatest potential source of PCBs within aquatic ecosystems. In experimental systems PCBs originating from sediments are taken up by zooplankton and planktivorous fish. PCBs are released from sediment to water, from where they are taken up directly as a result of water/fat partitioning processes in animals (Larsson, 1986). Bottom-living fish, such as eels, accumulate greater amounts (Larsson, 1984a). Benthic invertebrates are also directly exposed to PCBs. Many of these eventually emerge as adults into terrestrial environments and it has been estimated that 20 μg PCBs m^{-2} yr^{-1} are transferred by chironomid midges from the aquatic to the terrestrial environment (Larsson, 1984b). Chironomids, of course, are extensively eaten by insectivorous birds.

The accumulation of PCBs in fish is a function of size, lipid concentration and trophic position. Figure 2.26 illustrates the concentration of PCBs in flesh of roach collected at various stations down the River Seine, above and below Paris. The catchment above Paris is largely rural. Note the fourfold increase in PCBs in fish from below Paris, where the river is receiving discharges of waste. At the lowermost site PCB concentrations had again declined.

Although fish may accumulate high concentrations of PCBs there is as yet little evidence of any adverse effects on wild populations, though such effects are hard to demonstrate. In farmed Atlantic salmon fed on pellets made from Baltic herring (*Clupea harengus*) several physiological functions were disturbed, including effects on bone metabolism, steroid synthesis, the immune system and the induction of liver enzymes. As contaminant levels in farmed salmon were lower than in wild salmon, it was considered that the long-term survival of this stock should give cause for concern (Andersson *et al.*, 1993). Zebrafish (*Danio rerio*) fed a PCB mixture showed a reduction in both egg production and survival of larvae (Örn *et al.*, 1998). In the Great Lakes there was a strong correlation between PCB concentration and the hatching success of lake trout (*Salvelinus namaycush*) (Mac *et al.*, 1993). This has been demonstrated in the laboratory by injecting various concentrations of lake trout extract into eggs of rainbow trout, which showed dose-related embryotoxicity (Wright and Tillitt, 1999). In barbel (*Barbus barbus*) from the polluted River Meuse in Belgium, PCB concentrations were strongly correlated with several biomarker enzymes and with an altered liver ultrastructure: the effects these changes would have on metabolism may explain the decline in populations of this species (Hugla *et al.*, 1995).

The Great Lakes region has supplied much of the evidence for the effects of PCBs on wild populations of higher vertebrates. In Green Bay, Lake Michigan, a population of the fish-eating Forster's tern (*Sterna forsteri*) had impaired reproduction. The terns showed an extended incubation period compared with birds from other localities. Fewer eggs hatched, the chicks had a lower body mass, with an increased liver size and the occurrence of oedema. There was also a high incidence of congenital deformities. The parent terns were inattentive in nesting, which further reduced reproductive output.

A detailed toxicological analysis and the calculation of toxic equivalents concluded that those PCB congeners which induce the enzyme arylhydrocarbon hydroxylase (AHH) were the only contaminants present in sufficient amounts to cause the observed effects on eggs and chicks. Two pentachlorobiphenyls (PCBs with chlorine atoms attached to five of the six positions on the benzene ring) accounted for more than 90 per cent of the estimated TCDD equivalence. PCBs in general were the only contaminants causing the behavioural abnormalities in adults (Kubiak *et al.*, 1989).

Table 2.4 lists a number of effects and deformities reported in fish-eating birds from the Great Lakes. As well as Forster's tern, affected species include common tern (*Sterna hirundo*), Caspian tern (*Sterna caspia*), herring gull (*Larus argentatus*), bald eagle (*Haliaeetus leucocephalus*), night heron (*Nycticorax nycticorax*) and double-crested cormorant (*Phalacrocorax auritus*) (Giesy *et al.*, 1994; Grasman *et al.*, 1998; Ryckman *et al.*, 1998; Bowerman *et al.*, 2000). An example of a beak deformity in a cormorant is illustrated in Fig. 2.27. The prevalence of malformed cormorant chicks was 52 per 10 000 in Green Bay, much higher than in other regions of the Great Lakes (Fox *et al.*, 1991). The symptoms listed in Table 2.4 have been observed only over the past 20 years although the compounds have been in the environment for much longer. It seems likely that, in earlier years, they were masked by the effects of DDE, which thinned eggs to such an extent that they did not survive long enough for symptoms to be expressed. It appears that all of the

Table 2.4 Adverse effects in embryos and chicks of fish-eating birds of the Great Lakes (from Giesy *et al.*, 1994).

Effects	
Eggshell thinning	Cardiovascular haemorrhage
Deformities	Hormonal changes
Tumours	Enzyme induction
Behavioural changes	Metabolic changes, wasting syndrome
Immune suppression	Depletion of vitamin A
Oedema	Porphyria
Deformities	
Crossed bill	Eye
Clubbed foot	Brain
Hip dysplasia	Skull bones
Dwarf appendages	Gastroschisis
Ascites oedema	

populations of fish-eating birds in parts of the Great Lakes area are currently displaying symptoms of exposure to PCBs, dioxins and furans at the biochemical level and these cause deaths and deformities in all populations examined. It is only in the more contaminated areas, such as Green Bay, however, that the effects are sufficient to cause population declines (Giesy *et al.*, 1994). A large number of compounds can cause these effects, which are expressed through a common mechanism of action via the Ah-receptor protein. There may be synergistic and antagonistic interactions between the various compounds, including individual PCB congeners. In the case of the deformed cormorants, 67 per cent of the total TEQ was composed of two pentachlorobiphenyl (coplanar) PCBs, those associated with reproductive effects in Forster's terns, emphasizing the threat posed by these two compounds to wildlife in parts of the Great Lakes (Yamashita *et al.*, 1993; Ryckman *et al.*, 1998).

Effects are not confined to birds. Beluga whales (*Delphinapterus leucas*) in the St. Lawrence estuary downstream of the Great Lakes have failed to increase in numbers following a cessation of hunting. Necropsies carried out on 45 belugas revealed a high incidence of tumours (45 per cent) and of lesions to the digestive tract (53 per cent) and to the mammary glands of females (45 per cent). There was tooth loss and evidence of immunosuppression, while one individual was a hermaphrodite (Béland *et al.*, 1993). No such symptoms were found in belugas from the Arctic. A wide range of contaminants at high concentrations was found in St. Lawrence belugas. Many of the pathological observations were similar to those found in seals from the Baltic Sea, which were related to the effects of PCBs on the reproduction, endocrine and immune system (Bergman and Olsson, 1985). Not

Fig. 2.27 Cosmo, a double-crested cormorant from Lake Michigan. She was picked up by Dr James Ludwig and lived with him for two years. Her appearance on television in Washington and Japan, and in front of a Congressional Committee, did much to promote awareness of the environmental problems of the Great Lakes (photograph and information courtesy of Mike Gilbertson).

only has the seal population stopped breeding but immunosuppression makes it easier for infections that would otherwise have no adverse effects to become established.

In the Great Lakes region, mink (*Mustela vison*) and otter (*Lutra canadensis*) populations have also declined (Wren, 1991). Experiments with mink have shown that their reproduction is especially sensitive to non-*ortho* and mono-*ortho* PCBs, though the presence of di-*ortho* PCBs is also required to give expression to the effects (synergism) (Kihlström *et al.*, 1992). The relationship between declines in European otter populations and environmental contaminants was described earlier (p. 6).

In laboratory experiments PCBs have a wide range of effects on mammalian reproduction. Are there any observable effects on humans? As usual it has been extremely difficult to prove and is controversial. Studies do, however, give rise to concern. Mothers giving birth in hospitals close to Lake Michigan were asked about their history of eating fish from the lake. Blood was taken from the umbilical cords of their newborn babies and analysed for PCBs. A sample of 242 mothers ate moderate quantities of lake fish (two or three salmon or trout meals per month) and 71 mothers ate no fish. Effects on health observed among those in the highly exposed categories included increased anaemia, oedema and susceptibility to infectious disease. Babies of those mothers who had eaten fish weighed 160 to 190 g less than controls and their head circumferences at birth were disproportionately small in relation to their age and weight. At age 4 years there was still a relationship between mass and PCB measured in the umbilical cord at birth – the most highly exposed children weighing, on average, 1.8 kg less.

Babies also displayed behavioural defects including increased startle reflexes and were classified by physicians within the 'worrisome' neonatal category. At 7 months there was a more than 10 per cent decline in visual recognition memory among the most exposed babies, suggesting effects of PCBs upon the brain. At age 4 years and 11 years there were still significant deficits in short-term memory and speed of information processing (Jacobson and Jacobson, 1993; Lonky *et al.*, 1996; Carpenter, 1998). Although the deficits were small they could have a significant impact on the ability of a child to master basic reading and arithmetic skills.

It was calculated that an infant nursed for a year by a mother eating an average amount of Lake Michigan fish would have been exposed to 6.22 mg of PCB, of which 5.29 mg would be retained in the body. Infants nursed by mothers who had eaten fish from the lake exceeded the recommended consumption guidelines for *adults* of 1 $\mu g\,kg^{-1}$ body weight per day for their entire breast-feeding experience, some by as much as 25 times each day (Swain, 1988). Swain predicted that, if exposure to PCBs could be stopped now, the contaminants will be transmitted in measurable concentrations through transplacental passage of PCBs and postpartum exposure to breast milk for at least five generations beyond the original mother.

Behavioural changes have been detected in laboratory rats fed on a 30 per cent diet of Lake Ontario salmon over 20 days. The rats became hyper-reactive to negative and positive events in tests compared with controls. Similar hyper-reactivity was also present in their offspring, who had never eaten lake fish, and persisted into adulthood. The experimental group grew normally and showed no signs of illness, so that it seems that behavioural changes may occur at doses lower than those needed to produce physical illness (Daly, 1993). These results have relevance to observations on birds in the Great Lakes (inattentiveness to nesting etc.) and to the children described above.

Restrictions on the use of PCBs have led to a gradual decline in environmental levels in some areas (e.g. Bignert *et al.*, 1993; Jones *et al.*, 1992). There is, however, a large quantity of PCBs already in the environment and an even larger quantity in older electrical equipment, toxic dumps and storage. Lake sediments may present particular problems. There is evidence that microbes in anaerobic sediments are capable of dechlorinating some PCB congeners (Beurskens and Stortelder, 1995). PCBs can also be degraded aerobically (Focht, 1995). Genetic engineering holds much potential in the bioremediation of areas contaminated with PCBs (Beil *et al.*, 1999; Rapp and Timmis, 1999; Timmis and Pieper, 1999) but it will be some time before such methods move beyond the experimental stage. An alternative approach is to remove contaminated sediments. This has been achieved at Lake Järnsjön in

Sweden. Some 150 000 m^3 of sediment, holding 394 kg of PCBs, were dredged from the site and transferred to a contained landfill, from where no leaching has occurred. The PCB load in the lake was reduced by 97 per cent and concentrations in the water fell by 69 per cent within two years, and in fish by 52 per cent (Gullbring *et al.*, 1998).

Because many of these organochlorines are potentially so toxic, they have stringent safety guidelines for consumption. The total daily intake of dioxins, for example, should not exceed 14–37 pg kg^{-1} body weight d^{-1}, and even background levels of exposure could be exerting subtle health effects (van Leeuwen *et al.*, 2000). Unless great efforts are made to prevent further discharges, significant pollution with PCBs and related compounds is likely for the foreseeable future. Transfer across the generations and the persistence of PCBs within living tissues indicate that levels in top carnivores and humans are unlikely to decrease significantly for some years (Harrad *et al.*, 1994). In view of the apparent potential consequences for human health at very low levels of long-term exposure, this should be a matter of great concern.

CHAPTER 3

ORGANIC POLLUTION

Organic pollution occurs when large quantities of organic compounds, which act as substrates for microorganisms, are released into watercourses. During the decomposition process the dissolved oxygen in the receiving water may be used up at a greater rate than it can be replenished, causing oxygen depletion and having severe consequences for the stream biota. Organic effluents also frequently contain large quantities of suspended solids which reduce the light available to photosynthetic organisms and, on settling out, alter the characteristics of the river bed, rendering it an unsuitable habitat for many invertebrates. Toxic ammonia is often present.

A simple measure of the potential of biologically oxidizable matter for de-oxygenating water is given by the biochemical oxygen demand (BOD). The BOD is obtained in the laboratory by incubating a sample of water for five days at 20°C and determining the oxygen used. It estimates the pollution potential of a wastewater containing an available source of organic carbon by measuring the amount of oxygen used by the indigenous microorganisms in a standard sample. BOD provides a broad measure of the effects of organic pollution on a receiving water. Effluents with high BODs can cause severe problems in watercourses that receive them.

Organic pollutants consist of proteins, carbohydrates, fats and nucleic acids in a multiplicity of combinations. Raw sewage is 99.9 per cent water, and of the 0.1 per cent solids, 70 per cent is organic (65 per cent proteins, 25 per cent carbohydrates, 10 per cent fats). Organic wastes from people and their animals may also be rich in disease-causing (pathogenic) organisms.

ORIGINS OF ORGANIC POLLUTANTS

Organic pollutants originate from domestic sewage (raw or treated), urban run-off, industrial (trade) effluents and farm wastes.

Sewage effluent is the greatest source of organic materials discharged to freshwaters. In England and Wales there are almost 9000 discharges releasing treated sewage effluent to rivers and canals and several hundred more discharges of crude sewage, the great majority of them to the lower, tidal reaches of rivers or, via long outfalls, to the open sea. It has always been assumed, certainly incorrectly, that the sea has an almost unlimited capacity for purifying biodegradable matter. About 80 per cent of some 11 000 million litres of sewage effluent produced each day in England and Wales is of domestic origin. In England and Wales 96 per cent of the population is served by public sewers and 80 per cent of the sewage produced receives at least secondary treatment. This removes 95 per cent of the polluting load before effluent is released to rivers, or 50 per cent before it is discharged to coastal waters. Similar levels of treatment are typical of much of the developed world. In the economically developing world, however, where the capital for sewage treatment facilities is not available, the release of crude sewage into watercourses is still a major problem. Almost one-third of the disease in the world is the result of poor water supply and sanitation. The provision of wholesome water to populations of many millions of people remains an essential priority.

In urban areas, the run-off from houses, factories and roads can result in severe pollution, especially in storm conditions after periods of dry weather. This urban run-off may be routed through the sewage works (combined sewerage system), or it may be separately sewered and flow directly into rivers and streams. In the former case sewage treatment works may be severely overloaded during storm conditions and the effluent discharged may be of a much lower quality than under normal operating conditions, though the dilution is greater when the receiving river is in flood. Where urban run-off drains directly into rivers, pollution can be severe because drainage from the hard surfaces is so rapid that it may reach a river that has not yet increased above dry weather flow, so that the dilution will be minimal. Even in rural areas the volume of traffic on many highways is so great that run-off can cause significant local pollution in watercourses (Hamilton and Harrison, 1991; Maltby *et al.* 1995).

With urban run-off there is often a first flush of highly polluting water and much of this may come from the catchpits of roadside drains, which have thick bacterial scums and are usually anoxic. The constituents of urban run-off are obviously very variable, the effluent being rich in dog faeces, suspended material, heavy metals and, seasonally, chloride from road-salting operations. The quality of the run-off is also highly variable, but BODs as high as $7700\,mg\,l^{-1}$ have been recorded. It is considered that, with the clean-up of sewage effluents and factory wastes, 70 per cent of chemicals in San Francisco Bay, California, are nowadays derived from the everyday activities of the population, oil and metals from cars, chemicals used on lawns etc. being washed untreated into watercourses through storm drains. Ellis (1989), Walesh (1989) and Viessman and Hammer (1993) provide more details on surface water run-off from towns.

Industrial effluents are a further source of organic pollution. These may be routed via the sewage treatment works or they may be released, with or without treatment, directly into a waterway. Among the industries producing effluents containing substantial amounts of organic wastes are the food processing and brewing industries, dairies, abattoirs and tanneries, textile and paper making factories. Many trade effluents can be effectively treated by mixing with domestic sewage but some can inhibit the microbiological activity in the treatment works, resulting in a poor-quality effluent, and these must be treated on the industrial site.

Farm effluents have become an increasing pollution problem, particularly with the intensification of livestock production in recent years. Over the decade from 1979 the average size of dairy herds in England and Wales increased from 46 to 320 and of pigs from 200 to 345. The catchment of the River Torridge, a dairying area in southwest England, supports some 84 000 cattle, producing as much waste as 600 000 people. The human population of the catchment is little more than 15 000, most of them, in contrast to the cattle, connected to the sewerage system. Animal slurries have a BOD of some 20 000 mg l^{-1}, compared with an average of 350 mg l^{-1} for untreated human sewage. A substantial part of the diet of cattle, especially in winter, is silage, which is partially fermented grass. Silage liquor is two hundred times more polluting than untreated sewage. About half of all farm pollution incidents are caused by silage effluent.

Over the period 1978–85, the number of reported farm pollution incidents in England and Wales doubled. Over the period 1988–92, organic farm wastes accounted for 20 per cent of all pollution incidents in the United Kingdom. More stringent regulations introduced in 1991 have led to a reduction in major farm incidents, though pollution caused by run-off from the land has increased during the period (Fig. 3.1). Nevertheless incidents still have severe impacts on the river environment. For example, in August 1994, slurry from a dairy farm killed 100 000

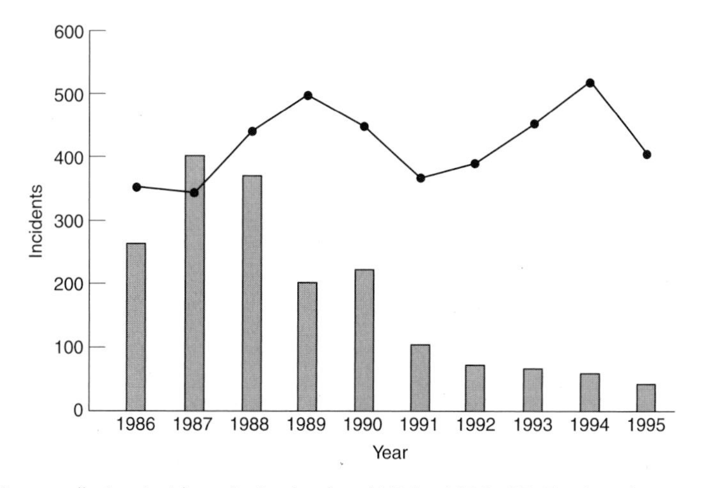

Fig. 3.1 Farm pollution incidents in England and Wales 1986–95. The bar chart represents major farm incidents, the line incidents of run-off from farmland.

mature trout in a fish farm taking its water from the River Camel in Cornwall. The more intensive and mechanized ways of dealing with vegetable crops may also cause pollution. The mechanical vining of peas, for example, and especially the drainage from pea haulm silage can cause severe, though seasonal, problems. One such effluent had a BOD of 15 000 $mg\,l^{-1}$.

Farms have, therefore, led to an increase in the chronic pollution of rivers and streams over the last two decades, often in areas where water was previously of a high quality. There has also been an increase in accidents (episodic events) having major effects on fish and other river life. As farms are distributed over wide areas and are often remote, pollution incidents are difficult to detect and police.

Fish farming may also result in a deterioration of water quality downstream. Farms mostly rear trout, which require clean water, but fish are kept at high densities. Effluents leaving the farm are high in suspended solids, BOD, ammonia and nutrients from faeces and excess food. The effluent may also contain compounds used in fish husbandry, such as antibiotics, algicides and fungicides. It has been suggested that the effluent from a trout farm producing 200–250 t of fish per year is equivalent to the untreated sewage from 1400–5000 persons (Pillay, 1992).

In general the origins of organic effluents and their composition are extremely diverse and we can expect a similar diversity in the effects they have on receiving waters.

PATHOGENS

Diseases contracted from water kill some 25 million people, most of them children, each year, while millions more are debilitated by waterborne diseases. Faecal contamination of water can introduce a variety of pathogens into waterways, including bacteria, viruses, protozoa and parasitic worms (Table 3.1). Waterborne diseases remain a major hazard in many parts of the world. Four classes of water-related diseases have been recognized (Cairncross and Feacham, 1983). Class 1, the true waterborne diseases, are contracted by drinking water that contains pathogenic organisms, usually because of faecal contamination. Examples include cholera, typhoid and hepatitis A. Class 2 diseases are indirect infections associated with a lack of personal hygiene (e.g. hand washing) which can be reduced by providing adequate amounts of water for bathing and washing. To control such diseases it is necessary to provide people with sufficient water of reasonable quality; achieving a high bacteriological quality is a secondary consideration (Ellis, 1989). This second class includes all the diseases in class 1, together with various other diarrhoeal diseases, some infections of eyes (e.g. trachoma) and skin (e.g. ringworm) and infections carried by lice and mites. Class 3 are diseases caused by helminths which spend part of their life cycle in water, while class 4 are diseases that require a water-related insect vector (e.g. yellow fever, malaria, river-blindness, filariasis), though

Table 3.1 Some water-related diseases and their causative organisms

Causative organisms		Disease or symptoms
Bacteria	*Salmonella typhi*	Typhoid fever
	S. paratyphi	Paratyphoid fever
	Salmonella spp.	Gastroenteritis
	Shigella spp.	Bacterial dysentery
	Vibrio cholerae	Cholera
	Escherichia coli	Gastroenteritis
	Leptospira icterohaemorrhagiae	Weil's disease
	Campylobacter spp.	Intestinal infections
	Francisella tularensis	Chills, fever, weakness
	Mycobacterium	Tuberculosis
Viruses	Enteroviruses	Many diseases, including poliomyelitis, respiratory diseases, meningitis and infectious hepatitis
	Rotaviruses	Diarrhoea and enteritis
Protozoa	*Entamoeba histolytica*	Amoebic dysentery
	Giardia lamblia	Diarrhoea, malabsorption
	Naegleria fowleri	Amoebic meningoencephalitis
	Cryptosporidium spp.	Diarrhoea
Helminths	*Diphyllobothrium latum*	Tapeworm infection
	Taenia saginata	Tapeworm infection
	Schistosoma spp.	Bilharzia
	Clonorchis sinensis	Trematode infection
	Dracunculus medinensis	Guinea worm

these are not necessarily associated with polluted waters. Hunter (1997) provides a detailed account of waterborne diseases.

Bacteria

Bacteria of the genus *Salmonella* can cause acute gastroenteritis (i.e. food poisoning), with diarrhoea, fever and vomiting. *Salmonella typhi* is the most virulent of the genus, responsible for typhoid fever. Some 200 immunologically distinguishable types (serotypes) of *Salmonella* are known to be pathogenic to humans and there are many more that infect animals, including livestock. Cross-infection between people and animals can occur via water pollution. The spreading of untreated sewage wastewater on land and its use for the irrigation of crops can also be a source of infection. In Morocco, *Salmonella* infected 32.6 per cent of a group exposed to this practice, compared with 1.1 per cent of a control group. Boys under 10 years old were most at risk, presumably because they were most likely to play in areas where wastewater was distributed (Melloul and Hassani, 1999).

Currently there appears to be an increase in the spread of *Salmonella* and this has been related to modern living conditions, such as mass food production (e.g. of poultry) and communal feeding. Salmonellae in rivers can survive for some distance downstream of the source. Large concentrations of gulls, feeding at rubbish tips and roosting on water-supply reservoirs, may excrete high numbers of *Salmonella* and faecal streptococci into the water.

Shigella species, especially *S. dysenteriae* and *S. sonnei*, are among the most commonly identified causes of acute diarrhoea in the United States. *Shigella* appears to be almost exclusively associated with humans and it is transmitted by the consumption of faecally contaminated water or food. If untreated, up to 15 per cent of the poorest populations may die of shigellosis.

Leptospirosis is an acute infection of the liver, kidneys and central nervous system and is caused by a group of spiral-shaped, motile bacteria, of which more than 100 serotypes are known. The primary hosts are rodents, which carry the organisms in their kidneys, and humans may become infected by wading or swimming in water contaminated with the rodents' urine. Most cases occur from infection through skin abrasions or from accidentally falling into contaminated water. Weil's disease is a particularly serious form of leptospirosis and is most likely to be caught in sewerage systems and rivers receiving sewage effluent where rats gather. All people sampling freshwaters or involved in angling and water sports should be aware of the potential risk. If 'flu-like symptoms develop after contact with potentially polluted water, a doctor should be consulted. Weil's disease can be fatal.

Gastroenteritis and diarrhoea, especially in young children, may be caused by four different groups and various serotypes of *Escherichia coli*. Most urinary infections of adults are also caused by pathogenic *E. coli*, though urinary infection is usually by spread of *E. coli* from the person's own intestinal flora, rather than from water supplies. People taking holidays abroad often suffer from travellers' diarrhoea (Montezuma's revenge or Delhi belly), which is commonly caused by enterotoxogenic strains of *E. coli*, or serotypes not previously encountered by the victims.

Since the late 1970s *Campylobacter* species have been increasingly recognized as one of the commonest causes of outbreaks of an intestinal infection which mimics *Shigella* dysentery. In young children it causes diarrhoea with obviously bloody faeces. Isolation of the organism, *C. jejuni*, requires a special medium, which is why it was not discovered until relatively recently. The natural hosts are birds, particularly poultry, and the disease can occur in young dogs and cats. During the 1980s in the United Kingdom, notified cases of food poisoning caused by campylobacters more than tripled and the problem may be 100 times worse than the reported cases suggest (Jones and Telford, 1991). Waterborne outbreaks have also been reported in the United States.

Vibrio cholerae causes cholera, an acute intestinal disease that can result in death within a few hours of onset. The El Tor vibrio lives on the gut wall and produces a protein exotoxin consisting of two subunits A and B. The B subunit binds to receptors on the surface of cells of the gut wall, while the A subunit acts as a hormonal mimic, causing the cells greatly to increase their output of water, and of sodium, bicarbonate and potassium ions. This fluid then leaves the body as 'rice-water' diarrhoea, which helps to disseminate the vibrios, and also results in severe dehydration which can be fatal if untreated. Cholera is largely under control in countries where widespread sewage treatment is practised, but it is still rife in many parts of the world and outbreaks occur when disasters happen, such as famine, earthquakes and floods. An outbreak occurred in Peru in 1991 where the provision of clean water and disposal of sewage had not kept pace with the population increase. The outbreak brought tourism to a halt while the export market of fruit and vegetables ceased. This cost the economy of Peru US$1 billion in just ten weeks. The total economic cost to the country of the epidemic was more than three times the total investment in improvements to water supply and sanitation during the previous decade.

Cholera is rife in over-populated and flood-prone Bangladesh. The vibrios are most numerous in the guts of copepods, whose populations are greatest in the plankton blooms of spring and autumn. A plankton bloom is invariably followed by an outbreak of cholera. While boiling drinking water to kill both copepods and vibrios would seem an obvious solution, fuel is too expensive. It was found that filtering drinking water through four layers of sari material removed more than 99 per cent of the bacteria, dramatically reducing the incidence of the disease. If the material is hung out to dry in the sun, the trapped bacteria are killed within two hours, and sari material is affordable by even the poorest of families (Coles, 1996).

Mycobacterium tuberculosis is responsible for tuberculosis. Transmission by water appears to be uncommon, but this may in part be due to the long time between infection and the appearance of symptoms, which makes the origins of the disease difficult to trace in any particular incident. Tubercle bacilli are able to survive in water for several weeks.

Legionnaire's disease is so named after an outbreak in Philadelphia in 1976 when 183 delegates to a large military convention developed pneumonia and 29 of them died. Six months of intensive epidemiological and microbiological investigation was needed before the causative organism, *Legionella pneumophila*, a previously unknown bacterium, was grown and the source of infection identified. The Legionellaceae occur naturally in pond water but the pathogenic variety colonizes the water in air-conditioning cooling towers, condensers, bathroom shower-heads and other man-made appliances that allow it to flourish. It becomes an opportunistic pathogen when minute infective water droplets from these sources are inhaled. It can also survive and multiply within amoeboid protozoa, which may then act as vectors for the disease. The disease can be prevented if water systems in large buildings are regularly cleaned, the water continuously chlorinated and hot water supplies maintained above 55 °C. About one-third of the 200 cases diagnosed annually in Britain are contracted abroad.

Viruses

Viruses seem to survive in water for much longer than faecal bacteria and there is concern over their occurrence and control in sewage effluents and in water supplies, though the evidence suggests that modern water treatment processes, which rely on chlorination, are effective in safeguarding public health (Cartwright, 1997).

Infectious hepatitis (jaundice) is endemic in many countries and, of the two main kinds, hepatitis A is transmitted by contaminated water. Like cholera, it thrives during disasters, such as floods, or in refugee camps. Hepatitis A virus is relatively resistant to the chlorine used in the water treatment process. Some 87 per cent of all waterborne viral outbreaks in the United States are caused by hepatitis A.

Poliomyelitis virus may occasionally be carried by water but most transmission is by personal contact. Enteritis can be caused by Coxsackieviruses and echoviruses and outbreaks may develop very rapidly if water supplies are contaminated with untreated sewage.

Contaminated water can be involved in the transmission of rotaviruses, which are a major source of diarrhoea in both adults and children. It is thought that up to 50 per cent of cases of acute diarrhoea in children under 2 years old are caused by rotaviruses, resulting in some 6 million deaths annually in developing countries. They may also have a role in travellers' diarrhoea in adults.

A review of viruses in water has been provided by Gerba *et al.* (1995).

Protozoa

Amoebic dysentery, caused by the parasitic protozoan *Entamoeba histolytica*, is endemic in the tropics and subtropics and the contamination of drinking water by amoebic cysts from faeces is a major transmission route. The trophozoite, a stage in the life cycle which causes the disease, may also get into the lymph and blood vessels to be carried to the liver, lungs and brain. It may cause bowel ulcers and liver abscesses. Chronic infection results in recurrent episodes of diarrhoea, accompanied by blood and mucus, alternating with constipation. Other protozoans causing waterborne disease include *Giardia lamblia*, which results in diarrhoea and a reduced absorption of nutrients and vitamins across the gut wall, and a pathogenic strain of *Naegleria fowleri*, contracted while swimming in the warm waters of small lakes, swimming pools or polluted estuaries, which causes a rare, but fatal, amoebic meningoencephalitis.

A pathogenic protozoan which is on the increase is *Cryptosporidium*. The organism multiplies in the intestine, causing stomach pains, vomiting, diarrhoea and fever. The illness usually abates within three weeks but certain groups, including children, the elderly and immuno-compromised individuals suffering from diseases such as leukaemia and AIDS, may be affected much more severely, and indeed fatally. At present there is no known cure.

The parasite lives in the gut of livestock and passes with the faeces into the environment where the spores (oocysts) can survive for up to 18 months. The intensification of livestock rearing, with attendant problems of waste disposal, may be one cause of the increase of *Cryptosporidium* and abattoir wastes are another significant source. A single oocyst may be sufficient to initiate an infection (Blewett *et al.*, 1993). A recent survey in the United States revealed that up to 77 per cent of surface waters examined contained the pathogen. *Cryptosporidium* caused the largest waterborne disease outbreak in the history of the United States, with 400 000 people falling ill in Milwaukee in April 1993, while earlier outbreaks in Georgia and Oregon infected thousands of people (Miller, 1997). In Britain the number of cases increased 60-fold between 1983 and 1986, with several major outbreaks since 1988. It is a serious public health concern.

Cryptosporidium is resistant to chlorine and standard tests used by water authorities prior to 1990 to monitor drinking water could not detect it. To provide a sample, 1000 litres of water need to be centrifuged and the organism can then be identified using genetically engineered polyclonal and monoclonal antibodies. Recent approaches to identifying and studying this organism are described by Smith (1997). Current expenditure in the UK to remove *Cryptosporidium* from drinking water by filtration is about £8 million per year but it is estimated that this could rise to £66 million per year to comply with new regulations (Pretty *et al.*, 2000).

Reviews of *Cryptosporidium* are presented by O'Donoghue (1995) and Fayer (1997).

Parasitic worms

The parasitic worms include the phyla Nematoda (roundworms) and Platyhelminthes (tapeworms and flukes). The ova of these parasites, passed out in faeces and urine, are often resistant to sewage treatment processes. A tapeworm can produce up to one million eggs in a day. The majority of life cycles involve an intermediate host. In the case of the beef tapeworm, *Taenia saginata*, cattle, the intermediate host, may become infected by grazing on pastures sprayed with sewage sludge or by drinking water contaminated with sewage.

Dracunculus medinensis, the guinea-worm, is prevalent in tropical regions. The female worm can be up to 1.3 m long, lives under the skin of the legs and produces a blister near the ankle which, when immersed in water, bursts to release large numbers of larvae. These infect copepods and the cycle is completed when people drink unfiltered water containing the copepods.

Schistosomiasis (bilharzia) in humans is mainly caused by one of three species of blood fluke: *Schistosoma mansoni*, particularly in Africa and South America, *S. haematobium* in much of Africa, and *S. japonicum* in the Far East. The last species also occurs in animals. The disease is of particular importance since 300 million people, mainly in the tropics, are infected. It is a debilitating illness which, though not often fatal, causes general weakness in the sufferer and results in an annual economic loss of hundreds of millions of dollars. Eggs of the flukes pass out with human faeces or urine and, if they reach freshwater, develop into miracidia larvae which infect snails. Cercariae develop in the snails which, on leaving those hosts, penetrate the skin of humans wading in water. After extensive migration in the body each parasite settles and produces ova in the venous system which the particular species frequents (e.g. *S. haematobium* in the bladder and urinary tract, *S. mansoni* in the portal blood system of the gut and liver). The late clinical features of these three pathogens depend on the site and injurious effects of the different types of ova. A single infected person who is passing parasite eggs daily in faeces or urine could contaminate a whole river if there are plenty of appropriate snails to perpetuate the disease.

The incidence of bilharzia appears to be influenced by changes in land use. With the filling of Lake Nasser behind the Aswan Dam in Egypt the surrounding land is now irrigated by canals and ditches rather than by seasonal flooding. This has led to the build-up of large populations of snail hosts, which now have a permanent rather than a seasonal habitat, and the incidence of bilharzia has greatly increased. Forest clearance and agricultural development in Africa have led to the appearance of *Schistosoma haematobium*, because the new conditions allow its snail host *Bulinus rohlfsi* to spread.

Bitton (1999) provides a review of the health risks associated with sewage discharges.

MONITORING WATER FOR PATHOGENS

Water used for drinking must be free of pathogens but the microbial contamination of bathing waters, situated close to discharges of poorly treated sewage, is also responsible for widespread outbreaks of gastrointestinal diseases, as well as infections of skin, ears and respiratory tract. The pathogens themselves may occur only in very low numbers, or intermittently, so other microbiological indicators of faecal pollution must be used. An indicator of faecal pollution must satisfy certain criteria (Bitton, 1999):

- it should be a member of the intestinal microflora of warm-blooded animals;
- it should be present when pathogens are present and absent in uncontaminated samples;
- it should be present in greater numbers than the pathogen for ease of detection;
- it should be at least equally resistant as the pathogen to environmental conditions and to disinfection in water and wastewater treatment plants;

- it should not multiply in the environment so that the ratio of pathogen to indicator in a sample is constant;
- it should be detectable by means of easy, rapid and inexpensive methods;
- it must be non-pathogenic.

The most frequently used indicators of faecal pollution are the coliform bacteria (e.g. *Escherichia coli*), faecal streptococci and *Clostridium perfringens*.

Coliform bacteria are Gram-negative, oxidase-negative, non-sporing rods, which are able to ferment lactose at 44.5 °C. In the past the total coliform count has been used as an indicator of faecal pollution, but it fails to satisfy the criteria for indicators of potential pathogens given above, because several types of coliform bacteria are non-faecal in origin. *Citrobacter* and *Klebsiella*, for example, are found in unpolluted soil and *Enterobacter aerogenes* and *E. cloacae* may be found on vegetation. Furthermore pathogens have been found in water when coliforms were absent and, conversely, some coliforms can multiply in clean waters.

The presence in water of faecal coliforms, especially *E. coli*, is a better indicator of sewage contamination than total coliforms. There are always high numbers of *E. coli* in human faeces and they can be readily distinguished from other coliforms. Direct relationships have been found between *E. coli* counts and the numbers of pathogenic organisms, such as *Salmonella*. There are also direct relationships between gastrointestinal illnesses (vomiting, diarrhoea, stomach ache and associated fever) in swimmers and the density of indicator bacteria in bathing waters (Fig. 3.2).

Caution should be exercised in using *E. coli* as a specific indicator of faecal pollution in hot climates for it can multiply in warm waters. Certain non-faecal

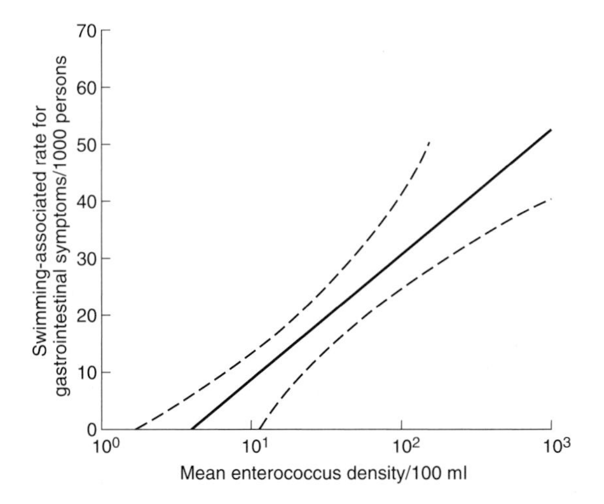

Fig. 3.2 Relationship between swimming-related gastrointestinal symptoms and the density of indicator bacteria in bathing waters. Dotted lines illustrate the 95 per cent confidence limits of the regression line (adapted from Cabelli *et al.*, 1983).

bacteria can also grow at 44°C and produce indole from tryptone, criteria used to identify *E. coli*.

Faecal streptococci are also used as indicators of faecal pollution and it is possible to distinguish between streptococci from human and animal sources. *Streptococcus faecalis* is the predominant streptococcus in the human gut, but is relatively rare in the gut of other animals. The ratio of faecal coliforms to faecal streptococci (FC/FS ratio) can be used to indicate the origin of contamination of water. A ratio greater than 4 is indicative of contamination of human origin, and a ratio below 0.7 indicates pollution derived from animals.

The anaerobic spore-forming bacterium *Clostridium perfringens* has been suggested as an alternative or accompaniment to *E. coli*, particularly where remote or old pollution is being examined. *Clostridium perfringens* is more resistant to toxic pollution than *E. coli*. Sorensen *et al.* (1989) showed that the spores of *Cl. perfringens* could be detected in decreasing concentrations for distances greater than 10 km below the discharge of effluent from a municipal wastewater treatment plant. Coliforms and faecal streptococci, in contrast, varied widely because of the presence of non-point sources, particularly livestock and animal feeding facilities in the catchment. It was therefore suggested that *Cl. perfringens* is a sensitive indicator of municipal wastewater discharges even when agricultural non-point sources of faecal indicator bacteria are present.

Techniques for estimating bacterial populations are described by Pepper *et al.* (1995), Bitton (1999) and Scragg (1999). Watkins and Jian (1997) discuss some recent rapid screening methods. The presence of faecal bacteria in a sample of water gives information on the degree of contamination of the water by humans or animals but, as Ellis (1989) stresses, it is only an indication. The relationship between indicator bacteria and pathogenic organisms is often not direct, so careful interpretation of results is essential.

More than 100 types of enteric viruses are excreted in human faeces, where as many as 10^6 may be present in a single gram. They represent a considerable potential health hazard in water used for potable supply and recreation, especially as some viruses survive the sewage treatment process well. Many infections caught while swimming in poor-quality water may be caused by viruses rather than bacteria. Viruses are present in water in much lower numbers than bacteria and large volumes of water must be examined to detect them but they also cause infections at lower levels of contamination than bacteria. Virus levels in river water are in general dependent upon virus input through domestic sewage.

Arguments in favour of using human enteric viruses as indicators are that they appear more tolerant of the aquatic environment than bacterial indicators and survive sewage treatment better. Because viruses can initiate infection in humans at low environmental levels, a reliable indication of their presence may not be provided by bacteria. A routine test procedure, however, must be easy to perform and current techniques for enumerating viruses are both technically more complex and more expensive than routine tests for bacteria. It has been suggested that coliphages and bacteriophages (viruses that are obligate intracellular parasites of bacteria) may be reliable indicators of recreational water quality, for numbers in samples are directly correlated with numbers of enteric viruses and faecal coliforms (Palmateer *et al.*, 1991). A discussion of viruses in water is provided by Gerba *et al.* (1995), Pepper *et al.* (1995) and Cartwright (1997).

SEWAGE TREATMENT

Historical

Until the last 200 years or so the deterioration of watercourses due to organic pollution was not a serious problem because a relatively small human population lived in scattered communities. The natural self-purification properties of rivers could cope with the waste dumped into them. Introduced pathogenic organisms, by contrast, were a severe problem, though the link between disease and faecal contamination had not, of course, yet been made.

Water pollution became a severe problem with industrialization, coupled with the rapid acceleration in population growth. Industrialization led to urbanization, with people leaving the land to work in the new factories. Domestic wastes from the rapidly expanding towns and wastes from industrial processes were all poured untreated into the nearest river.

The historical pollution of the River Thames has been described by Wood (1982). Though pollution in the Thames at London was recorded as early as the thirteenth century, it was not until the eighteenth century that problems became acute. The population of London doubled between 1700 and 1820 to 1 250 000. Sewage was collected in individual or communal cesspits, which were periodically emptied to fertilize the surrounding land. In 1843 main sewers were laid and 200 000 cesspits were abolished. The sewage drained directly into the Thames, causing gross pollution and epidemics of cholera, while noxious odours rising from the river periodically disrupted the work of Parliament and the Law Courts, 1858 being known as the 'Year of the Great Stink' (Fig. 3.3).

In 1865 a scheme designed by Sir Joseph Bazalgette diverted the sewage, via three main sewers, to outfalls situated 10 miles (16 km) below London Bridge, where it was discharged untreated on the ebb tide. The river through London showed some improvement, though sewage and industrial effluents from many other discharges continued to add to the pollution. The situation around the outfalls downstream of the capital was atrocious.

During the latter part of the nineteenth century a Rivers Pollution Committee recommended various treatment processes and in 1882 a Royal Commission on Metropolitan Sewage Disposal was set up and eventually concluded that the suspended solids in sewage should be separated from the liquid before it was discharged into the river. At this time Mr W. J. Dibdin put before the Commission the idea that sewage could be treated biologically, though this was not taken further at the time. The solids in sewage were separated at the outfalls from 1889 and the resulting sludge was dumped at sea.

FARADAY GIVING HIS CARD TO FATHER THAMES;

And we hope the Dirty Fellow will consult the learned Professor.

Fig. 3.3 The 'Year of the Great Stink', 1858 (from the magazine *Punch*).

Dibdin's ideas on bacteriological treatment of sewage were researched further throughout the 1890s and the first treatment plant was put into commission in 1914 – though in Manchester, not London. Three experimental plants for biological treatment were installed downstream of London in 1920 and a major plant was erected in 1928, with greatly expanded facilities provided since then. A marked improvement in the quality of the water in the River Thames became apparent during the 1960s and a diverse flora and fauna, absent for many years, began to colonize the river (Attrill, 1998).

Similar stories could be told of cities throughout the world.

The basic treatment process

There are three objectives in the treatment of sewage: to convert sewage into suitable end products, that is an effluent that can be satisfactorily discharged into the local watercourse and a sludge that can be readily disposed of; to carry this out without nuisance or offence; and to do so economically and efficiently.

There are four stages in the treatment of sewage but, depending on the quality of the effluent required, not all the stages may be used:

- **preliminary treatment**, which involves screening for large objects, maceration and removal of grit, together with the separation of storm flows;
- **primary treatment (sedimentation)**, where the suspended solids are separated out as sludge;
- **secondary (biological) treatment**, where dissolved and colloidal organics are oxidized in the presence of microorganisms;
- **tertiary treatment**, which is used when a very high-quality effluent is required. It may involve the removal of further BOD, bacteria, suspended solids, specific toxic compounds or nutrients.

The amount of treatment that is provided will depend to some extent on the amount of dilution that is available in the receiving water and on the quality

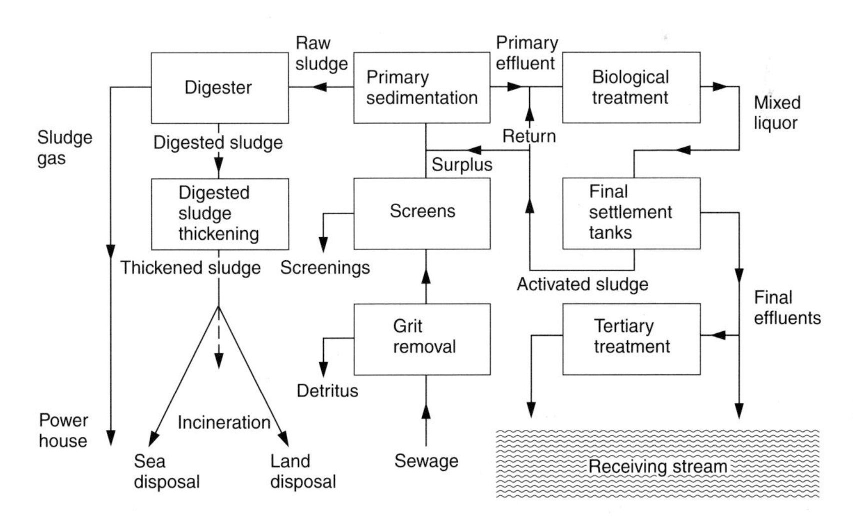

Fig. 3.4 Flow diagram of a typical sewage treatment works (from Wood, 1982).

objectives for that water. Preliminary treatment only may be given for effluents being discharged out to sea, though this is increasingly considered to be unacceptable. Tertiary treatment may be required if water is abstracted for potable supply downstream of a discharge. In Britain the minimum requirement for an effluent, which can be achieved with secondary treatment, is the Royal Commission Standard, allowing no more than 30 mg l^{-1} of suspended solids and 20 mg l^{-1} BOD (a 30:20 effluent). The effluent should be diluted with at least eight volumes of clean river water, having a BOD of no more than 2 mg l^{-1}, to ensure that there is minimal impact on the environment. Such a dilution may not always be available so that a more stringent standard than 30:20 may be required for the effluent. Ammonia is also often now included in any discharge consent, a concentration of 12 mg l^{-1} being a typical target value.

The basic organization of a sewage works is shown in Fig. 3.4.

Preliminary treatment

The sewage is initially passed through screens (rows of iron bars, with a spacing of 40–80 mm) which remove large debris, such as wood, paper and bottles. The screens are operated automatically and the screenings are either incinerated or buried. Grit and small stones are removed by passing the sewage either along a constant velocity channel or through a grit chamber.

Primary treatment

In the primary treatment or sedimentation process, sewage is passed slowly and continuously through tanks to remove as much solid matter as possible by sedimentation. The raw sludge can then be passed to the sludge digestion tank, the supernatant liquid (primary effluent or settled sewage) being given secondary treatment. There are a variety of designs of sedimentation tank but the ones most frequently installed are of a shallow, radial design, equipped with mechanical gear to remove the sludge. The sewage is retained for several hours and about 50 per cent of the suspended solids settle out as primary sludge. Sedimentation is cheaper than biological treatment in terms of unit removal of pollution so that the tanks need to be operated at their maximum efficiency.

Secondary (biological) treatment

Secondary treatment involves the oxidation of dissolved and colloidal organic compounds in the presence of microorganisms and other decomposer organisms. The aerated conditions are obtained usually either by trickling filters or activated sludge tanks, while in warmer climates oxidation ponds may be used. The secondary sludge which results from biological treatment is combined with the primary sludge in sludge digestion tanks, where anaerobic breakdown by microorganisms occurs.

There are advantages and disadvantages in the methods of biological treatment. Filters are often installed in small sewage works, serving populations of less than 50 000. They tend to be higher in capital cost, but lower in running costs, than activated sludge plants. Filters take up a greater amount of land than activated sludge plants, but they require less skill and active control in functioning. Trickling filters may breed flies, which can cause a nuisance to local residents, but activated sludge plants are noisy. Filter beds oxidize more nitrogen than activated sludge plants but the final effluent carries more suspended solids. Oxidation ponds are much cheaper than other methods of treatment to construct and maintain and they produce a good-quality effluent. However, they only function in warm, sunny climates and require large areas of land. The effluent is turbid, owing to large populations of algae and the ponds can be breeding grounds for noxious insects, especially mosquitoes.

Trickling (percolating) filters

A cross-section of a typical trickling filter is illustrated in Fig. 3.5. Trickling filters are circular or rectangular tanks, some 1–3 m high and filled with a packed bed of mineral or plastics. The mineral may be broken rock, gravel, clinker or slag but it must be uniformly graded so as to give a large proportion of spaces (voidage). The size range is usually 3.8–5.0 cm, with a specific surface area of 80–110 m^2 m^{-3} of volume and a proportion of spaces of 45–55 per cent of the total volume. Plastics are nowadays widely used instead of minerals because a high voidage can be obtained. The sewage from the primary settling tank is applied to the bed from above, either from rotating arms (circular tanks) or from pipes that travel backwards and forwards over the bed (rectangular tanks). The effluent flowing from the base of the bed contains suspended matter (humus) which is settled out and may be added to the primary settling tank. The clarified effluent may be recycled through the filter so as to dilute the incoming wastewater. The main factors influencing the

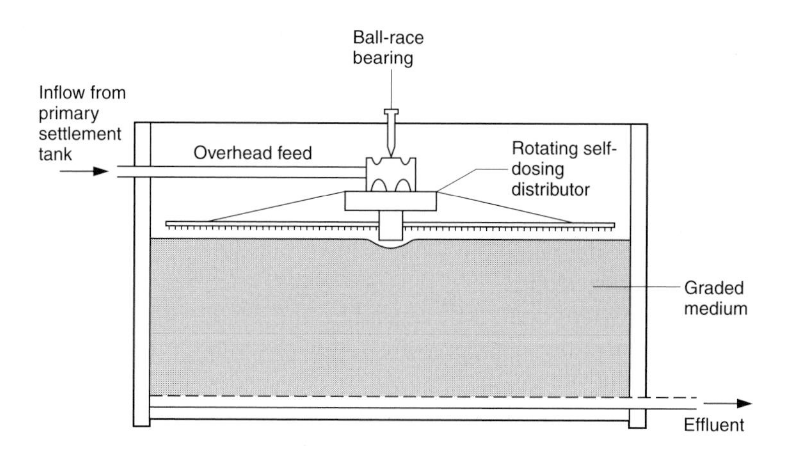

Fig. 3.5 Section through a trickling (percolating) filter.

rate of removal of BOD are the specific surface area of the filtration medium, the hydraulic loading and the temperature of the sewage.

With the application of settled sewage to the surface of the filter a biological community is gradually established as a slime (*biofilm*) on the surface of the material and its constituent organisms oxidize the pollutants in the wastewater. Bacteria are the most numerous organisms and form the base of the food web. A very wide range of bacteria has been recorded but the dominant aerobic genera appear to be the Gram-negative rods *Zoogloea*, *Pseudomonas*, *Achromobacter*, *Alcaligenes* and *Flavobacterium*. Fungi are normally outnumbered 8 : 1 by bacteria and are most abundant in the top 15 cm. The growth of fungal hyphae helps to transfer oxygen to the lower depths of the biofilm. These heterotrophic bacteria and fungi are responsible for the primary oxidation of the effluent. Autotrophic bacteria tend to be more predominant in the lower layers of the filter, with *Nitrosomonas* oxidizing ammonium to nitrite and *Nitrobacter* oxidizing nitrite to nitrate.

The filters hold large populations of algae (e.g. *Chlorella*) and cyanobacteria (e.g. *Oscillatoria*, *Ulothrix*) but they seem to play only a minor role in the purification process, whereas in large numbers they may reduce the efficiency of the filter.

Protozoa are present in proportions similar to fungi and over 200 species have been identified, of which ciliates predominate. They include species of *Carchesium*, *Chilodonella* and *Colpoda*. The major role of protozoa is to feed on bacteria that are freely suspended in the liquid film so that the effluent is clarified.

A diverse grazing fauna is present in percolating filters. It consists of rotifers, nematodes and annelid worms (e.g. *Eiseniella*, *Dendrobaena*), larval and adult flies (Diptera such as *Anisopus*, *Psychoda*, *Metriocnemus*) as well as beetles (Coleoptera) and springtails (Collembola). Filters with macroinvertebrates produce a much better effluent than those without. This is partly due to the grazing activity of the animals, which prevents the accumulation of too much film and increases oxygen diffusion. The film material is made more settleable by passing through the animals' guts. Material that has been converted to the chitin of the moulted exuviae (skins) of larval and pupal *Psychoda* settles quickly and a considerable proportion of the humus solids in percolating filters may consist of chitin. The respiration of the macroinvertebrate community makes a valuable contribution to the purification of sewage in biological filters.

To function efficiently, the percolating filter requires a continuous inoculation of microorganisms in the sewage. To assist in the establishment of an active film, a new filter bed may be seeded with humus sludge or activated sludge solids. Establishment is most rapid in the summer but it may be up to two years before the community is fully developed.

The community developing will depend on the depth within the filter bed, the time of the year and the composition of the waste to be treated. The rate of accumulation of the film is determined by the difference between the rate of growth and rate of removal and its control is important in the efficient functioning of a filter bed. A thick film impedes the flow of water through the bed and the growth rate of bacteria may be reduced because the rate of diffusion of nutrients through the slime decreases. Recirculation of effluent after it has passed through the filter dilutes the incoming effluent, thus reducing the strength of feed. It also increases the hydraulic load and together these reduce the rate of growth of the film. Alternate double filtration is frequently used, in which two filters are operated in series. The wastewater is applied at a relatively high rate to the primary filter and its effluent, after settlement, is passed to the second filter. The filters are switched at daily or weekly intervals. The primary filter quickly grows a thick film, but when it is switched to the secondary position the film rapidly shrinks. The overall costs are generally lower than for a single filtration. Lester and Edge (2001), Horan (1998) and Bitton (1999) provide detailed accounts of these fixed film reactors.

Activated sludge process

In this process the settled sewage is mixed with a flocculent suspension of microorganisms and aerated in a tank for 1 to 30 hours, depending on the treatment required. The medium is rich in dissolved and suspended nutrients, rich in oxygen and is violently agitated. The suspended and colloidal matter adsorbs on to the microbial flocs, bacterial extracellular polymers being involved in the process. The flocs settle and microbial metabolism breaks them down largely to carbon dioxide, water, ammonia and nitrate. The sludge, which increases by 5–10 per cent during the process, is removed from the purified effluent in a sedimentation tank and returned to the inlet of the aeration tank. Any excess is returned to the inlet of the primary sedimentation tank.

By contrast to the percolating filter, the activated sludge is a truly aquatic environment. The community lacks the higher links in the food web because turbulence within the tank produces unsuitable conditions for macroinvertebrates. The amount of microbial mass in the system is controlled by withdrawing excess sludge, whereas in the filter excess film is removed chiefly by biological agencies. In the activated sludge tank the microbial community is initially associated with the untreated waste and finally with the purified effluent, whereas in the filter bed a succession of communities is established at different depths in the bed and is associated with different degrees of effluent purification.

Most bacteria present in the activated sludge belong to Gram-negative genera such as *Pseudomonas*, *Zoogloea* and *Sphaerotilus*. Fungi are not usually dominant, though they may grow profusely if bacteria are inhibited, often when industrial effluent is present in the sewage. Fungi may cause ‘bulking’ in the sludge, in which loose-flocculent growths of microorganisms impede settling.

Over 200 species of protozoans are associated with activated sludge, some 70 per cent of them ciliates. Activated sludge is therefore more species-rich than percolating filters, but a much greater proportion of species are ciliates. Densities of protozoa are of the order of 50 000 cells ml^{-1} and a generalized succession can be observed. Rhizopods and flagellates, able to use dissolved and particulate wastes, initially predominate but later, as bacteria increase, predatory flagellates and free-swimming ciliates are dominant. Crawling and attached ciliates predominate in the later stages of floc formation. The role of protozoa is, as in the filter, to clarify the effluent. They improve the quality of the effluent by substantially reducing chemical oxygen demand, organic nitrogen, suspended solids and viable bacteria (Lester, 1996). Small numbers of nematodes and rotifers are also present in activated sludge.

For efficient operation the activated sludge process requires that the concentrations of substrate and microorganisms should be low. The basic activated sludge system has been developed in a number of ways, giving the process versatility and enabling it to be adapted to a wide range of operational circumstances. It is nowadays the most widely used biological process for the treatment of organic and industrial wastewaters. Detailed reviews of the activated sludge process are provided by Lester and Edge (2001), Horan (1998) and Bitton (1999), while Sell (1992) describes its use in dealing specifically with industrial wastes.

Oxidation ponds

Oxidation (or stabilization) ponds are used in warm climates to purify sewage, either as secondary treatment or to polish the final effluent before discharging it to the environment. They are a very ancient system of waste treatment. The most frequent type are facultative ponds, treating waste by both aerobic and anaerobic processes and involving an interaction between bacteria and algae (Fig. 3.6). The ponds are shallow lagoons, with an average depth of 1 m. Settled sewage passes through them in one to four weeks, but raw sewage may be retained for up to six months. The bacteria in the ponds decompose the biodegradable organic matter to release carbon dioxide, ammonia and nitrates. These are used by the algae, together with sunlight, and the photosynthetic process releases oxygen, enabling bacteria to break down more waste. A layer of organic sludge settles on the bottom of the ponds and anaerobic decomposition results in the release of methane. Oxidation ponds are used not only for treating sewage, but also for treating wastes

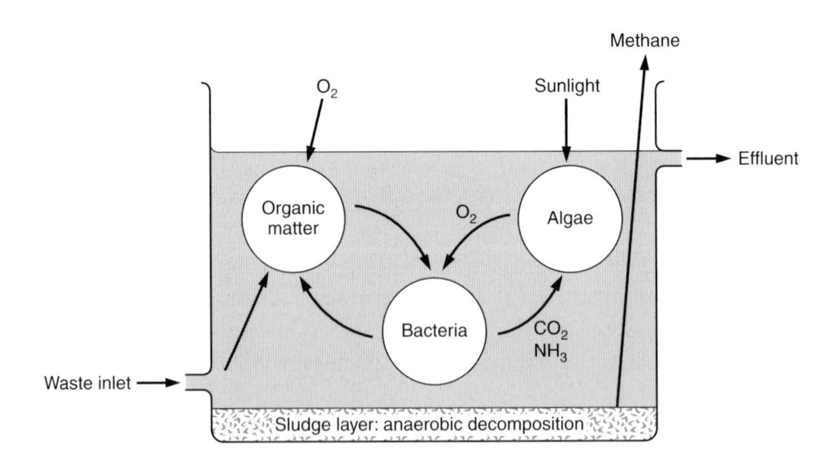

Fig. 3.6 Summary of processes in an oxidation pond.

from food processing, paper production and other industries. Bitton (1999) provides details of the processes involved in oxidation ponds.

Oxidation ponds not only purify wastes but they can also be used to provide energy. Algae have a solar conversion efficiency of 3–5 per cent and average production is 70 $t\,ha^{-1}\,yr^{-1}$ (35 $t\,ha^{-1}\,yr^{-1}$ algal protein). When harvested they may be fed directly to animals, used as fertilizer, as a source of chemicals, fermented to produce methane or burnt to produce electricity (Polprasert, 1989; Laliberté *et al.*, 1994). Fish also grow extremely fast in stabilization ponds, yields as high as 14 600 $kg\,ha^{-1}\,yr^{-1}$ being claimed, though in the tropics yields of 3000–8000 $kg\,ha^{-1}\,yr^{-1}$ seem more typical, with lower yields (generally less than 1000 $kg\,ha^{-1}\,yr^{-1}$) in temperate waste ponds (Polprasert, 1989). If fish are used for human consumption, care must be taken to avoid transferring pathogens and helminths.

Duckweed (*Lemna*) has been used in oxidation lagoons instead of algae (Hancock and Buddhavarapu, 1993). This simple, higher plant floats on the surface of the water, doubles its mass in 24 hours and is easily harvested and dried. It can be used directly to feed livestock or transferred to adjacent fish-ponds that are producing *Tilapia* (Cave, 1991).

Sludge digestion and disposal

The sludge produced by primary and secondary treatment processes is passed to sludge digestion tanks, where it is decomposed anaerobically. Alternatively, it may be stored for later disposal. The sludge amounts to some 50 per cent of the initial organic matter entering the sewage works. The role of anaerobic bacteria is to convert the raw sludge into a stable and disposable product, which neither gives rise to offensive smells nor attracts harmful insects or rodents. The major constituents of raw sludge are proteins, fats and polysaccharides and they are degraded in three processes. Hydrolysis involves the formation of long chain fatty acids, amino acids, monosaccharides and disaccharides. The second process, acid formation, results in the production of a range of fatty acids, alcohols, aldehydes and ketones, together with ammonia, carbon dioxide, hydrogen and water. The third process, methanogenesis, results in methane, carbon dioxide and water. There is also, of course, an increase in bacterial biomass.

Sludge digestion often takes place in two stages. The first stage involves digestion in closed tanks, heated to a temperature of 27–35 °C for 7–30 days; most of the gas evolution occurs here. Some of the gas evolved is used to heat the tanks. In the second stage, further digestion occurs for 20–60 days in open tanks at ordinary temperatures and the sludge consolidates and dries with the separation of a supernatant liquor, which is then returned to the sewage inlet for treatment with the sewage.

The sludge, after digestion, has been reduced in volume by two-thirds and its disposal presents some problems. It is no longer legal to dump it at sea, a frequently used option in the past. It may be incinerated, a process that could yield usable energy to produce electricity, or it may be used as land-fill, while it is also a potentially valuable fertilizer. Sludge may be composted or sprayed directly on to agricultural land. In England and Wales, some 473 000 t of sewage sludge are added to agricultural land each year, at an application rate of some $100\,m^3\,ha^{-1}\,yr^{-1}$ (Try and Price, 1995). Sludge may, however, contain toxic materials, such as metals or fluoride, which will gradually build up in cultivated soils. Metal availability to crops increases at lower pH and may affect productivity or accumulate in those parts eaten by humans (Smith, 1994a,b). Liming of soils will reduce the uptake of metals and, at pH 6 or more, current regulations provide high protection of the human food chain from this potential source of contamination. Whether natural food chains, for example via metal uptake by earthworms, are so protected is less certain. When applied to pastures, the sludge and associated metals adhere to the leaves of grass, taking a month or more to decline to control levels and presenting a potential risk to livestock (Aitken, 1997). There is some evidence that sheep grazed on pastures fertilized with sewage sludge have exceptionally high concentrations of cadmium in their livers and kidneys (Coghlan, 1997). There is also the risk that foodstuffs grown in sludge may be a potential source of viruses to consumers and the current science and regulations appear inadequate to preclude this risk. Sewage sludge is a good fertilizer, however, and improves soil aggregation. With the high cost of artificial fertilizers, it is economically attractive to farmers, particularly where transport costs are low or where these costs are met by the water industry.

The use of untreated sewage sludge on the land is being phased out in Great Britain. Treating sludge to a higher standard, including pasteurization, will cost the industry some £270 million in capital expenditure, with annual operating costs of £14 million.

Tertiary treatment

In many situations the dilution available to an effluent in the water that receives it is insufficient to prevent a deterioration of water quality. A higher-quality effluent must be produced and this effluent 'polishing' is known as tertiary treatment. The removal of phosphates and nitrates may also be necessary for environmental or public health reasons. Tertiary treatment is becoming the norm.

Nitrogen is usually removed from wastewater by biological processes, involving nitrification and denitrification and, when preceded by secondary treatment, over 90 per cent removal of total nitrogen can be achieved. Nitrification involves the oxidation of ammonia to nitrate, with nitrite as an intermediate:

$$2NH_4^+ + 3O_2 \rightarrow 2NO_2^- + 2H_2O + 4H^+$$
$$2NO_2^- + O_2 \rightarrow 2NO_3^-$$

The reactions are carried out by the bacteria *Nitrosomonas* and *Nitrobacter* respectively. Denitrification involves the conversion of nitrate to nitrogen gas and a number of facultative heterotrophs use nitrate instead of oxygen as the final electron acceptor during the breakdown of organic matter under anoxic conditions. With methanol as the organic carbon source the reaction is:

$$6NO_3^- + 5CH_3OH \rightarrow 3N_2 + 5CO_2 + 7H_2O + 6OH^-$$

Because nitrified effluent contains little carbon, a carbon source is normally added. This is frequently methanol because it is almost completely oxidized, thus producing less sludge for disposal, and it is relatively inexpensive. Reviews are provided by Robertson and Kuenen (1992) and Hardman *et al.* (1993).

The oxidation ponds described above may be used in conjunction with a standard sewage treatment works, secondary effluent being run into shallow lagoons to mature, resulting in a high-quality final effluent. Effluents can also be passed through rapid sand filters, or micro-strained, resulting in a final effluent with a BOD of less than $10\,mg\,l^{-1}$ and a similar level of suspended solids.

Fluidized beds

These systems treat wastes to a high quality and they are also efficient at removing nitrates. The fluidized bed combines features of both the trickling filter and activated sludge processes. Wastewater is passed upward through a reaction vessel that is partially filled with a fine-grained medium. The velocity is sufficient to fluidize the bed (i.e. impart motion to it) and a microbial community develops on the surface of the medium. Activated carbon and sand are typical media and offer large surface areas per volume of reactor. As the particles are in fluid motion there is no contact between them, allowing the entire surface to be in contact with the wastewater, the microbial biomass attained being several times that of the trickling filter or activated sludge process. The required degree of treatment can therefore be achieved in a much smaller reactor volume, saving considerable amounts of land. Operating costs are, however, greater because pure oxygen must be injected into the incoming wastewater stream to support the increased microbial biomass. The sludge produced is more concentrated (10 per cent solids compared with 2 per cent for activated sludge). If the main aim is the removal of nitrates by denitrification then the expensive oxygen, of course, is not required.

Constructed wetlands

Secondary effluent may also be passed through beds of aquatic macrophytes to remove wastes, the pollutants being immobilized and degraded by the normal physical and biological processes that operate in a wetland ecosystem (Hardman *et al.*, 1993). Macrophytes may be submerged (e.g. Canadian pondweed, *Elodea canadensis*), floating (e.g. water hyacinth, *Eichhornia crassipes*) or emergent (e.g. reed, *Phragmites australis*). Water hyacinth has a very high productivity in the tropics and is being used in countries such as Brazil as a tertiary treatment system, removing nitrogen and phosphorus. The biomass is removed frequently to sustain maximum productivity and to remove incorporated nutrients. Water hyacinths can also be used as integrated secondary and tertiary treatment systems, removing BOD as well as nutrients, both decomposition of organic matter and the microbial transformation of nitrogen proceeding simultaneously; in this system harvesting is only carried out for maintenance purposes and performance with respect to phosphorus removal is poor (Brix and Schierup, 1989). Polprasert (1989) provides a review.

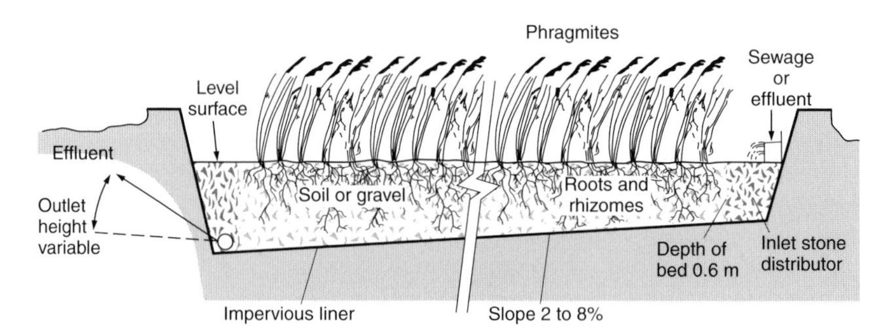

Fig. 3.7 Section through a typical reedbed system (from Cooper *et al.*, 1989).

There is much current interest also in reedbed systems, especially for the cost-effective secondary treatment of wastes from small communities and for tertiary treatment. They may also be useful in treating the run-off from major roads prior to its discharge to watercourses (Ellis *et al.*, 1994). A typical reedbed system is illustrated in Fig. 3.7. Reedbed designs are of two types, those that have a subsurface flow and those in which water flows over the surface. A bed is dug to a depth of 60 cm and lined with an impervious liner (clay or butyl rubber) and filled with gravel or sand (subsurface system) or back-filled with soil (surface system), and planted with reeds or other emergent macrophytes. In the subsurface system the rhizomes grow deep into the substrate, providing channels along which the water can run. Oxygen is transported to the roots, some of which diffuses out to allow aerobic treatment of wastes by bacteria in the rhizosphere. Anaerobic treatment takes place in the surrounding substrate. In the surface flow system water flows to a depth of about 10 cm and waste is decomposed aerobically in the above-ground layer of litter formed from dead leaves and stems. The reedbed system works on the same basic principle as trickling filters but the reedbed has a greater structural diversity and provides a large, vertical gradient in redox potential, which, it is believed, allows a greater range of chemical and biochemical processes to take place.

A specific example of a constructed wetland is illustrated in Fig. 3.8. This has been designed to cope with sewage waste created by visitors at the Wildfowl & Wetlands Trust centre at Slimbridge in England. The maximum daily waste is some 120 000 l but varies greatly both daily and seasonally. The sewage is first pumped into a settlement tank, where liquids and solids are separated. The solids are drawn off into a holding tank, from where they are passed on to two, sludge-drying vertical flow reedbeds. The sludge dries to form a friable compost, while the liquid draining from the bottom joins the main system to receive treatment with the liquid from the separation tank. This flows into a vertical subsurface flow reedbed (1 in Fig 3.8), constructed on sands and gravels, which filters and breaks down organic matter and oxidizes ammonia to nitrate. It is considered that the quality of effluent leaving this wetland is already as good as that achieved by many conventional sewage treatment works. The effluent then passes through a settlement pond (2) and on to a horizontal subsurface flow wetland (3 and Fig. 3.9),

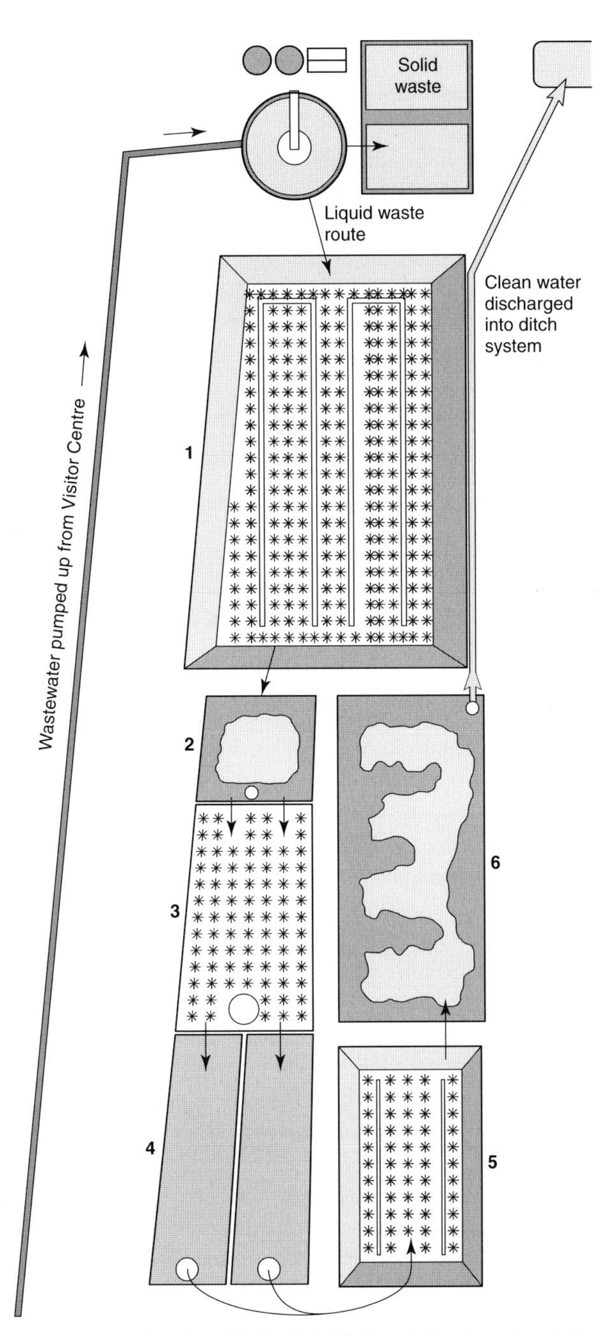

Fig. 3.8 The constructed wetland at Wildfowl & Wetlands Trust centre at Slimbridge. See text for explanation. This system is open to public view.

Fig. 3.9 A view of the constructed wetland at Slimbridge (photograph by Matthew Millett, Wildfowl & Wetlands Trust).

planted with a mixture of yellow iris (*Iris pseudacorus*) and sedges (*Scirpus* spp.). It is here that denitrification (see p. 102) takes place to convert nitrates to nitrogen gas. A large surface flow reedbed (4) acts as a security system for periods when the wetland is being used at maximum capacity. The next stage (5) is a reedbed constructed on clay to bind phosphates and polish the effluent before it passes to a wildlife lake (6), where the success of the system in producing a clean effluent can be demonstrated to visitors. From there it flows via a ditch system to the river.

In general, reedbed systems appear to be very efficient at removing pollutants. On average they remove 73 per cent of BOD, 69 per cent of suspended solids, 44 per cent of ammonia, 64 per cent of total nitrogen and 55 per cent of total phosphorus (Stauffer, 1998). From an analysis of a large database of natural and artificial wetlands in North America, Richardson and Qian (1999) have proposed a 'One Gram Assimilative Capacity Rule'; wetlands are able to assimilate $1\ g\ P\ m^{-2}\ yr^{-1}$ without significant ecosystem change or elevated phosphorus output. Constructed wetlands may also be useful at removing pesticides and heavy metals from wastewaters (Odum, 2000).

The advantages of constructed wetlands include low capital costs, simple construction, low maintenance costs, robust operation and a consistent effluent quality. The macrophytes within the system stabilize the surface of the wetland, reduce flow velocity, insulate the surface, take up nutrients, provide attachment surfaces for microbes, transfer and release oxygen, and provide an aesthetically pleasing wildlife habitat (Brix, 1994). They require, however, relatively large areas of flat land for construction and may take up to three years to reach maximum efficiency. Very many wetlands have been constructed over the last 20 years, treating wastes of a single household up to towns with populations of 100 000 (Stauffer, 1998). They should, however, still be considered as experimental, as their performance over long timescales is unknown. It must be stressed that systems for purifying wastewater using macrophytes should be custom-built; natural wetlands are too scarce and vulnerable a resource to be used for waste disposal. Reviews of constructed wetlands are provided by Merritt (1994), Mitsch (1994), Kadlec and Knight (1996), Worrall *et al.* (1997), Cole (1998), Stauffer (1998), Campbell and Ogden (1999) and Nuttall (1999).

Removal of pathogens

The reduction in the numbers of pathogens during the sewage treatment process is governed by the length of retention time during treatment, chemical composition of the wastes and their state of degradation, antagonistic forces in the biological flora, pH and operational temperature, together with other less well-understood factors. Secondary treatment removes 90–99 per cent of bacteria and viruses. Before release to the environment, effluent may be disinfected with chlorine, ozone or ultraviolet radiation. With further considerable dilution in receiving waters the sewage treatment process can be seen as an efficient way of reducing the incidence of pathogens. Greater potential problems are associated with home septic tanks, which are often inefficiently maintained and overloaded, and also with storm drainage.

Treatment of farm wastes

Wastes produced by intensive livestock units present considerable problems of disposal and pollution incidents caused by farm wastes are now of great concern. The wastes can, however, be treated in small, on-farm units that can be designed to suit the needs of the individual farming operation. Anaerobic digestion is the basis of many such systems. Digesters vary in capacity from $10\,m^3$ to $1000\,m^3$. In a typical system, the anaerobic digester tank is heated using methane produced in the digestion process, and stirred. Material progresses through the digester in 10–20 days and sufficient excess methane is produced to provide energy for heating and power generation on the farm. The polluting power of the waste is reduced by as much as 80 per cent. The digested waste can be split into liquid and solid components. The liquid may be pumped through small-bore pipes to be used as a fertilizer when and where it is required on the farm, while the solid waste is easy to handle and can be used on the farm or even sold as a high-quality compost. The treatment of animal wastes is discussed by Polprasert (1989), Fallowfield *et al.* (1992) and Freitas and Burr (1996).

Overview of waste treatment

This section has largely been concerned with the treatment of sewage, but the principles apply equally well to the treatment of industrial organic wastes. Many factories have small-scale effluent treatment plants to deal with their organic wastes before these are discharged into rivers.

Sewage treatment has been described in detail because it demonstrates how biological systems can be applied effectively to industrial processes. A study of these processes can in turn help in the development of new fundamental biological theory, for instance in understanding the dynamics of ecological communities. The discussion has also shown that there are many potential by-products of economic importance to be gained from the treatment process, such as energy, organic compounds and food, either directly as fish or indirectly as animal feed. The exploitation of this potential will involve the skilled manipulation of biological systems. There has been considerable recent interest in methods to recover or produce compounds from organic and industrial wastes. These include the direct extraction of proteins, fats and other organic compounds, the microbial conversion of carbohydrate to protein and the recovery of metals. The development of genetically engineered microorganisms (GEMs) will enable many valuable materials to be recovered or synthesized from domestic or industrial wastes.

Microorganisms can also be used to break down recalcitrant compounds, which are not broken down when released to the environment. The majority of these are synthetic organic molecules, including detergents and pesticides. Some microorganisms can metabolize particular compounds. The bacteria *Alcaligenes*, *Azotobacter* and *Flavobacterium* use aromatics as a substrate, for example, and the fungus *Trichosporon cutaneum* uses phenols.

Microbial cultures to degrade these compounds can be made by enrichment culture or gene manipulation. In enrichment culture an inoculant of microorganisms is taken from a habitat already exposed to the compound of concern and is cultured and exposed to increasing concentrations of the compound until it becomes a source of essential nutrient. Gene cloning techniques have recently been used to

Fig. 3.10 A poor-quality effluent discharged from a sewage treatment works (photograph by Sheila Macdonald).

produce bacteria that can break down a wide range of recalcitrant organics and some of these microorganisms have been patented. Packaged microbes are also now available for dealing with wastes that are difficult to treat. Packaged microbes can, for example, be used to improve BOD removal in treatment plants, or to improve methane generation in digesters, or for degrading spilled oil (p. 227). Further discussion on the use of microorganisms in pollution biotechnology can be found in Hardman *et al.* (1993), Scragg (1999) and Madigan *et al.* (2000).

In an ideal world, all the materials in sewage would be economically recovered and water alone would be returned to the hydrological cycle. The world, of course, is not ideal. Even where sewage is receiving secondary treatment, the effluent produced may be substandard because the design capacity of the works may not be sufficient to cope with the population it is now having to serve and resources are not available to expand the plant (Fig. 3.10). Similarly, many plants are old and inefficient or become overloaded during unusual conditions such as storms. Sewage is also changing with time (e.g. increased amounts of oils and plastics) and these may present problems with which old plants are not designed to cope. In England and Wales effluents from sewage treatment works are still one of the main causes of deterioration in river water quality, responsible for some 23 per cent of all reported pollution incidents. In 1999, five water companies topped the table of prosecutions for pollution offences caused by sewage discharges. In addition to sewage, other forms of organic pollution may be affecting the receiving water.

Many parts of the world lack any effective waste treatment facilities. In Latin America, 98 per cent of sewage is untreated when discharged to rivers, while 60–70 per cent of the daily flow of the river passing through Manila, the capital of the Philippines, consists of untreated sewage. In Portugal, Greece and Turkey, 79.1, 88.6 and 93.7 per cent respectively of the population are not served by sewage treatment plants (Stauffer, 1998). The next section will outline the effects of organic pollution on the receiving stream.

EFFECTS OF ORGANIC EFFLUENTS ON RECEIVING WATERS

The oxygen sag curve

When an organic polluting load is discharged into a river it is gradually eliminated by the activities of microorganisms in a way very similar to the processes in the sewage treatment works. This *self-purification* requires sufficient concentrations of oxygen, and involves the breakdown of complex organic molecules into simple inorganic molecules. Dilution, sedimentation and sunlight also play a part in the process. Attached microorganisms in streams play a greater role than suspended organisms in self-purification. Their importance increases as the quality of the effluent increases since attached microorganisms are already present in the stream, whereas suspended ones are mainly supplied with the discharge.

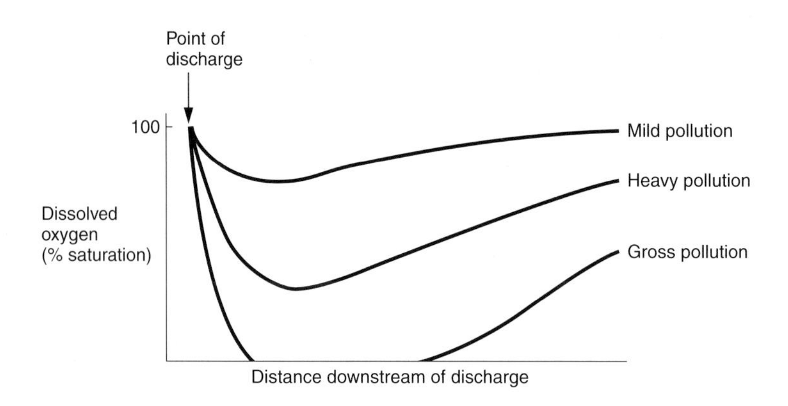

Fig. 3.11 Effect of an organic discharge on the oxygen content of river water.

The de-oxygenation of a river as a result of organic wastes is generally slow so that the point of maximum de-oxygenation may occur considerably downstream of a discharge. The degree of de-oxygenation depends on a number of factors such as the dilution that occurs when the effluent mixes with the stream, the BOD of the discharge and of the receiving water, the nature of the organic material, the total organic load in the river, temperature, the extent to which re-aeration occurs from the atmosphere, the dissolved oxygen in the stream and the numbers and types of bacteria in the effluent.

Figure 3.11 illustrates generalized curves, known as *oxygen sag curves*, which are obtained when dissolved oxygen is plotted against the time of flow downstream. The curves show slight, moderate and gross pollution. A poor-quality effluent contains a high concentration of ammonia and the eventual oxidation of this to nitrite and nitrate may cause a further sag in the oxygen curve.

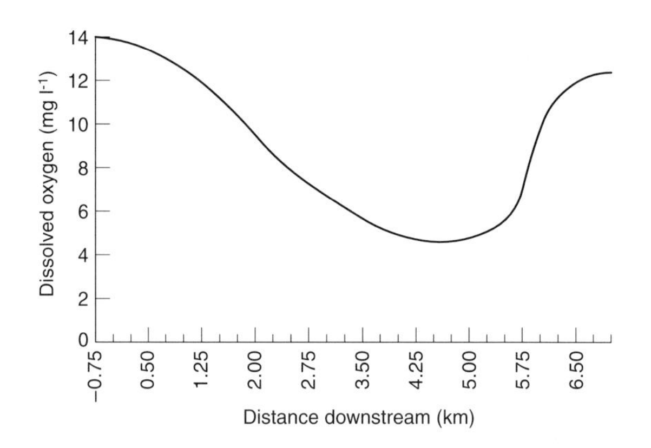

Fig. 3.12 Oxygen sag curve in the River Deben, eastern England, in July 1999 (data from Lynn Parr).

Figure 3.12 illustrates a real example of an oxygen sag curve, that measured below a discharge of treated sewage effluent in the River Deben in eastern England in summer. The effluent met the standard required but in this dry year there was very little flow in the river, allowing the growth of abundant macrophytes and algae, which further added to oxygen uptake from the water.

If the oxygen demand of a sewage effluent is studied over a number of days, it is found that oxidation proceeds quite rapidly to begin with, but then slows over a period of 15–20 days. There are often, however, two further stages in oxidation, which may account for a large proportion of the total oxygen consumption. The oxygen demand in the first 20 days is caused by the oxidation of organic matter (carbonaceous BOD), while later demand involves the oxidation of ammonia to nitrite and then nitrate.

To increase the rate of aeration and speed up self-purification below discharges, weirs are often built into rivers to considerable effect. To combat the disastrous effects of accidental pollution incidents, techniques have been developed to pump pure oxygen directly into threatened rivers. It is essential that the pollution be spotted in time, but the technique can be especially valuable where prime fisheries are at risk.

Effects on the biota

Organic pollution affects the organisms living in a stream by lowering the available oxygen in the water. This causes reduced fitness, or, when severe, asphyxiation. The increased turbidity of the water reduces the light available to photosynthetic organisms. Organic wastes also settle out on the bottom of the stream, altering the characteristics of the substratum. The general effects of fairly severe organic pollution are illustrated in Fig. 3.13. The top part of the figure (A) shows the oxygen sag curve, together with a massive increase in BOD, salts and suspended solids at the point of discharge, followed by a gradual decline in these parameters as self-purification occurs. The peak of ammonia (B) is replaced by a peak of nitrate as nitrification proceeds and both are gradually diluted as the polluting load travels downstream.

There is a large increase in the number of bacteria (C) immediately below the outfall and these gradually decline as their substrate is depleted. There is also a large increase in sewage fungus (see later), which disappears as the stream re-oxygenates. The protozoa are chiefly predators on bacteria and increase in response to bacterial increases, subsequently to decline as bacterial numbers fall. Algae, especially *Cladophora* (blanket weed, a large filamentous green alga) increase in numbers as recovery begins, light conditions improve and nutrients are released from the oxidizing organic matter. They decrease as nutrients are used up or diluted.

The stream animals are shown at the bottom of Fig. 3.13. The clean water fauna is eliminated at the point of discharge of the pollutants, unable to tolerate the lowered oxygen tension. Sludge worms (Tubificidae) may be the only

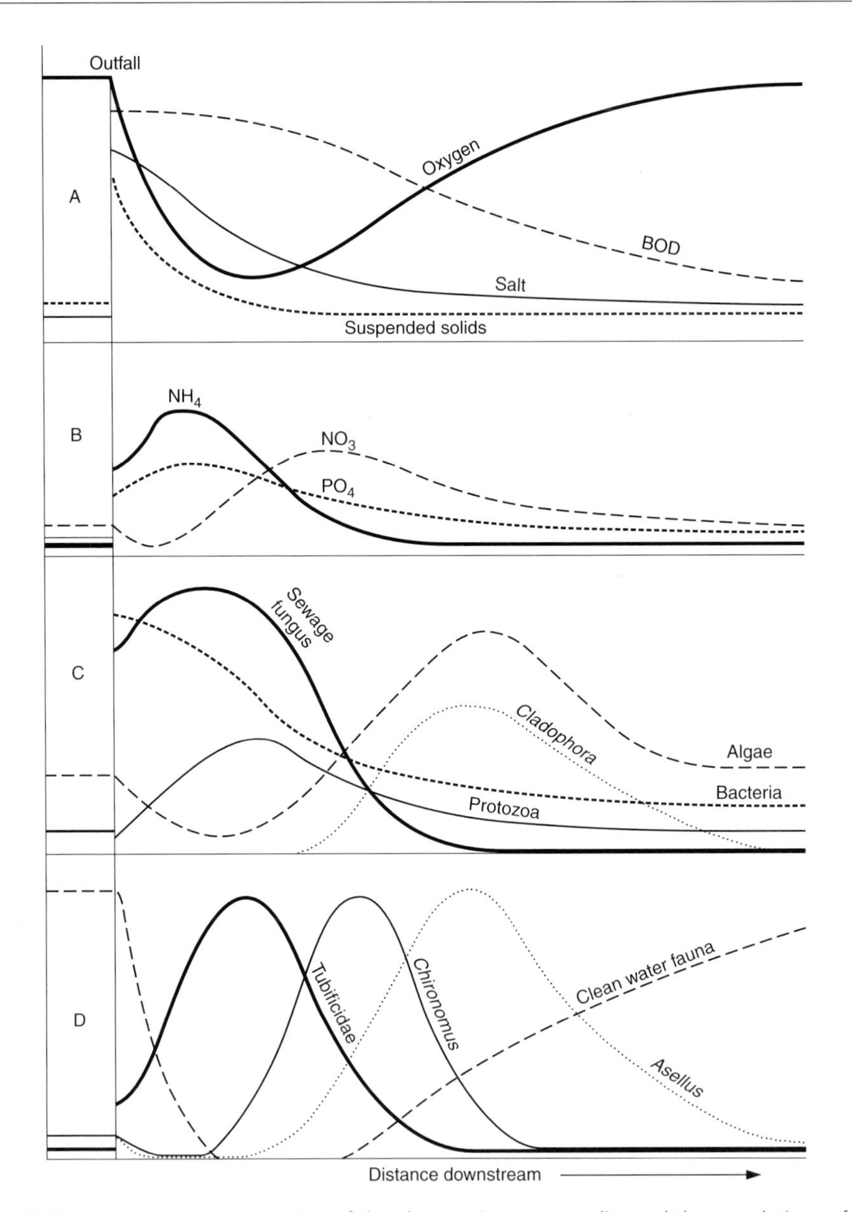

Fig. 3.13 Schematic representation of the changes in water quality and the populations of organisms in a river below a discharge of an organic effluent (from Hynes, 1960). A, physical changes; B, chemical changes; C, changes in microorganisms; D, changes in macroinvertebrates.

macroinvertebrates present immediately below the discharge and, in some cases of very severe pollution, even these may be absent. As conditions improve, blood worms, larvae of the midge *Chironomus*, become abundant and, as further amelioration occurs, large populations of the isopod *Asellus* build up. As the stream gradually re-oxygenates, the clean water fauna increases in numbers and diversifies.

The absence of a particular species or group from a river may not be indicative of pollution because not all reaches are suitable for them. Stoneflies, for example, are largely confined to eroding substrata with fast currents and do not occur in the slow-moving, silty, lowland reaches of rivers, even where these are free of pollution. Various zones can be recognized along a river and these have characteristic animals and plants associated with them (Hawkes, 1975).

Microorganisms

The primary effect of organic pollution is to increase the numbers of bacteria that use the waste as a substrate. Bacteria increase markedly below the sewage outfalls of large cities, with direct counts of up to 36×10^6 ml^{-1}. Most of the bacteria below an outfall are suspended, about 10 per cent being attached to plant surfaces. Viruses may also be present in significant numbers, the majority of them deriving from sewage effluents (Walter *et al.*, 1989).

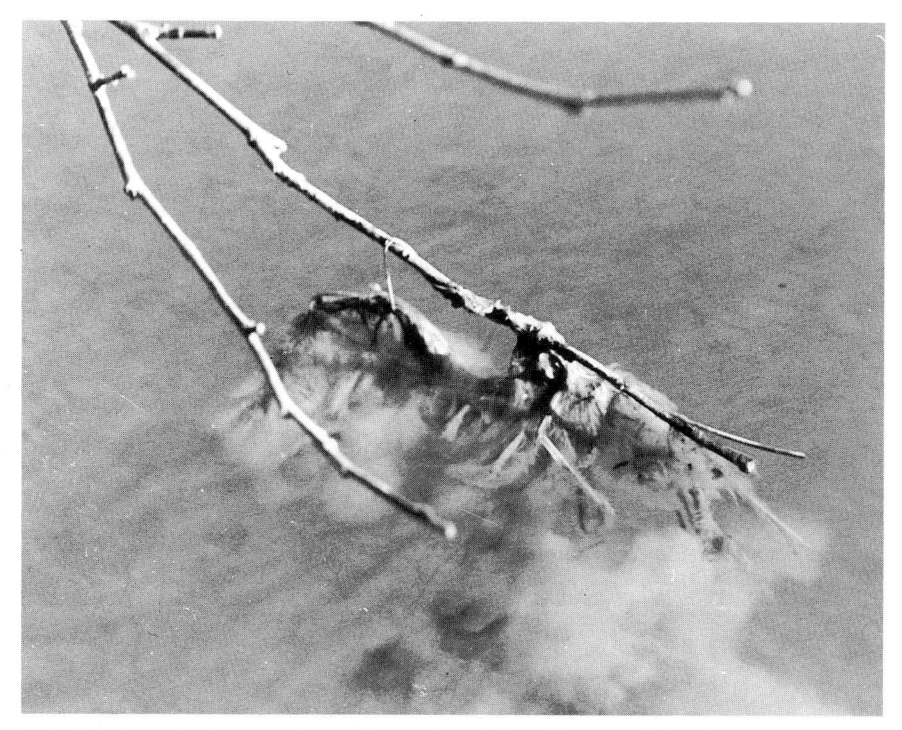

Fig. 3.14 Growth of sewage fungus in a polluted river (photograph by the author).

Table 3.2 Typical organisms of the sewage fungus community (from Curtis and Curds, 1971)

Bacteria	*Sphaerotilus natans*
	Zoogloeal bacteria
	Beggiatoa alba
	Flavobacterium sp.
Fungi	*Geotrichum candidum*
	Leptomitus lacteus
Protozoa	*Colpidium colpoda*
	Colpidium campylum
	Chilodonella cucullulus
	Chilodonella uncinata
	Cinetichilum margaritaceum
	Trachellophyllum pusillum
	Paramecium caudatum
	Uronema nigricans
	Hemiophrys fusidens
	Glaucoma scintillans
	Carchesium polypinum
Algae	*Stigeoclonium tenue*
	Navicula spp.
	Fragilaria spp.
	Synedra spp.

Under fairly heavily polluted conditions a benthic community of micro-organisms, known as sewage fungus, develops. Sewage fungus (Fig. 3.14) is an attached macroscopic growth, which may form a white or light brown slime over the surface of the substratum, or may exist as a fluffy, fungoid growth, with long streamers. It is usually bacteria, not fungi, that dominate the sewage fungus community. The organisms making up sewage fungus are listed in Table 3.2, and some organisms are illustrated in Fig. 3.15. The dominant species are usually *Sphaerotilus natans* and zoogloeal bacteria.

Sphaerotilus natans consists of an unbranched filament of cells, enclosed in a sheath of mucilage. The sheath enables the bacterium to attach to solid surfaces, as well as protecting the organism from parasites and predators. These bacteria use a wide variety of organic compounds as substrates and they can also use inorganic nitrogen sources though growth is less luxuriant than with organic nitrogen. When glucose and acetate were added as carbon sources to artificial channels, slime did not form below a substrate concentration of $1\,mg\,l^{-1}$, while above this concentration slime formation was proportional to oxygen concentration (Curtis *et al.*, 1971). The bacteria are aerobes, but they can survive in oxygen concentrations down to $2\,mg\,l^{-1}$. For typical growth, a certain amount of flow is also necessary.

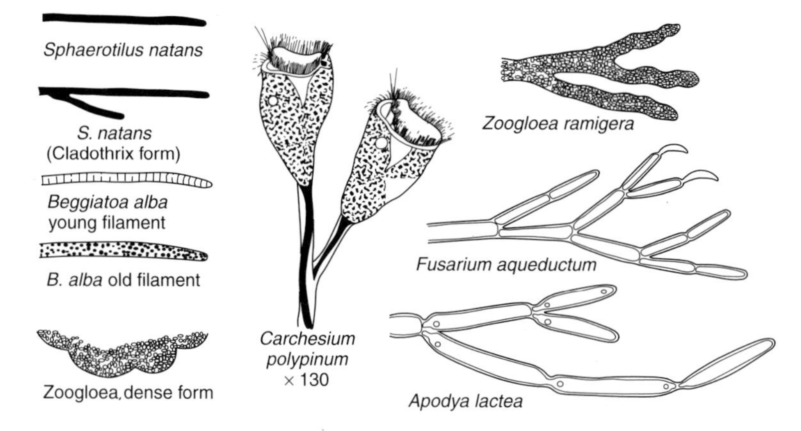

Fig. 3.15 Some of the characteristic species in the sewage fungus community (× 330, except *Carchesium*) (from Hynes, 1960).

The sewage fungus community extends further downstream during the winter than in summer, because the oxidation of organic matter in the effluent proceeds more slowly, such that the pollution plug extends further downstream. Also the sewage fungus bacteria are able to compete more effectively with other heterotrophic bacteria at lower temperatures.

Eleven protozoan species occur in very large numbers in the sewage fungus community (Table 3.2). *Carchesium*, a stalked, ciliated protozoan, tends to be most abundant at the lower end of the sewage fungus zone, where oxygen conditions are somewhat improved. It is a predator, feeding mainly on the large populations of bacteria present in sewage effluent. *Carchesium*, when present in large numbers, can cause silting of the river bed because it flocculates suspended matter. *Colpidium colpoda* and *Chilodonella cucullulus* are very abundant in the sewage fungus. *Colpidium* feeds mainly on bacteria, while *Chilodonella* is a more generalized feeder on bacteria, diatoms, filamentous growths and flagellates. Gray (1985) provides a review of sewage fungus.

Algae

With heavy organic pollution, de-oxygenation and low light, algae are eliminated from rivers, but there is a gradual reappearance as conditions improve and populations and growth are stimulated by the large concentrations of nutrients present. Algae have been investigated by immersing glass slides in organically enriched rivers. The filamentous *Stigeoclonium tenue* is common in the zone immediately below the region of gross pollution. *Stigeoclonium* also occurs in the sewage fungus community (Table 3.2). The characteristic species of the recovery zone are the diatoms *Nitzschia palea* and *Gomphonema parvulum*. The cyanobacterium *Chamaesiphon* sp., the green alga *Ulvella frequens* and the diatom *Cocconeis placentula* appear when the pollution has dispersed. A long distance is required for the complete recovery of the algal community in grossly polluted rivers.

The diversity of algal species in clean waters can be very variable. Heavily polluted environments always have communities low in numbers of species because sensitive species are gradually eliminated as the pollution load increases. At low levels of organic pollution, however, tolerant species tend to increase, while conditions are not sufficiently severe to cause a great loss of sensitive species. Mildly polluted rivers can thus have high diversity. The filamentous green alga *Cladophora* becomes abundant in the recovery zone and forms dense blankets over the substratum, where it provides both cover and food for invertebrates. The ecology of the plant has been reviewed by Dodds and Gudder (1992). Its dense growth in the recovery zone is linked to an increase in nutrients, and especially to phosphate but other factors are probably also involved. *Cladophora* may be so prolific in organically enriched streams that it de-oxygenates the water at night, causing fish to die. It also interferes with river recreation, such as angling and boating.

Higher plants

Macrophytes are also adversely affected by organic pollution and may be eliminated by severe pollution events. The load of suspended solids reduces light and, by settling out, may render the bed of the river unstable. When conditions improve downstream, the rich supply of nutrients can result in a very dense growth of some species. Haslam (1987) has described the effects of organic pollution on plants in rivers. She recorded a decrease and loss of those species most sensitive to pollution, with a decrease in overall species richness, and an increase in any species favoured by pollution. The only species whose range appears to be increased by organic pollution is the pondweed *Potamogeton pectinatus*, which is very tolerant. Five species of higher plant were described as tolerant, being common in both clean and polluted streams, namely *Mimulus guttatus* (monkey flower), *Potamogeton crispus* (curled pond-weed), *Schoenoplectus lacustris* (club-rush) and the bur-reeds, *Sparganium emersum* and *S. erectum*. A further seven species were fairly tolerant.

Figure 3.16 illustrates the distribution of some macrophyte species in the River Mersey catchment in northwest England. The Mersey has three main tributaries, the Goyt, the Tame and the Etherow. The River Goyt receives the greatest number of industrial discharges in the catchment, including effluent from textile firms. Pollution begins at the head of the river and, although there are some improvements downstream, water quality is still poor at its confluence with the Mersey. The River Tame is bounded on both sides by industrial and urban development throughout its length and receives effluents from six sewage works, all of which also treat industrial effluent, and several sewer overflows. The lower stretches of the river have continuous growths of sewage fungus. The River Etherow has a higher water quality than its tributaries but it deteriorates at the confluence with the Goyt, and the Mersey remains of poor quality until it discharges into the estuary.

The aquatic vegetation in the headwaters of the catchment is generally rich and

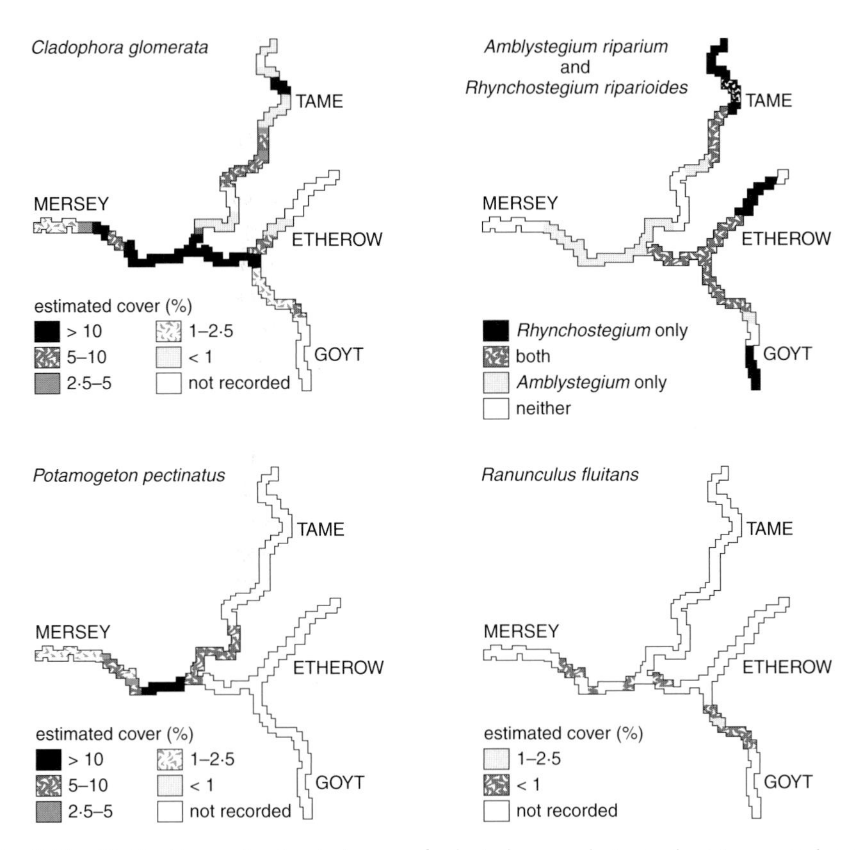

Fig. 3.16 Distribution and estimated cover of selected macrophytes in the Mersey catchment (from Holland and Harding, 1984).

dominated by bryophytes (mosses and liverworts). For example, the top of the Goyt supports a diverse and dense macroflora but in its middle reaches the river receives a severely polluting effluent from a textile works and all aquatic plants are absent. There is then a progressive increase in diversity and cover downstream. Plant communities undergo rapid changes in distribution, as evidenced by the discontinuous pattern of distribution and abundance shown in Fig. 3.16. The relative abundance of the mosses *Rhynchostegium ripariodes* and *Amblystegium riparium* appear to be related to organic pollution. In waters free of pollution neither species occurs. As enrichment increases, *Rhynchostegium* appears, to be joined by *Amblystegium* with further enrichment. As pollution becomes more severe, *Rhynchostegium* declines, followed by *Amblystegium*, and neither species is present in conditions of gross pollution. The patchy distribution of *Cladophora* is related to heavy metal contamination rather than organic pollution.

Overlying the effects of pollution on macrophytes are many natural characteristics of rivers, such as substratum type, flow regime, current speed, water chemistry and shading. These factors make it difficult to relate a particular plant community to pollution conditions, unless neighbouring streams with similar characteristics, but different loads of pollutants, are compared.

Invertebrates

Protozoa dominate the animal community in polluted rivers where oxygen is severely limiting and some species have already been mentioned in the discussion of sewage fungus. The response of protozoan communities to pollution is difficult to evaluate because large numbers are washed into streams from the surrounding land and in the effluents from the secondary sewage treatment process, making it difficult to distinguish between natural and intrusive members of the fauna.

There has been a great deal of work on the effects of organic pollution on macroinvertebrate communities. Heavy pollution affects whole taxonomic groups of macroinvertebrates, rather than individual species. Specific differences only become important in cases of mild pollution. In general those organisms associated with the silted regions of rivers are the most tolerant of organic pollution, while species associated with eroding substrata and swiftly flowing water are the most sensitive. Siltation of the river bed as solids settle out may clog the gills of species associated with eroding substrata. For example, the mayflies *Rhithrogena semicolorata* and *Ephemerella ignita* are particularly sensitive to siltation (Scullion and Edwards, 1980). Invertebrates from swiftly flowing waters also generally have higher metabolic rates than those from slow flowing waters and they would hence

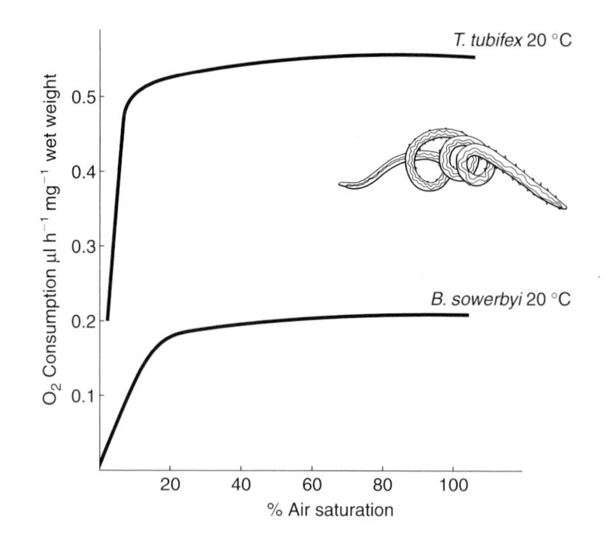

Fig. 3.17 Rate of oxygen consumption in relation to dissolved oxygen concentration, as a percentage of air saturation, in two species of tubificid worms, *Tubifex tubifex* and *Branchiura sowerbyi* (adapted from Aston, 1973).

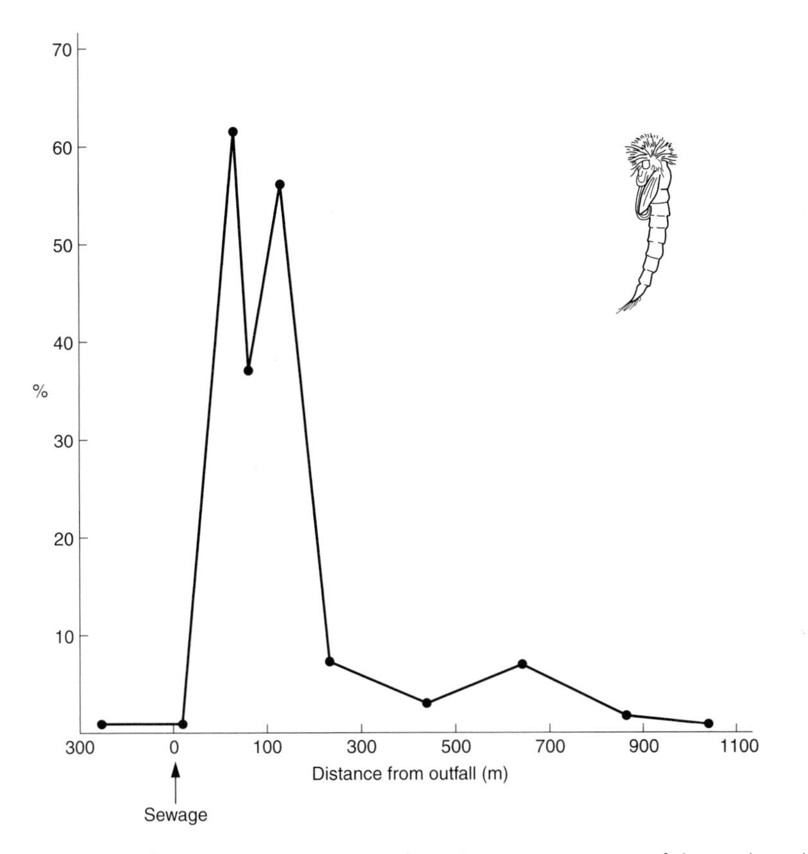

Fig. 3.18 Number of *Chironomus riparius* pupal exuviae as a percentage of the total number of chironomid exuviae at stations above and below a sewage outfall on the River Chew in southwest England (adapted from Wilson and McGill, 1977).

tend to be more sensitive to decreases in oxygen content in the water. Respiratory adaptations in invertebrates are discussed by Williams and Feltmate (1992) and Giller and Malmqvist (1998).

In heavily polluted waters tubificid worms are frequently very abundant, often forming monocultures with densities of over 10^6 m^{-2}. Successive species of tubificids are eliminated as conditions become more severe and in very severe conditions only *Limnodrilus hoffmeisteri* and/or *Tubifex tubifex* remain. For these animals the effluent provides an ideal medium for feeding and burrowing, while the absence of predators allows populations to increase largely unchecked.

The success of tubificids in these environments is due to their ability to respire at very low oxygen tensions. Figure 3.17 shows that the respiration rate of *Tubifex tubifex* and *Branchiura sowerbyi* is almost unaffected by dissolved oxygen concentrations down to 20 per cent of air saturation. Tubificids contain the pigment haemoglobin, which has a high affinity for oxygen. The haemoglobin in *T. tubifex* exhibits a negative Bohr effect, which means that oxygen can be taken up at low pH, when the carbon dioxide content of the water is high, a frequent occurrence in organically polluted waters. The pigment functions in the transport of oxygen, but it appears not to store oxygen for use during prolonged periods of anoxia. Tubificids are known to survive in anaerobic conditions for up to four weeks and they may metabolize glycogen anaerobically at this time. They can also feed and lay eggs at low oxygen tensions. Naidid worms may also respond to organic pollution by large increases in numbers, especially on the stony substrata that are favoured by the group. *Nais elinguis* appears particularly tolerant of pollution and may increase in numbers 20-fold below a sewage outfall.

As the water below a discharge of organic material becomes more oxygenated, tubificids decrease in abundance and are replaced by the midge larva (blood worm) *Chironomus riparius*, which cannot withstand oxygen conditions as low as those tolerated by tubificids. The chironomid zone contains species in addition to *C. riparius*, including the carnivorous Tanypodinae, which feed on tubificids and small chironomids. Chironomid densities may exceed 50 000 m^{-2} downstream of an organic discharge. Pupal exuviae (the 'skins' that remain floating on the surface of the water after the adult midges have hatched) have been used to characterize the distribution of chironomids in streams. Figure 3.18 shows how *C. riparius* dominates the chironomid community below a sewage outfall, being absent in the clean water immediately above. As the water self-purifies downstream the proportion of *C. riparius* decreases until it is absent 1 km below the discharge. Chironomid pupal exuviae can be used to detect point sources of organic pollution (Wilson, 1994).

The life cycle of *C. riparius*, with many broods each year, together with the selection of sites by ovipositing females, gives the species numerous opportunities during a year to invade a site that has become suitable as a result, for example, of temporary pollution caused by organic run-off from a farmyard.

Chironomus, like tubificids, has haemoglobin in its blood and the content may reach up to 25 per cent of the value for human blood. The haemoglobin has a molecular weight half that of mammalian blood and the pigment content of the blood increases when the water is poorly aerated. It acts as a carrier mainly when the oxygen tension of the water is low, at a time when the amount of oxygen required by the animal cannot be supplied by physical solution. *Chironomus riparius* lives in a tube, which is kept oxygenated by the undulatory movements of the animal's body. This activity extends the penetration of oxygen into the sediments and increases their rate of oxidation.

Below the chironomid zone, the isopod crustacean *Asellus aquaticus* becomes numerous (Fig. 3.13D, p. 112), especially where beds of *Cladophora* occur. The body length and mass of *Asellus* were found to be significantly lower in polluted sites, while females produced fewer eggs (Tolba and Holdich, 1981). Leeches, molluscs and the alder fly *Sialis lutaria* are also often abundant in this zone. The amphipod crustacean *Gammarus pulex* is very much more sensitive than *Asellus* to organic pollution, being killed within five hours at oxygen concentrations in water of $1\,mg\,l^{-1}\,O_2$. Its distribution in rivers was found to be largely determined by the number of hours during the night that the oxygen concentration fell below this critical concentration (Hawkes and Davies, 1971).

As the self-purification process continues, other species of invertebrates appear in the community. The species most sensitive to organic pollution are the stoneflies (Plecoptera) and to a lesser extent the mayflies (Ephemeroptera), which are often absent even at mild levels of pollution. *Amphinemura sulcicollis* is more tolerant than other stoneflies to organic pollution and *Baetis rhodani* and *Caenis horaria* are more tolerant than other mayflies.

The distributions of selected invertebrate species in the chronically polluted River Mersey catchment (p. 116) are shown in Fig. 3.19. The upper reaches of the Goyt support a rich invertebrate fauna, including stoneflies and mayflies, and in the absence of pollution, physical conditions would be suitable for them for most of the

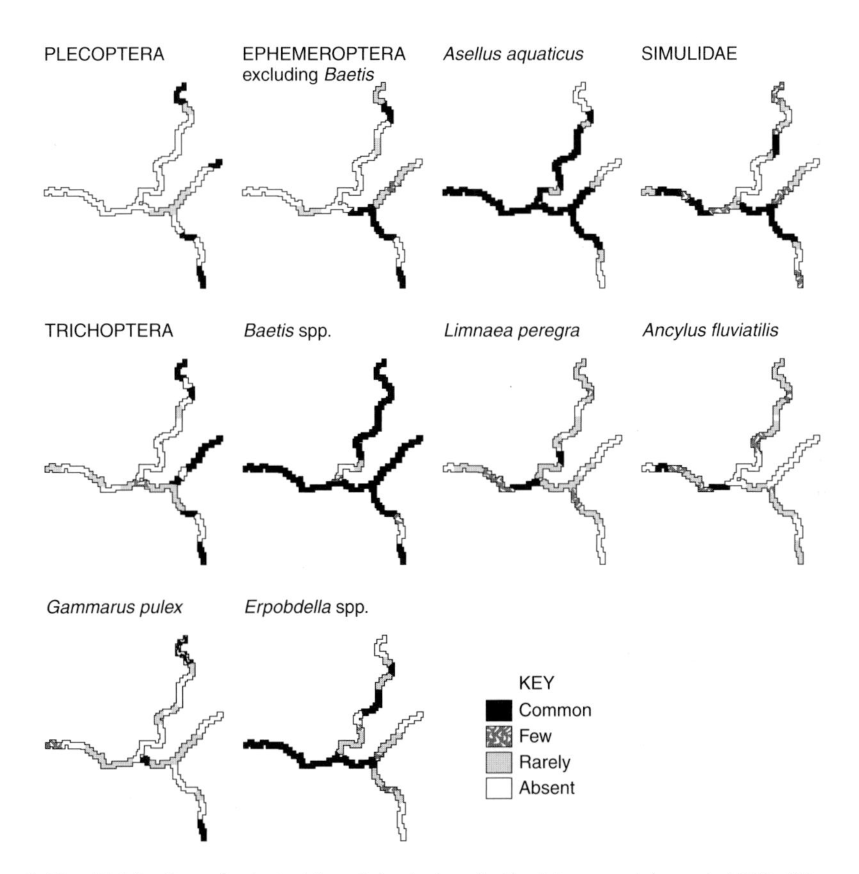

Fig. 3.19 Distribution of selected invertebrate taxa in the Mersey catchment, 1978–86 (from Holland and Harding, 1984).

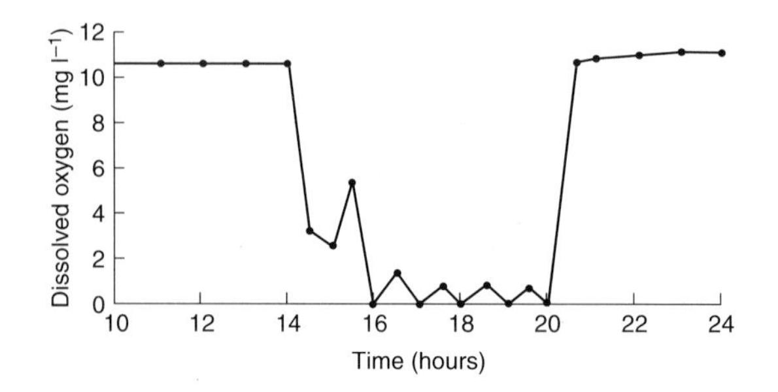

Fig. 3.20 Dissolved oxygen in stream water following treatment with sodium sulphite to mimic an organic pollution episode (from Edwards *et al.*, 1991).

length of the river. Below the discharge of effluent from a textile factory however, only oligochaete worms and chironomids occurred. Aeration provided by the steep gradient of the river allowed species such as *Asellus* and *Baetis* to recolonize, and lower down small numbers of other mayflies and stoneflies survived by colonizing from a clean tributary. The invertebrate fauna of the River Tame steadily deteriorated downstream as effluents from various sewage works were added to the river.

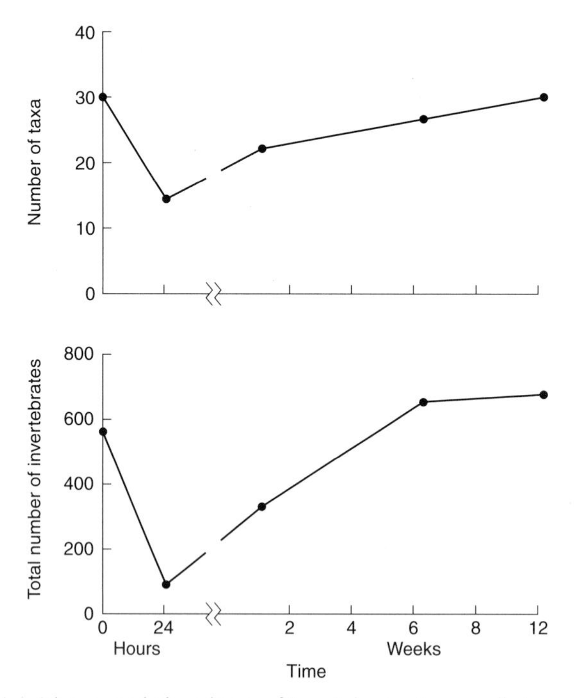

Fig. 3.21 Species richness and abundance of invertebrates in a small stream following a pulse of deoxygenation (from Edwards *et al.*, 1991).

The River Etherow had a richer fauna and that of the Mersey has shown improvement with the upgrading of sewage works but the lowest stretch of the river, just above the tidal limit, supported only tubificids and chironomids.

These impacts of organic pollution described for temperate watercourses also occur in tropical streams. However, evidence suggests that tropical streams are rather more sensitive than their temperate counterparts to organic pollution (Jacobsen, 1998).

Field experiments have been conducted to investigate the effect of episodes of de-oxygenation on stream invertebrates, events associated, for example, with accidents at farms. Sodium sulphite was added to a stream to reduce oxygen to around $1\,mg\,l^{-1}$ for a period of six hours (Fig. 3.20). Drift densities of invertebrates, normally less than 0.5 individuals per m^3 during daylight, increased within one hour to 43 individuals per m^3, mainly of mayflies and stoneflies. A later peak included *Gammarus*, blackfly larvae (Simulidae) and chironomids. The invertebrate taxa in the stream decreased from 30 to 14 but recovery was relatively rapid, with the initial species richness and numbers being achieved within 6 to 12 weeks (Fig. 3.21).

Fish

Fishes are considerably more mobile than any of the organisms discussed so far and they can sometimes avoid pollution incidents. The oxygen consumption of a fish is independent of the oxygen concentration of water down to the incipient lethal level, below which survival becomes difficult. When the oxygen supply becomes deficient, a fish breathes more rapidly and the amplitude of the respiratory movements increases. Oxygen lack and carbon dioxide excess, both of which occur with organic pollution, increase the ventilation volume of fishes and, at lower levels of oxygen, the cardiac output is reduced. This lowers the rate of passage of blood through the gills, so allowing a longer period of time for the uptake of oxygen, and also conserves oxygen by reducing muscular work. The percentage volume of red corpuscles in the blood may also be increased as the fish increases its urine flow to reduce the volume of the blood. When the oxygen tension in the water falls so low that the homeostatic mechanisms of the fish are no longer able to maintain the oxygen tension in the blood, the basic metabolic rate begins to fall.

Many factors, both natural and relating to pollution, affect the abilities of fishes to take up oxygen, so that their reactions to organic pollution in field situations are complex. Ammonia, often present in organic effluents and highly toxic to fish, is a frequent problem. Severe organic pollution renders rivers devoid of fish for considerable distances downstream of the discharge. Fish begin to reappear in the *Cladophora/Asellus* zone, the stickleback *Gasterosteus aculeatus* often being the first species to occur.

Organic pollution tends to be most severe in the lower reaches of rivers and in estuaries and this can cause particular problems for migratory fish such as salmon *Salmo salar* and sea trout *Salmo trutta*, which have high oxygen requirements.

These fishes may be prevented by severe pollution downstream from reaching their breeding grounds in the headwaters of rivers, even though conditions are perfectly satisfactory there.

The improvement in water quality of the River Thames in London following upgrading of sewage treatment works has already been described (p. 93). The Thames used to support an abundant community of fish, including salmon and, even as late as 1828, some 400 fishermen earned their living from the tidal reaches of the river. Fish disappeared in the middle years of that century but, following the first attempts at sewage treatment, they began to re-establish from 1890. A further deterioration took place from about 1915 and for around 45 years the tidal Thames was devoid of fish over a stretch of at least 60 km. Further improvements in sewage treatment facilities in the 1960s and 1970s have resulted in a marked improvement in oxygen and a sharp decline in metals (Power *et al.*, 1999). By the end of 1991, 112 species of fish had been recorded (Thomas, 1998). Attempts are being made to re-establish migratory stocks of salmon. Several hundred salmon now return to the river to spawn each year but most of these are hatched in nurseries. The aim is to have a self-sustaining salmon population with at least a thousand fish returning to breed each year. Evidence suggests that a self-sustaining population is becoming established.

Organic pollution may have an impact on species not directly associated with water but that use the river corridor. The foraging of pipistrelle bats (*Pipistrellus pipistrellus*) was found to be significantly reduced downstream of discharges of treated sewage effluent compared with upstream stretches, whereas Daubenton's bats (*Myotis daubentonii*) foraged significantly more downstream of discharges. It was suggested that Daubenton's bats may feed on insects tolerant of organic pollution, such as chironomids, whereas pipistrelle bats feed on pollution-sensitive species (Vaughan *et al.*, 1996).

CONCLUSION

The organic contamination of freshwaters was the first type of pollution to be recognized and studied and the first for which treatment facilities were developed, dating back to the nineteenth century. Where effort is put into providing or upgrading sewage treatment facilities the improvements in river water quality can be dramatic. Figure 3.22 shows the recovery of salmon and sea trout populations on the River Tyne in northeast England following improvements to both sewage and industrial discharges.

Nevertheless, poor maintenance and poor operational management at sewage works results in many pollution events in Great Britain and elsewhere each year, while many rivers remain grossly polluted, even in the developed world. Organic pollution becomes most acute in the developing world, where the treatment both of sewage effluents and drinking water supplies is often rudimentary or non-existent, faecal contamination of the latter leading to chronic problems of disease and

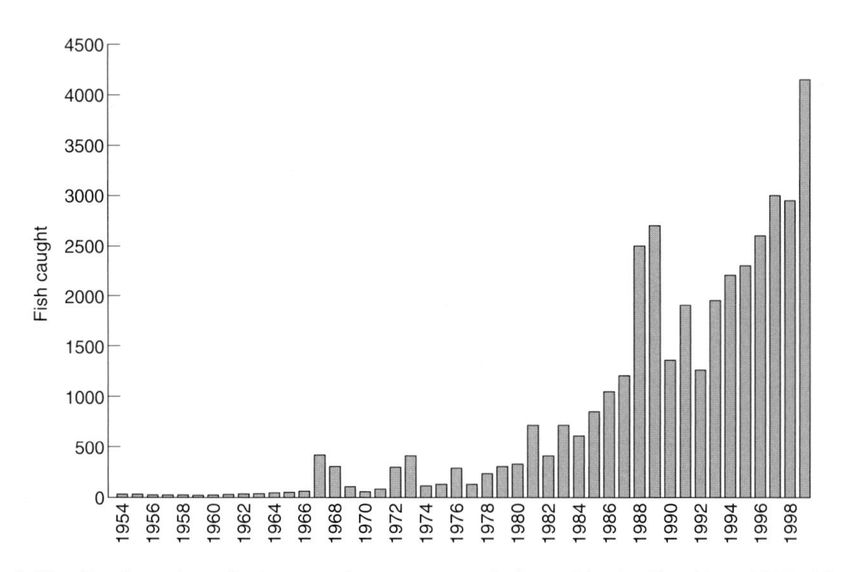

Fig. 3.22 Total number of salmon and sea trout caught by rod in the River Tyne 1952–98 (data from the Environment Agency).

periodic epidemics. But organic pollution and the diseases associated with it are the consequence of a lack of financial resources or a lack of political will, or both, rather than of a lack of scientific and technological know-how.

CHAPTER 4

EUTROPHICATION

The terms oligotrophic and eutrophic were originally used in the early years of the twentieth century to describe nutrient conditions in the development of peat bogs. When applied to limnology, lakes were classified as having oligotrophic water if they were clear in summer and eutrophic water if they were turbid owing to the presence of algae. Considerable confusion over these terms has developed as other definitions have arisen. In particular, oligotrophic lakes have been defined as having a low productivity and eutrophic lakes as having a high productivity. Although in general terms high nutrient levels and high productivity go together, this may not always be the case. Some lakes (marl lakes) have highly calcareous, nutrient-rich water (eutrophic), with a heavy precipitation of carbonates. Under these conditions phosphates and many micro-nutrients form insoluble compounds which precipitate to the bottom of the lake. Thus algal productivity is low and marl lakes might be considered as oligotrophic.

Eutrophication can be defined as the *enrichment of waters by inorganic plant nutrients*. These nutrients are especially nitrogen and phosphorus and result in an increase in primary production. *Artificial* or *cultural* eutrophication results from an increase in nutrients due to human activities, *natural* eutrophication results from an increase caused by a non-human process, such as a forest fire. From our definition of pollution in Chapter 1 (p. 1) only artificial eutrophication will concern us here.

The general characteristics of oligotrophic and eutrophic waters are given in Table 4.1. Lakes with intermediate characteristics are described as *mesotrophic*. A fourth general category of lake includes those that receive large amounts of organic matter of terrestrial plant origin – dystrophic or brown-water lakes, so-called because of the heavily stained water. They usually have a low productivity of plankton.

SOURCES OF NUTRIENTS

A number of compounds and elements (e.g. silicon, manganese, vitamins) may at times be limiting to algal growth but it is only excesses of

Table 4.1 The general characteristics of oligotrophic and eutrophic lakes

	Oligotrophic	**Eutrophic**
Depth	Deeper	Shallower
Summer oxygen in hypolimnion	Present	Low, sometimes absent
Algae	High species diversity with low density and productivity, often dominated by Chlorophyceae	Low species diversity with high density and productivity, often dominated by cyanobacteria
Blooms	Rare	Frequent
Plant nutrient flux	Low	High
Animal production	Low	High
Fish	Salmonids (e.g. trout, char) and coregonids (whitefish) often dominant	Coarse fish (e.g. perch, roach, bream) often dominant

nitrogen and phosphorus that give rise to algal nuisances. In the majority of freshwaters phosphorus is normally the limiting element because the relative proportions of biologically available phosphorus in the water compared with other elements are often less than the relative proportions of these elements in algae. An increase in phosphorus will therefore result in an increase in productivity. In many oligotrophic waters, however, it appears that nitrogen is limiting. If nitrogen becomes limiting in eutrophic conditions some cyanobacteria (also known as blue-green algae or cyanophytes) are able to fix atmospheric nitrogen and will grow provided phosphorus is not limiting. Cyanobacteria can also store large amounts of phosphorus within their cells which allows them to survive when phosphorus may become limiting in late summer (Thompson and Rhee, 1994).

The majority of polluting nutrients enter watercourses and lakes in effluents from sewage treatment works, in untreated sewage, from farming activities and from precipitation. Local eutrophication may also occur where livestock are given access to the water for drinking or where large congregations of waterfowl gather. Sources might be discrete, such as a specific sewage outfall, or diffuse, such as from farmland within the catchment.

Urban sources

Nutrients from urban sources may be derived from domestic sewage, industrial wastes and storm drainage. The contribution of nitrogen and phosphorus per person per day averages 10.8 g N and 2.2 g P, though there is a considerable range. In the 1940s detergents were developed containing sodium tripolyphosphate, which softens water by neutralizing calcium and keeps dirt in suspension once it has washed off the clothes. Between 1950 and 1970 detergent consumption increased more than five times in the United States and more than seven times in Great Britain and detergent is now an important source of phosphorus in domestic sewage, often making up more than half the total phosphorus in effluents. Some 54 per cent of the soluble reactive phosphorus in the inflowing rivers to eutrophic Lough Neagh in Northern Ireland was from sewage treatment works (Smith, 1993). Sewage effluents contributed 18–84 per cent to the total annual phosphorus loading of four UK rivers (Mainstone *et al.*, 2000).

Industrial sources of nutrients may be locally important, depending on the type of industry, the volume of effluent and the amount of treatment it receives. For instance the brewing industry releases about 11 000 m^3 effluent each day into rivers in England and Wales, containing some 156 $mg\,l^{-1}$ N and 20 $mg\,l^{-1}$ P. Food processing generally and businesses such as the woollen industry, which require substantial washing procedures, are likely to produce effluents containing high concentrations of nitrogen and phosphorus.

There is a considerable increase in the export of phosphate as forest is converted to agricultural use and agricultural land is subject to urban development. The relative importance of sources of nutrients will vary with the type of land-use in the catchment.

Rural sources

Rural sources of nutrients include those from agriculture, from forest management and from rural dwellings, the first of which is the most universally important. Rural dwellings often dispose of their sewage into septic tanks, which may cause local pollution. Summer houses, which frequently have primitive sewage disposal facilities, are often built on the banks of lakes.

Nutrients are lost from farmland in three ways:

- by drainage water percolating through the soil, leaching soluble plant nutrients;
- by run-off of animal manure applied to the land;
- by the erosion of surface soils or by the movement of fine soil particles into subsoil drainage systems.

The use of artificial fertilizers has vastly increased. In the United States the total amount of nitrogen fertilizer used has increased from 0.5 million tonnes in the 1940s to 10 million tonnes in the 1990s, while phosphorus use has doubled. The total inorganic fertilizer used is equivalent to some 40 $kg\,person^{-1}\,yr^{-1}$. Nitrogen and phosphorus behave differently in soils. The nitrate anion is fairly mobile because of the predominantly negative charge on soil particles so that it is readily leached if it is not taken up by plants. By contrast, phosphate is precipitated as insoluble iron, calcium or aluminium phosphate and then released only slowly. However, recent data has shown that, in soils that have been heavily fertilized with phosphorus, significant leaching of phosphorus can occur (Sims *et al.*, 1998; Johnes and Hodgkinson, 1998; Turner and Haygarth, 1999). Artificial fertilizers made up 7–22 per cent of the annual loadings of total phosphorus to four UK rivers (Mainstone *et al.*, 2000).

The solubility of nitrate means that agriculture is a major contributor of it to freshwaters and often up to half the nitrogen applied to crops is lost to groundwater. Agriculture accounts for 71 per cent of the mass flow of nitrogen in the River Great Ouse in the English Midlands, compared with only 6 per cent for phosphorus, the balance being from sewage effluents. The concentration of nitrate in rivers follows closely that of river flow. Levels are low during summer, even when fertilizer is being added, because the growing plants utilize nitrogen as soon as it becomes available. There is also little net downward movement of water in the soil during the summer because of high rates of evaporation and transpiration. With a decrease in transpiration and evaporation in autumn and winter, nitrate is leached from the soil and levels in rivers rise. The rate of loss declines again in late winter because soluble nitrate reserves are depleted and low temperatures reduce the rate of nitrification. The annual nitrate level in rivers closely follows the annual levels of fertilizer application within the catchment and this has been steadily increasing (Fig. 4.1). For instance in eight stretches of river in the English Midlands annual increases ranged from 0.07 to 0.22 mg l^{-1} yr^{-1}, the greatest rates of increase occurring in the most urbanized stretches of river (José, 1989). However in the rural and intensively arable eastern region of England, rates of increase are higher still, of the order of 0.25–0.28 mg l^{-1} yr^{-1} (Warn and Page, 1984).

The loss of phosphate from agricultural land is largely by erosion. Arable farming increases the natural rate of erosion of land because the soil is often bare for several months of the year. Average losses of total phosphorus are of the order of 3 kg ha^{-1} yr^{-1} but may be up to ten times that in undrained clay-rich soils.

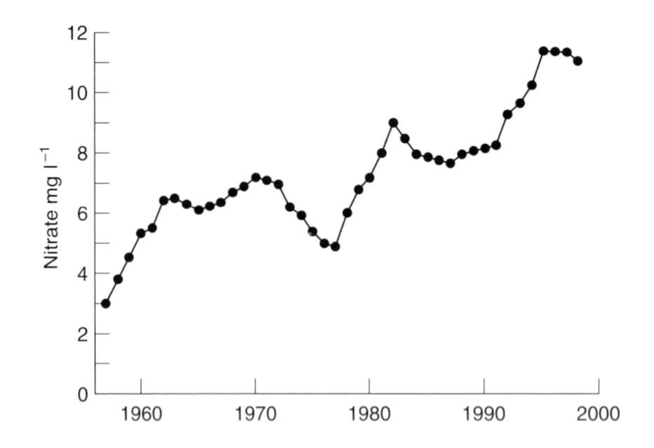

Fig. 4.1 Trend (1956–98) in the concentration of nitrate in the River Deben, eastern England, a typical lowland river (data provided by Lynn Parr from Environment Agency files).

Losses from grassland average 0.5 kg ha^{-2} yr^{-1}. Losses of phosphorus to water equivalent to 60 per cent of the fertilizer applied to the land have been reported. Much of this phosphorus is initially tightly bound to the soil particles and is not immediately biologically available when it reaches freshwaters, but the solubility of phosphate is enhanced when the soil particles become incorporated into anaerobic sediments.

The other chief source of nutrients from the agricultural industry is livestock farming, especially where this involves intensive rearing. Livestock currently contribute 7–15 per cent of the total annual phosphorus loadings to four UK rivers (Mainstone *et al.*, 2000). The increase in size of livestock units and the problem of waste disposal has already been described (p. 83). Labour costs for handling farmyard manure are high and areas of intensive livestock rearing are often distant from areas of arable farming, where the wastes could be used on the land. The amount of phosphorus excreted by British livestock each year is four times that excreted by the human population. The outdoor rearing of pigs, the preferred method on grounds of animal welfare, can, if too intensive, lead to severe soil erosion and high levels of nutrients find their way to nearby rivers. In Northern Ireland increases in diffuse sources of phosphorus to rivers and lakes from agriculture have negated the reductions in phosphorus achieved by improved treatment of sewage effluents (Foy and Lennox, 2000).

The management of forests may have local effects on the nutrient loading of rivers. Experiments in the Hubbard Brook Watershed in the United States, where a forest was cut and left on site, with regrowth prevented by the application of herbicides, showed that nitrate increased some 50 times compared with uncut controls (Likens *et al.*, 1970). Most forest practices, of course, are nowhere near as extreme as this. The harvesting of conifer plantations in Wales led to an increase in nitrogen in streams up to 3.2 mg l^{-1} but concentrations fell to background levels (0.5 mg l^{-1}) after five years (Reynolds *et al.*, 1995). In some countries, forests are regularly fertilized and this may result in local eutrophication. Phosphate is added to newly established conifer plantations in Britain and with the increase in afforestation this could result in an overall increase in productivity in upland catchments. Naturally oligotrophic streams flowing from plantations frequently have dense growths of filamentous algae in spring, smothering stoney bottoms which may be important spawning grounds for salmonids.

A comparison of ten contrasting British catchments in 1931, 1970 and 1988 showed that there had been a marked loss of rough grazing to fertilized temporary and permanent grasslands, an increase in the stocking rate of livestock, an increase in the area of farmland under arable cultivation and a movement of humans into the catchments. All of these trends increased the flows of nutrients with a doubling of both nitrogen and phosphorus over the period. The wastes of livestock were the greatest contributors to the export of nutrients (Johnes *et al.*, 1996).

In the UK it is considered that natural phosphorus concentrations in river water are of the order of 0.03 mg P l^{-1}. Current concentrations are generally ten times higher than this (Mainstone *et al.*, 2000).

LAKE WASHINGTON – A CASE HISTORY

Before dealing with the general effects of eutrophication I want to describe conditions in Lake Washington, which has been studied in detail. Lake Washington (Fig. 4.2), in the northwestern United States, has a surface area of 87.6 km^2 and a maximum depth of 76.5 m. Water, low in nutrients, enters the lake via the Cedar River from the south and the outflow is the Sammamish River, which flows into Lake Sammamish. Lake Washington is connected to the Pacific Ocean at Puget Sound, but a series of locks holds the lake level higher. The city of Seattle (with a population currently of some three million) lies between Puget Sound and the western shore of the lake. The events in Lake Washington have been described by Edmondson (1969, 1970, 1972).

Early in the twentieth century raw sewage was disposed of directly into Lake Washington and, by 1926, a population of 50 000 was adding raw sewage to the lake through 30 outfalls. By 1936 a series of interceptors and tunnels had been constructed to divert Seattle's sewage to the sea at Puget Sound and this reduced pollution within the lake. However, Seattle began to expand along the edges of the lake and development took place on the east side. A series of sewage works was

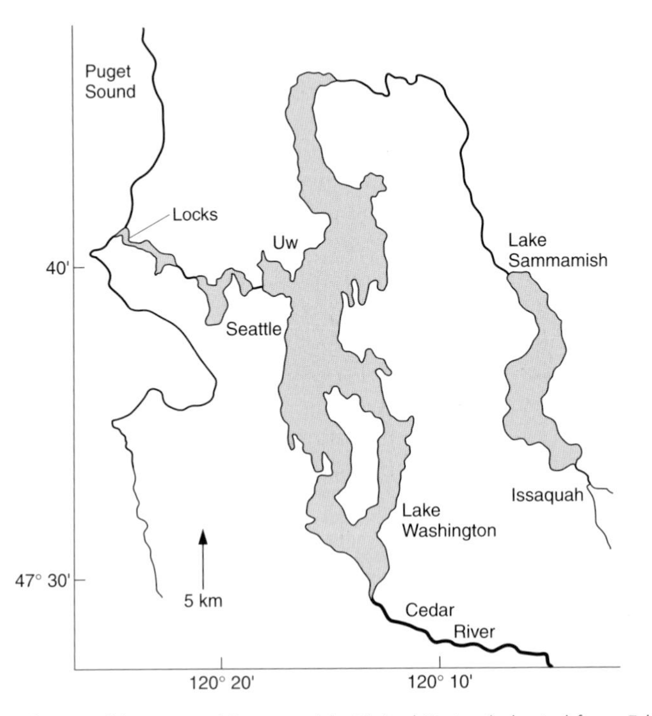

Fig. 4.2 Lakes Washington and Sammamish, United States (adapted from Edmondson, 1969).

built to cope with the expansion and discharged secondarily treated sewage into the lake, there being ten sewage works by 1954. In addition, feeder streams into the lake were being contaminated with drainage from septic tanks. Sewage was responsible for 56 per cent of the total input of phosphorus to the lake.

Work during 1950 showed that conditions in Lake Washington were very different from those revealed in an earlier survey in 1933, and during the next few years the situation deteriorated sufficiently to attract considerable public attention. There was a greater summer oxygen deficit in the hypolimnion in 1950 compared with 1933 while the winter phosphate concentration, which gives a good index of supply to the plankton, had shown a marked increase. The summer densities of phytoplankton increased several fold during the 1950s and in 1955 a dense bloom of the cyanobacterium *Oscillatoria rubescens* developed. A *bloom* can be defined as *an aggregation of plankton sufficiently dense to be readily visible*. Lakeside bathing beaches were periodically closed owing to pollution, but the situation was far worse in Puget Sound, which was receiving Seattle's untreated sewage.

The amenity value of Lake Washington, close to a large urban centre, had seriously declined and this gave rise to a vociferous movement to prevent further deterioration. It was decided to divert the majority of sewage from the lake to Puget Sound and at the same time to improve the quality of the effluent entering the Sound to reduce pollution there. Diversion began in March 1963

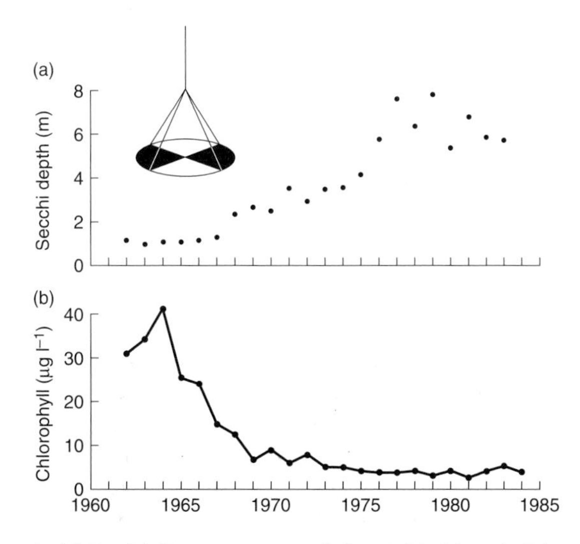

Fig. 4.3 Changes in (a) Secchi disc transparency (m) and (b) chlorophyll in surface water (μg l^{-1}) in Lake Washington 1962–84 (adapted from Gulati, 1989). Note: a Secchi disc is 20 cm in diameter and divided into four segments, two black and two white. It is lowered into the water on a line until it just ceases to be visible – the Secchi depth.

with one-third of the sewage being transferred to Puget Sound. Ninety-nine per cent of the sewage had been diverted by March 1967 and the project was completed a year later at a total cost of US$125 million.

Lake Washington responded quickly to the reduction in nutrients. Winter concentrations of phosphate began to decline rapidly after 1965. Levels of nitrogen declined more slowly because agriculture is a major contributor of nitrogen to the lake. Chlorophyll *a* concentrations had fallen by 1970 to one-fifth the level in 1963 and there were corresponding increases in the transparency of the water (Fig. 4.3). Edmondson (1991) described the lake as being in excellent condition in 1990 for a waterbody in an urban area. The transparency of the water was rarely less than 5 m and usually much greater, while, although small numbers of cyanobacteria occurred, phytoplankton were no longer considered a nuisance.

The observations on Lake Washington illustrate clearly the relationship between an increase in nutrients and an increase in undesirable biological productivity, but they also show that eutrophication can be reversed if a major source of nutrients is removed. Edmondson (1991) provides an overview of Lake Washington, including the social and political problems encountered during its restoration.

THE GENERAL EFFECTS OF EUTROPHICATION

We can now generalize about the impacts of an increase in the rate of nutrient input on waterbodies. The effects of eutrophication on the receiving ecosystem and the particular problems these cause to humans are summarized in Table 4.2.

Eutrophication causes marked changes in the biota. The Norfolk Broads area of eastern England can serve as an initial example. The Norfolk Broads are a group of relatively small, shallow lakes formed during medieval times by the flooding of peat diggings. They are calcareous and naturally eutrophic, but the water used to be generally clear and dominated by a rich flora of charophytes (stoneworts – large multicellular algae) and aquatic flowering plants (macrophytes). These plants support a diverse fauna of invertebrates, some of them rare in Britain. The area as a whole is outstanding for its wildlife interest, but at the same time maintains a large tourist industry, based mainly on boating holidays and angling. A detailed description of the region is provided by George (1992).

Conditions began to deteriorate in the Broads during the late 1950s but the area was not subject to any scientific study. A survey in the early 1970s of 28 broads showed that 11 were completely devoid of aquatic macrophytes or had only a poor macrophyte growth, consisting mainly of floating-leaved water lilies (Mason and Bryant, 1975). A later study of 42 broads found only four with clear water and a flora of submerged charophytes (phase 1). A further six had a flora dominated by submerged species which can grow rapidly to the surface and form mats (phase 2a). Such species include *Ceratophyllum demersum* and *Potamogeton pectinatus*, often

Table 4.2 The effects of eutrophication on receiving ecosystems and the problems to humans associated with these effects

Effects
1. Species diversity often decreases and the dominant biota change
2. Plant and animal biomass increases
3. Turbidity increases
4. Rate of sedimentation increases, shortening the lifespan of the lake
5. Anoxic conditions may develop

Problems
1. The water may be injurious to health
2. The amenity value of the water may decline
3. Increased vegetation may impede water flow and navigation
4. Commercially important species of fish may disappear
5. Treatment of drinking water may be difficult and supply may have an unacceptable taste or odour

associated with floating-leaved water lilies. At a later stage only water lilies remain (phase 2b). Thirty-two broads were devoid of aquatic plants (phase 3). These phases represent stages in the eutrophication process, phases 2b and 3 falling into a group of broads with high phosphorus, group 2a being intermediate (Moss, 1983). These changes are clearly illustrated in Fig. 4.4. With the loss of macrophytes permanent algal blooms developed and the transparency of the water in summer at one site, Barton Broad, measured with a Secchi disc was only 11 cm. Broads with no submerged macrophytes had a poorly developed benthic fauna dominated by tubificid worms and chironomids. A detailed comparison of the benthos of a culturally enriched broad with one still holding clear water (Mason, 1977a) showed that the latter (Upton Broad) supported 40 taxa in the benthos, 17 of them occurring commonly. The enriched site (Alderfen Broad) had 22 taxa, only 7 of them occurring commonly. In the Norfolk Broads there has been a change from a community dominated by macrophytes to one dominated by phytoplankton.

We can now consider in more detail the effects of eutrophication on the biota of freshwaters.

Plankton

The types of phytoplankton associated with lakes of different trophic status are listed in Table 4.3. Bacteria-sized picoplankton and desmids are particularly important in lakes low in nutrients, while cyanobacteria often dominate lakes with high nutrient concentrations. The diatom flora also changes. Oligotrophic lakes have a diatom flora which is often dominated by species of *Cyclotella* and *Tabellaria*,

Fig. 4.4 The Norfolk Broads: (a) a small broad (turf pond) dug some 20 years ago and isolated from river water (phase 2a) and (b) a broad, fed by the River Bure, from which macrophytes disappeared in the 1960s (phase 3) (photographs by the author).

while eutrophic lakes have *Asterionella*, *Fragilaria crotonensis*, *Stephanodiscus astraea* and *Melosira granulata* as dominants. It must be emphasized, however, that we are looking at a continuum, not a series of sharp divisions of trophic state, and the phytoplankton communities in any one lake must be viewed accordingly.

As an example, the seasonal changes in phytoplankton biomass of the

Table 4.3 Characteristic algal associations in lakes of increasing trophic status

	Algal group	Examples
Oligotrophic lakes	Picoplankton (often small cyanobacteria)	*Synechococcus*
	Desmid plankton	*Staurdesmus, Staurastrum*
	Chrysophycean plankton	*Dinobryon*
Mesotrophic lakes	Diatom plankton	*Cyclotella, Tabellaria*
	Dinoflagellate plankton	*Peridinium, Ceratium*
	Chlorococcal plankton	*Oocystis, Eudorina*
Eutrophic lakes	Diatom plankton	*Asterionella, Fragilaria, Stephanodiscus, Melosira*
	Dinoflagellate plankton	*Peridinium, Glenodinium*
	Chlorococcal plankton	*Scenedesmus*
	Cyanobacterial plankton	*Aphanizomenon, Anabaena, Microcystis*

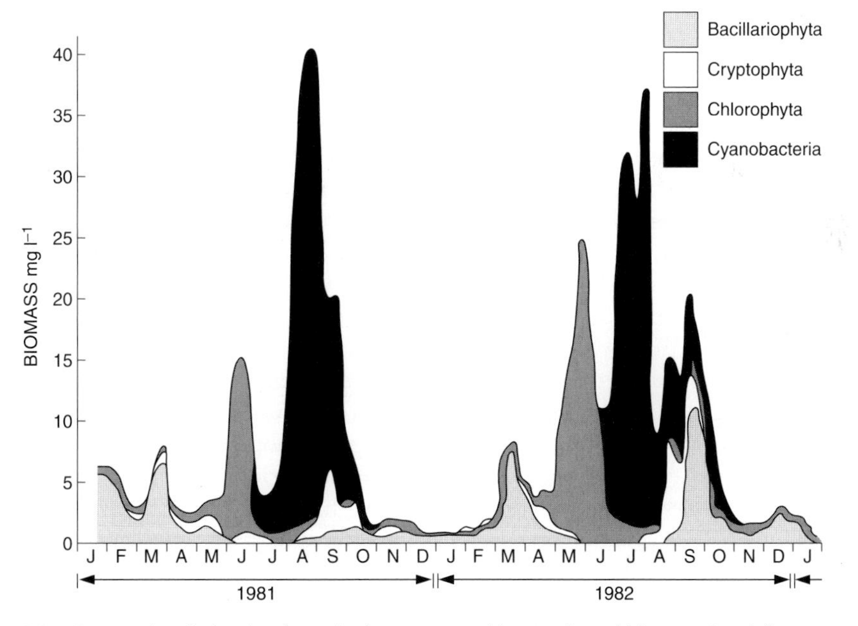

Fig. 4.5 Seasonal variation in phytoplankton composition and total biomass (mg l^{-1} wet weight) in Ardleigh Reservoir (adapted from Abdul-Hussein and Mason, 1988).

hypertrophic Ardleigh Reservoir, in eastern England, are shown in Fig. 4.5. The reservoir is shallow and is characterized by high concentrations of phosphorus, up to 0.78 mg l^{-1}. Diatoms (Bacillariophyta) dominated in winter and early spring, to be succeeded by a mixture of cryptomonads (Cryptophyta) and green algae (Chlorophyta), the latter group becoming dominant in late spring. In early summer the cyanobacteria *Anabaena* and *Aphanizomenon* joined the chlorophytes, while during high summer the cyanobacterium *Microcystis aeruginosa* dominated, the intense bloom of this species subsiding in early autumn to be replaced by cryptophytes and later by diatoms. *Microcystis* is unable to fix atmospheric nitrogen and is dependent on nitrate and ammonia released by the processes of decomposition in the waterbody. There was a significant relationship between ammonia and *Microcystis* in Ardleigh Reservoir. This seasonal change in the phytoplankton appears fairly typical of eutrophic waters. Turbulence is important in determining the dominant phytoplankton group, for cyanobacteria are outcompeted in turbulent waterbodies even though conditions would otherwise be suitable for them (Steinberg and Hartmann, 1988).

Blooms of cyanobacteria form where nutrient loading is high, where the ratio of nitrogen to phosphorus is low, where conditions in the water are warm and still, and where large-bodied grazers, such as *Daphnia*, are scarce (Elser, 1999). Once established, bloom-forming cyanobacteria influence their surrounding environment, physically, chemically and biologically, in ways that favour their continued persistence. Vincent (1989), for example, reported how a large bloom of *Anabaena* in Lake Rotongaio, New Zealand, restricted the euphotic zone to less than 1.5 m. *Anabaena*, with low maintenance energy requirements and an efficient light-harvesting mechanism, could cope with this poor light regime. *Anabaena* also strongly absorbed solar radiation, which raised surface water temperatures and further promoted cyanobacteria, for these have a high temperature optimum for growth. The thermal gradient that was created restricted the propagation of turbulence down through the water column, favouring buoyant species of flagellates at the expense of diatoms, which sank in the still water. It also reduced the upward transfer of ammonium ions from the hypolimnion, putting all but nitrogen-fixing species at a disadvantage. *Anabaena*, by contrast, was able to take up low concentrations of ammonium ions and ammonia over a broad range of pH, while the increased pH in the water induced by the *Anabaena* bloom may have additionally favoured cyanobacteria over other phytoplankton groups through decreased CO_2 availability at high pH. Cyanobacteria are largely resistant to grazing by zooplankton. They also produce specialized cells that can survive prolonged periods of unfavourable conditions. Therefore once nutrient conditions have developed that allow the establishment of cyanobacteria (in Lake Rotongaio, high dissolved inorganic phosphorus and low dissolved inorganic nitrogen concentrations), positive feedback mechanisms enable populations to grow rapidly and persist at high densities for long periods.

Diatoms have siliceous frustules (outer walls), which fall to the lake floor on the death of the cell. By examining a time series of sections through a sediment core it is possible to reconstruct the changes that have occurred in the waters of a lake. Moss (1972) examined a sediment core from Gull Lake, Michigan. When the earlier sediments were laid down, the diatoms *Cyclotella michiganiana* and *Stephanodiscus niagarae* were dominant. From 20 cm in the sediment upwards they were joined by *Cyclotella comta* and *Melosira italica*. At 15 cm in the sediment a marked change occurred, with those species already present becoming much more abundant and *Asterionella formosa* appearing. At 10 cm *Fragilaria crotonensis* appeared while the abundance of the other species continued to increase. The total diatom density near the surface of the sediment was 50 times greater than in the lower layers. The sediment record thus reflected the increase in nutrient supply to the lake both in terms of a change in diatom species and in an increase in diatom abundance.

Cores through the sediment of several of the Norfolk Broads have shown similar changes as eutrophication progressed. There is historical evidence for eutrophication from agricultural fertilizers and from dwellings with their requirement for effluent disposal (Moss, 1983). Increased algal growth has led to an increased rate of sedimentation. In Barton Broad the rate of sedimentation doubled in the 1950s (from 1.2–3.1 mm yr^{-1} to 5 mm yr^{-1}), doubled again in the 1960s, and increased to 12 mm yr^{-1} in the 1970s (Moss, 1980).

The concentration of chlorophyll *a* in the water is often taken as an index of the biomass of algae present and, together with total phosphorus and water transparency (Secchi depth) has been used to classify the trophic status of lakes (Table 4.4). Enrichment also affects the rate of primary production. The mean daily rates of primary production are 30–100 mg C m^{-2} d^{-1} for oligotrophic lakes and

Table 4.4 OECD boundary values for fixed trophic classification system (from Ryding and Rast, 1989, after OECD, 1982)

Trophic category	TP	Mean Chl	Maximum Chl	Mean Secchi	Minimum Secchi
Ultra-oligotrophic	< 4.0	< 1.0	< 2.5	> 12.0	> 6.0
Oligotrophic	< 10.0	< 2.5	< 8.0	> 6.0	> 3.0
Mesotrophic	10–35	2.5–8.0	8–25	6–3	3–1.5
Eutrophic	35–100	8–25	25–75	3–1.5	1.5–0.7
Hypertrophic	> 100	> 25	> 75	< 1.5	< 0.7

Explanation of terms:
TP = mean annual in-lake total phosphorus concentration ($\mu g\, l^{-1}$)
mean Chl = mean annual chlorophyll *a* concentration in surface waters ($\mu g\, l^{-1}$)
maximum Chl = peak annual chlorophyll *a* concentration in surface waters ($\mu g\, l^{-1}$)
mean Secchi = mean annual Secchi depth transparency (m)
minimum Secchi = minimum annual Secchi depth transparency (m)

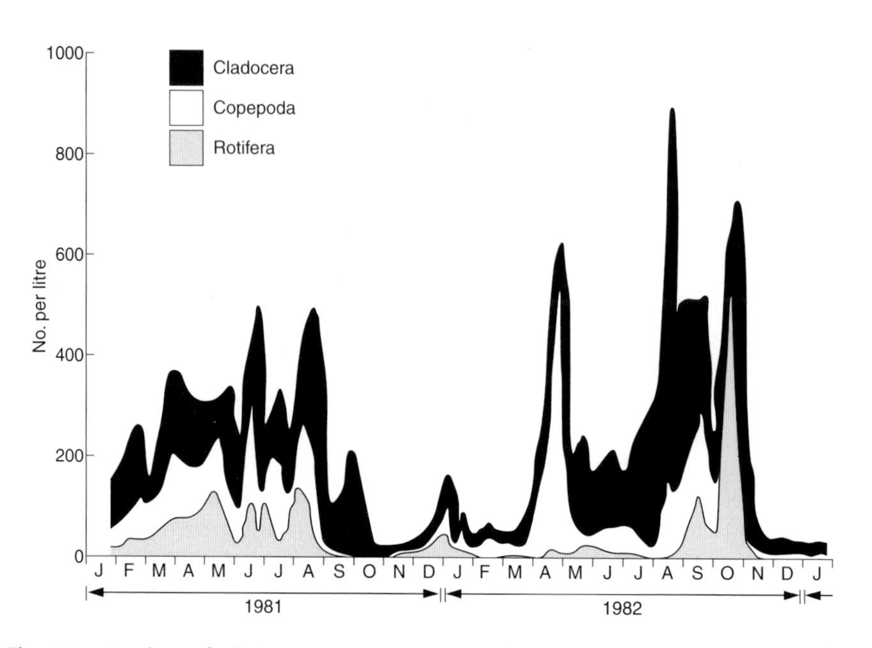

Fig. 4.6 Numbers of Cladocera, Copepoda and Rotifera in the zooplankton of Ardleigh Reservoir (unpublished data of M.M. Abdul-Hussein).

300–3000 mg C m^{-2} d^{-1} for eutrophic lakes (Rohde, 1969), being equivalent to annual rates of 7–25 g C m^{-2} and 75–700 g C m^{-2} respectively.

A typical pattern of seasonal change in the zooplankton is shown in Fig. 4.6. A spring increase in algae is followed by an increase in herbivores. They overeat their food supply and phytoplankton decline, followed by the zooplankton. There is usually a clear-water phase before algae again increase in numbers, followed by a further expansion in zooplankton numbers. With any rise in primary productivity we could reasonably expect an increase in the zooplankton. This does not necessarily occur, however, for different types of algae are utilized to different extents by zooplankton. Changes in the availability of suitable algal food are responsible for much of the summer fluctuation in zooplankton numbers apparent in Fig. 4.6. Both the quantity of food and its quality affect the ingestion rate of zooplankton. Cyanobacteria are assimilated very inefficiently compared with green algae and diatoms. The bottom-dwelling *Chydorus sphaericus* can eat cyanobacteria and its population often increases markedly during phytoplankton blooms. Larger cladocerans, such as *Daphnia*, filter much more efficiently than smaller plankton, such as *Bosmina*, and this greater feeding efficiency may suppress smaller zooplankton when larger animals are abundant. However, fish generally eat the larger forms, so that smaller zooplankton may dominate in the presence of these predators (see below).

Macrophytes

The loss of macrophytes from the Norfolk Broads was described above and similar changes have occurred in other lake systems, such as in the Netherlands (Van Liere *et al.*, 1984) and Lake Geneva on the boundary of France and Switzerland (Lehmann and Lachavanne, 1999). The loss of aquatic plants is often abrupt, luxuriant beds disappearing in only one or two seasons. The decline in charophytes and rooted macrophytes, such as *Najas marina*, is easiest to explain. There is a direct relationship between nutrient loading and increased growth of both phytoplankton and periphyton (the microflora and fauna attached to leaves and stems of aquatic plants) and also of non-rooted macrophytes, with a concomitant decline in rooted, submerged macrophytes (Hough *et al.*, 1989). The low-growing rooted macrophytes are effectively shaded out (Middleboe and Markager, 1997), and in the Norfolk Broads this group was the first to disappear. But why have species such as *Ceratophyllum* and *Myriophyllum*, at best only partially rooted and forming dense mats at the surface of lakes, also disappeared? Similarly those species, such as water lilies, which are rooted but have floating leaves, are eventually also lost.

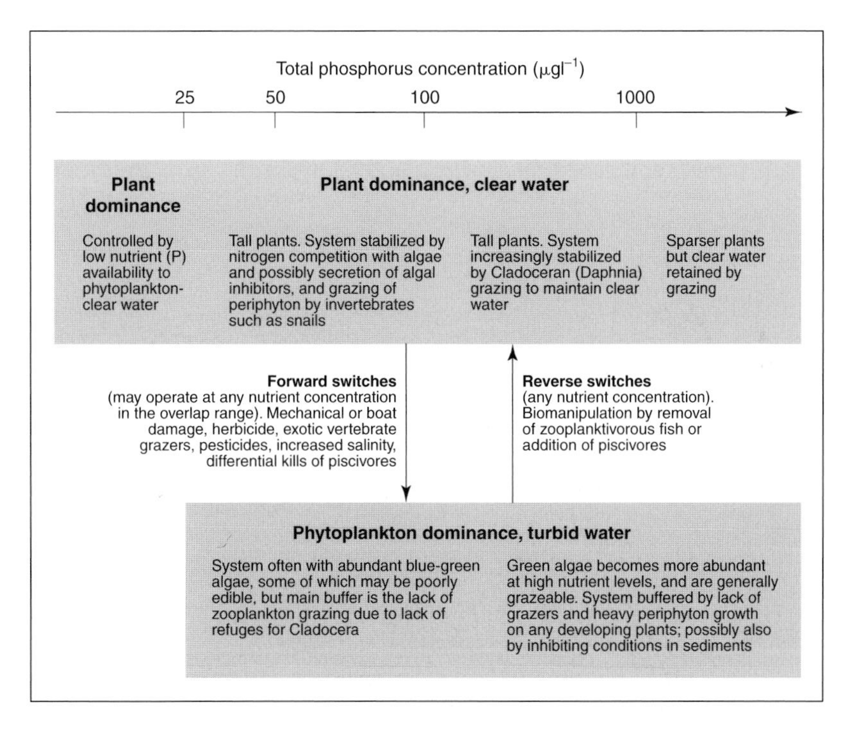

Fig. 4.7 Alternative stable states model for dominance by aquatic plants or phytoplankton in shallow lakes. The gradient of total phosphorus concentrations ranges from pristine to those which occur in highly eutrophic conditions (from Moss *et al.*, 1996).

Macrophyte-dominated or phytoplankton-dominated states of waterbodies may exist over a wide range of high nutrient concentrations (Irvine *et al.*, 1989) and, indeed, species such as *Ceratophyllum* and *Myriophyllum* have been proposed as suitable species for use in wastewater treatment lagoons, where nutrient levels will be extremely high (Polprasert, 1989). In experimental ponds there was no significant relationship between phosphorus loading or concentrations and the biomass of aquatic plants (Balls *et al.*, 1989).

It appears that, over a range of nutrient concentrations, shallow lakes may have alternative stable states (Fig. 4.7), either clear water and dominated by aquatic vegetation or turbid water characterized by a high algal biomass (Irvine *et al.*, 1989; Scheffer *et al.*, 1993; Scheffer, 1998). The macrophyte community may be stabilized by luxury uptake of nutrients (thus making nutrients unavailable to plankton), allelopathy, i.e. the secretion of chemicals that prevent plankton growth (Nakai *et al.*, 2000), shedding of leaves with heavy epiphyte burdens, sheltering of large populations of grazers that eat phytoplankton, and the provision of refuges for these grazers from predation by fish. The phytoplankton community may be stabilized by an early growing season (shading the later growth of macrophytes), easier acquisition of carbon dioxide, especially when pH is high late in the season, increasing the vulnerability of herbivores to predation in the unstructured environment and the production of large, inedible algae. In many lakes the stability of the macrophyte-dominated community has, however, been broken, with a switch to communities dominated by algae. Possible reasons for this will be discussed below.

Nutrient-rich lakes are surrounded by emergent macrophytes, such as reedmace or cat-tails (*Typha*) and reed (*Phragmites australis*). Over the same timescale as the

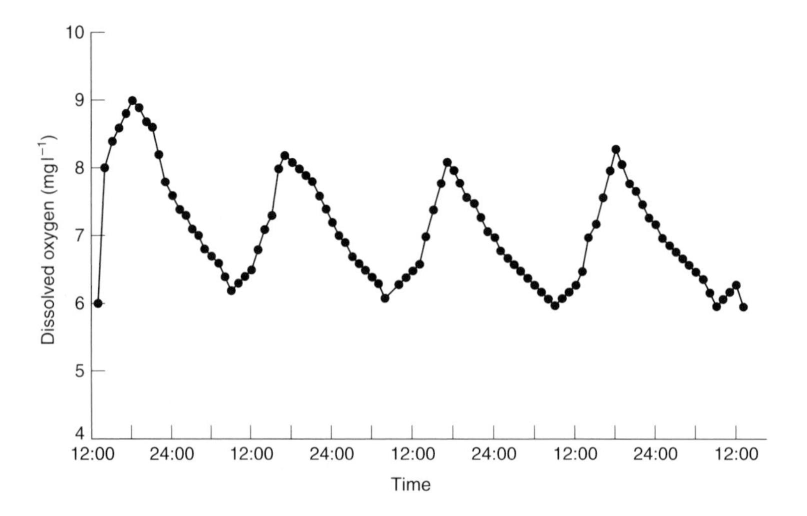

Fig. 4.8 Changes in dissolved oxygen concentrations in the River Brett, eastern England, over five days in August 1998 (24:00 is midnight). The site had abundant growth of aquatic macrophytes (from unpublished data of Lynn Parr).

loss of aquatic macrophytes, reedbeds have also regressed. In the Norfolk Broads Boorman and Fuller (1981) calculated that there were 216 ha of reedswamp in 1880, 121 ha in 1946 and only 49 ha in 1977. Declines in reed area have been described from more than 35 lakes in Europe (Ostendorp, 1989). For the Norfolk Broads Boar *et al.* (1989) have suggested that, at nitrate concentrations above $6\,mg\,l^{-1}$, which occur commonly in the water, there is a decrease in the rhizome to shoot ratio of reed. This may make floating reedbeds unable to withstand mechanical disturbances caused by wave action. Ostendorp (1989) has pointed out, however, that reed decline is not restricted to polluted (eutrophic) lakes but also occurs in mesotrophic lakes, while reeds can thrive in hypertrophic lakes and wastewater treatment plants (p. 103) with nitrate concentrations as high as $10\,mg\,l^{-1}$. There is some evidence that accumulated phytotoxins, such as organic acids and sulphide, produced from the death and decay of plants may be responsible for die-back, eutrophication causing excessive loading with organic matter (Armstrong *et al.*, 1996a,b). It may also be significant that, over the same timescale as the loss of reedswamp, large areas of intertidal salt marsh, a broadly similar habitat, have also regressed (Long and Mason, 1983).

In flowing waters it is not macrophyte loss, but excessive macrophyte growth that occurs. This may include the large, filamentous alga *Cladophora*, especially in waters rich in sewage effluents (see p. 116) and angiosperms such as species of *Ranunculus* and *Potamogeton*, or in the tropics, water hyacinth *Eichhornia crassipes*. Macrophytes add oxygen to the water during the day, when photosynthesizing vigorously, but their respiration depletes oxygen at night (Fig. 4.8). This may lead to complete de-oxygenation of the river water, especially when there is also a large oxygen demand in the sediments. Fine silts, with a high oxygen demand, settle around the roots of abundant macrophytes over substrates that, in the absence of excessive plant growth, would be gravels with a low oxygen requirement.

Benthos

The reduction in diversity that occurs in the benthic fauna with enrichment has already been referred to but there may also be changes in the seasonal pattern of occurrence, with consequences for organisms higher in the food chain. The benthos of the eutrophic, but unpolluted, Upton Broad mentioned earlier, was dominated by larvae of the midge *Tanytarsus holochlorus*, with lesser numbers of caddis larvae, mayfly nymphs and snails, whereas the culturally eutrophic Alderfen Broad was dominated by the tubificid worm *Potamothrix hammoniensis* and the midge larva *Chironomus plumosus*. The benthic populations in Upton Broad were highly seasonal, with a large peak in June, and there was little biomass present in the autumn and winter. Alderfen Broad was markedly less seasonal, with a substantial biomass of benthic animals in autumn and winter (Mason, 1977a). The fish in Upton Broad would probably have very little food from late summer to mid-spring and indeed the populations of fish were small and recruitment poor compared with high populations and good recruitment in Alderfen Broad.

Fish

The general changes in the fish fauna with eutrophication are summarized in Fig. 4.9, using information from 51 European lakes. Four stages in eutrophication are recognized from changes in fish yield. Coregonid fish dominate in oligotrophic lakes and their yield increases in the early stages of enrichment and then declines. There is some increase in the percids in the intermediate stages of eutrophication but they then decline. The yield of cyprinid fish increases sharply at intermediate stages of eutrophication and falls off sharply in highly eutrophic waters.

Fish kills may occur in shallow lakes when surfaces freeze in winter, decomposition continuing in the water beneath the ice, utilizing oxygen that cannot be replenished from the atmosphere. In the summer, large beds of macrophytes in rivers and lakes may remove the oxygen from the water by respiration during the night, especially in very warm conditions when little oxygen is dissolved, resulting in extensive fish mortalities.

In Dutch lakes, eutrophication has led to the dominance of bream (*Abramis brama*) in the fish community (Lammens, 1989). Bream grow rapidly compared with other fish species and quickly become too large to be eaten by the chief predatory fish, zander (*Stizostedion lucioperca*). They exert considerable grazing pressure on zooplankton and accelerate the eutrophication process. Large bream also stir and mix the sediments, which may cause further difficulties for the re-establishment of macrophytes.

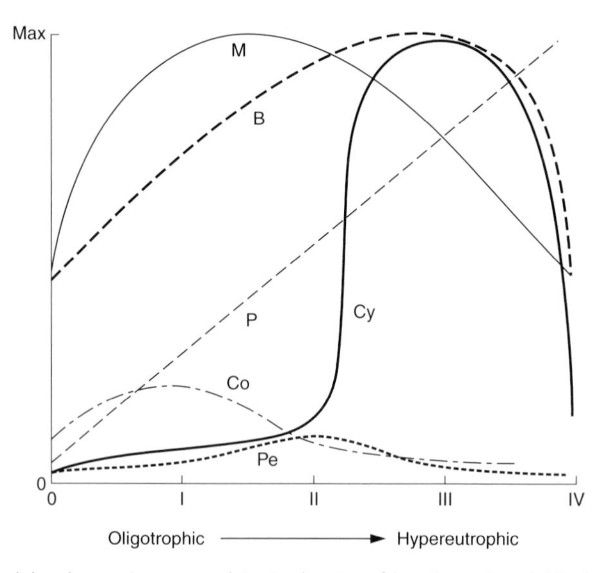

Fig. 4.9 Trends in lakes becoming eutrophic. B. density of benthos; Co, yield of whitefish (coregonids); Cy, yield of cyprinid fish; M, submerged macrophytes; P, phosphorus content of water, density of plankton and turbidity; Pe, yield of percid fish. 0–1V, stages of eutrophication (from Hartmann, 1977).

COMMUNITY INTERACTION AND EUTROPHICATION

A naturally eutrophic lake has an abundant and diverse community of aquatic macrophytes in the littoral zone, while in shallow lakes the plants will occupy much of the waterbody. The canopy provides a refuge for zooplankton, especially those larger species that would otherwise be vulnerable to predation by fish. Large zooplankton, such as *Daphnia*, are efficient grazers of phytoplankton so that water remains clear, providing ideal conditions for macrophyte growth. Fertilization experiments have generally failed to reduce the preponderance of macrophytes (other than low-growing, rooted species) in these diverse systems (Moss and Leah, 1982; Irvine *et al.*, 1989) showing that the ecosystem is very robust. Why then have these macrophyte-rich ecosystems collapsed in the Norfolk Broads and elsewhere?

Stansfield *et al.* (1989) have suggested that the switch was operated by a reduction in large-bodied zooplankton caused by the widespread introduction of organochlorine insecticides into agriculture in the late 1950s and 1960s, when chemicals were used with little control. Cladocera are rather susceptible to pesticides. Sediment core analysis indicated that *Daphnia* became scarce at this time. With the absence of *Daphnia*, large populations of phytoplankton and epiphytes developed to shade out macrophytes. Any later recovery in the zooplankton population would be prevented because, no longer having a macrophyte refuge, they were intensively preyed upon by fish. The community would remain dominated by small forms, such as *Bosmina*, which are less efficient grazers and feedback mechanisms would maintain the system in the phytoplankton stage. Daytime refuges, in the form of water lilies, have been shown to be essential for the survival of large-bodied Cladocera in the Broads (Timms and Moss, 1984).

Would the sudden elimination of zooplankton and the consequent algal blooms have been sufficient to stifle completely those vigorous macrophytes, such as *Ceratophyllum*, which grow rapidly through the waterbody to form surface mats of vegetation, and those such as water lilies which grow from a large rhizome to put out floating leaves, immune to shading by phytoplankton? Once rooted macrophytes become sparse, the sediment is more likely to be suspended by wind and by the activities of bottom-feeding fish, adding to the problems of shading (Barko and James, 1998). The loss of charophytes, which grow in dense carpets on the bottom of shallow lakes but which are the first to disappear with eutrophication, may be especially significant (Scheffer, 1998). Also, those waters that brought increased concentrations of nutrients and raised levels of novel insecticides into the Broads in the 1950s and throughout the 1960s would have additionally carried various new herbicide formulations, applied to control weeds on agricultural land and indeed to control water plants in drainage ditches within the region. A cocktail of herbicide residues may have reduced the vigour and hence the competitive ability of macrophytes at a time when conditions were optimal for algal blooms to develop. These herbicides may also have had a role in the decline of reedbeds and indeed of coastal salt marshes. Herbicides, however, unlike organochlorines, break down quickly in lake sediments, so any part that they may have played in macrophyte decline must remain conjectural.

PROBLEMS FOR HUMANS

The changes in nutrient levels and biology outlined above can directly affect human activities (Table 4.2) and these problems can be summarized into three main areas: those associated with water purification, supply and consumption; those affecting aesthetic and recreational activities; and those causing difficulties with the management of watercourses and lakes.

Water purification

A water treatment plant must produce a finished product of water of a consistently high quality, irrespective of the size of demand (Gray, 1994). The increase in phytoplankton with eutrophication frequently causes severe problems in water purification. Water is treated either by double slow sand filtration, preceded by passage through rapid sand filters or microstrainers, or by coagulation and sedimentation, followed by rapid sand filtration. In the first method large algae are removed at the primary filter, which may rapidly become blocked at high algal densities or high densities of zooplankton, whereas large quantities of small algae can overload the slow sand filters. Coagulation and sedimentation in the second method of treatment are more efficient at removing large numbers of small algae, so that the residual large algae may block the rapid sand filters. Blocked filters seriously reduce the throughput of water at the treatment works and occasionally a reservoir has to be taken out of service temporarily to clean the filters. Where the water supply for potable and industrial use in an area comes mainly from one large source, the blockage of filters in the treatment works can be potentially very serious.

The smallest algal cells often pass through the filters, producing a turbid final water. The cells may decompose in the distribution pipes and some breakdown products, notably mucopolysaccharides, chelate with iron and aluminium that is added to the treatment, leading to increased metal levels passing to supply. The decomposing algae also stimulate the growth of bacteria, and invertebrates will feed on these biofilms growing within the distribution network. The resulting water coming from the tap can have an unpleasant appearance, taste and odour and may contain chironomids and *Asellus*, leading to many complaints from consumers!

Nitrates

A water supply with a high nitrate level presents a potential health risk. In particular infants under six months of age may develop *methaemoglobinaemia* by drinking bottle-fed milk which is high in nitrates. Babies have gastric juices with a very low pH, which favours the reduction of nitrate ions to nitrite. Nitrite ions readily pass into the bloodstream, where they oxidize the ferrous ions in the haemoglobin molecules, reducing the oxygen-carrying capacity of the blood. Above 25 per cent methaemoglobin there is a blueing or cyanosis of the skin (hence the common name of blue-baby syndrome) and associated symptoms. Death occurs at levels of between 60 per cent and 85 per cent methaemoglobin. There have been several thousand cases notified worldwide, many of them fatal. It appears that methaemoglobinaemia only occurs when bacteriologically impure water with nitrate levels approaching 100 mg l^{-1} is supplied and the problem is non-existent with treated, piped water. The disease is largely associated with infants bottle-fed on milk made up with impure well-water containing high levels of nitrates (Gray, 1994). In the UK the last reported case was in 1972.

The European Health Standards for drinking water recommend that nitrate concentrations should not exceed 50 mg NO_3 l^{-1}. Water remains acceptable at nitrate levels of 50–100 mg NO_3 l^{-1} but is unacceptable at concentrations above this. Standards in the United States are more stringent, with nitrate levels higher than 45 mg NO_3 l^{-1} unacceptable. Water derived from river intakes in lowland England frequently has nitrate levels exceeding the maximum recommended level (100 mg NO_3 l^{-1}) and has to be mixed with water low in nitrates before it enters the public supply. Alternatively, bottled water, low in nitrates, is supplied to mothers who are bottle-feeding young infants.

High concentrations of nitrates (greater than 100 mg NO_3 l^{-1}) in water may result in the formation of nitrosamines in the stomach. Nitrosamines have caused stomach cancer in animal experiments, but epidemiological studies have, as yet, failed to find a link between exposure to nitrates and cancer.

Amenity value

The aesthetic and recreational value of eutrophic waters is often reduced (Fig. 4.10). There may be interference with angling, sailing and swimming as a result of the production of surface scums during algal blooms and of the growth of algae and macrophytes on shores. The smell resulting from decaying algae washed ashore is often highly offensive. The dense swarms of midges that emerge from eutrophic lakes can be a considerable nuisance to lakeside visitors and residents, while the insecticides used to control midges cause damage to other organisms, including fishes, in the lake (p. 70).

Management problems, other than water treatment, include excessive weed growth and fisheries. The biomass of aquatic macrophytes increases with nutrient input and plants spread over lakes, rivers and canals, making navigation difficult and hindering recreation. Large growths of macrophytes impede the flow of water and increase the risk of flooding during storms. Many waterways have to be manually cleared of weeds, often twice a year, at great expense.

It has already been noted (p. 144) that increases in enrichment cause changes in

Fig. 4.10 A bloom of *Aphanizomenon* (photograph by the author).

the fish community. Salmonids and coregonids, which are high-quality food fish, are replaced by cyprinids, which are of low quality, though their biomass is usually higher. This change in dominant species is due to de-oxygenation of the hypolimnetic water. Low oxygen conditions may develop in eutrophic waters when algae and macrophytes die and decompose and this will also affect the numbers and species of fishes present. An increase in pH, associated with increased growth of plants, may also kill fish.

Algal and cyanobacterial toxins

High densities of algae may produce toxins that are lethal to animals. *Prymnesium parvum*, for example, grows well in nutrient-rich, slightly brackish waters and produces a toxin that is very potent to fish. It has caused problems in commercial fishponds in Israel and has severely damaged a first-class recreational fishery at Hickling Broad, eastern England (Wortley and Phillips, 1987). Fish kills have occurred in Hickling Broad since 1969 and the high populations of *Prymnesium* appear to be related to an increase in total phosphorus in the water. Large populations of cells are necessary to produce sufficient toxins to kill fish and high toxin activity occurs when phosphorus becomes limiting, because the formation of phospholipids in cell membranes is disrupted, allowing leakage of the toxin.

Toxins produced by cyanobacteria may also pose risks to human health, livestock and wildlife. Species such as *Microcystis*, *Aphanizomenon* and *Anabaena* can produce potent poisons, as measured in mouse bioassays. They can induce rapid and fatal liver damage at low concentrations and may also be neurotoxic, causing paralysis and death in as little as five minutes. Algal toxins in water used for kidney dialysis were suspected of killing 43 people in Brazil in 1996 and there are a number of cases of algal toxins being implicated in illness. Worldwide there are many cases of mortalities of livestock and wildlife. In September 1989, at Rutland Water, which was experiencing a bloom of *Microcystis*, eight dogs and a number of sheep died rapidly after drinking water from the lake. Filtrates from algal scums collected from the lake contained the toxin microcystin at concentrations of 0.1–1.7 $\mu g\, l^{-1}$ and public access, including angling and watersports, was banned for several weeks. Toxins do not always occur in blooms of cyanobacteria, and concentrations, when they do occur, can be highly variable with time. Somewhat more than 50 per cent of blooms tested have contained toxins (Lawton and Codd, 1991) but the presence of toxins is difficult to predict, detect and monitor. Chorus and Bartrum (1999) provide a detailed review of toxic cyanobacteria.

Problems also arise with the bacterium *Clostridium botulinum*, which grows in the sediment of shallow, eutrophic lakes and releases a toxin during periods of hot weather. Birds are especially susceptible to botulism in shallow waters and serious losses of waterfowl periodically occur, especially in the United States (Eklund and Dowell, 1987).

These wide-ranging effects of increases in nutrient loadings on receiving waters show why eutrophication remains an important problem.

UNDERSTANDING EUTROPHICATION BY EXPERIMENT

Much of the information concerning eutrophication has been gained by comparing observations made on polluted and unpolluted waters or by comparing data collected from one site at successive intervals. It is often difficult, however, to determine the cause–effect relationships in such a complex situation where many factors may vary simultaneously. The opposite approach is to conduct laboratory experiments (bioassays), where cause–effect relationships can be verified by direct test, but it is often dangerous to apply results obtained under precise laboratory conditions to field situations. The manipulation of waters in the field lies between these two approaches and has proved very useful in elucidating aspects of the eutrophication process.

Bioassay

Bioassay tests using algae are used to evaluate the nutrient status of a waterbody, for distinguishing between total and biologically available nutrients, and for determining the potential effects of changing water quality on algal growth. Algal bioassays can be done in the field but most involve laboratory studies under carefully defined conditions.

The Standard Bottle Test (American Public Health Association, 1989) has been widely used and is a measure of either maximum specific growth rate or maximum standing crop. The specific growth rate (μ), defined as the number of logarithmic units of increase in cell number per day during the exponential phase of growth, is:

$$\mu = \frac{\log_e N_2 - \log_e N_1}{t_2 - t_1}$$

where N_1 is the initial cell density at time t_1, and N_2 is the final cell density at time t_2. The maximum standing crop is the maximum algal biomass achieved during incubation. Absorbance, chlorophyll measurements or total cell carbon can be substituted for cell counts or biomass determinations. Standard test algae are normally used, namely the green *Selenastrum capricornutum*, the diatom *Asterionella formosa* and the cyanobacteria *Microcystis aeruginosa* and *Anabaena flos-aquae*. *Selenastrum* is easiest to culture and use. The algae initially present in the test water must be removed, by membrane filtration or autoclaving, which may alter the quality of the water. Nutrient depletion and the build-up of metabolic wastes also occur during the test period in the closed bottle.

Figure 4.11 shows the effect of adding primary and secondary treated sewage effluents to water from two Californian lakes, the oligotrophic Lake Tahoe and the eutrophic Clear Lake. Both effluents resulted in a large increase in the populations of *Selenastrum* in Tahoe water but there was no effect in Clear Lake water. The effect of tertiary treatment was examined in bioassays by using water from

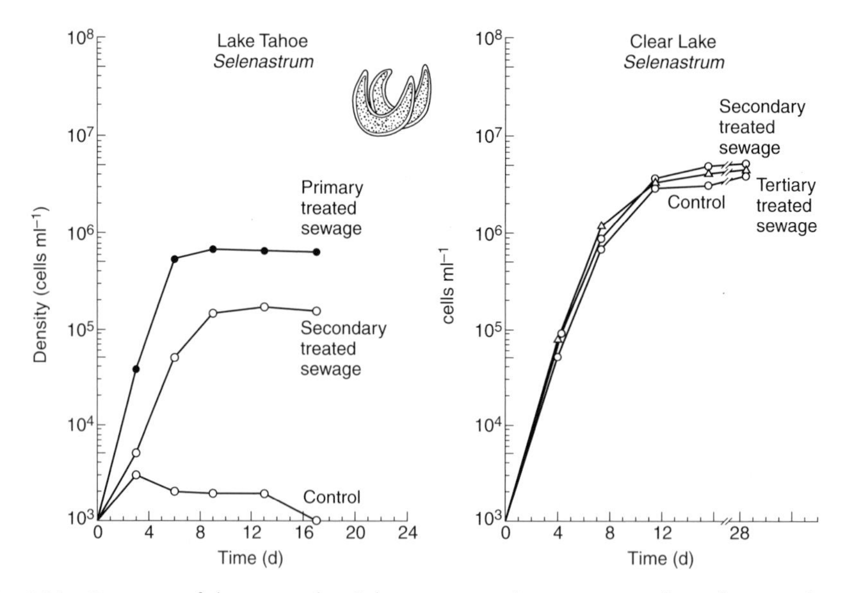

Fig. 4.11 Response of the green alga *Selenastrum capricornutum* to spikes of sewage in oligotrophic Lake Tahoe and eutrophic Clear Lake waters (from Payne, 1975).

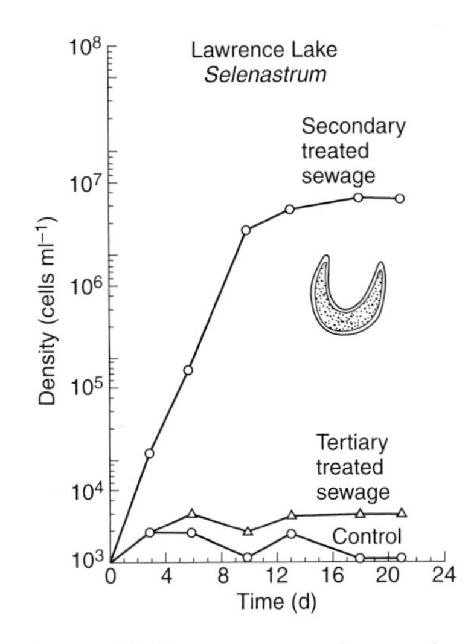

Fig. 4.12 Effect of secondary and tertiary sewage spikes on the growth of *Selenastrum capricornutum* in water from Lawrence Lake, Michigan (from Payne, 1975).

Lawrence Lake, Michigan (Fig. 4.12), where spiking with tertiary treated sewage (from which nutrients had been removed) produced no significant growth, compared with the large growth when the lake water was spiked with secondary treated sewage. The tertiary treatment of sewage (p. 102) can therefore significantly reduce the eutrophication of waterbodies.

An alternative to bottle tests is the continuous flow or chemostat method, in which nutrient medium is added continuously so that its constituents are maintained at the required concentrations. A steady state is established, at which the test organism maintains a constant specific growth rate and cell concentration in a constant substrate concentration.

Mesocosms within lakes

Part of a lake may be isolated and nutrients added, the water outside the enclosure acting as a control, or there may be a series of experimental and control enclosures (Ravera, 1989). Some enclosures may be very large. For example, Lund Tubes, used in the Lake District of England, were made of butyl rubber, with a diameter of 45.5 m and a depth of 15 m (Lund, 1978). Such enclosures are too large and expensive to replicate. Smaller enclosures are generally made of polyethylene. One problem with enclosures is that the walls develop a microflora and fauna which can markedly alter the nutrient balance within, such problems increasing in smaller enclosures with large surface to volume ratios.

In eutrophic Lake Stigsholm, Denmark, 18 enclosures of area 100 m^2 were constructed to examine the impact of submerged macrophytes on the inter-relationships among phytoplankton, zooplankton and small planktivorous fish. When macrophytes were absent, the biomass of zooplankton was low and dominated by cyclopoid copepods, irrespective of fish density, while algal biomass was high. When plant cover exceeded 15–20 per cent and fish density was low, the biomass of cladoceran zooplankton was high and phytoplankton biomass was low. At high fish densities, the zooplankton again became dominated by copepods and the phytoplankton density increased markedly, irrespective of the density of macrophytes. Provided fish density was not high, the macrophytes provided refuges for the larger zooplankton and algal biomass was reduced (Schriver *et al.*, 1996). Perrow *et al.* (1999) suggest that a plant cover of 30–40 per cent will give refuge to zooplankton, provided predation is not excessive.

The refuge effect of macrophytes has also been demonstrated in smaller mesocosms (e.g. Beklioglu and Moss, 1996; Tüzün and Mason, 1996). In the latter study a complete exclusion of fish from mesocosms resulted in the development of a population of *Neomysis* shrimps and these suppressed *Daphnia* populations.

In the Norfolk Broads mesocosm studies have been undertaken at Hoveton Great Broad (Irvine *et al.*, 1991; Phillips and Kerrison, 1991). The broad (Fig. 4.13) is fairly large (31 ha), shallow (1–1.5 m) and largely devoid of macrophytes. Pens of 4 m^2 were constructed, which allowed for free movement of water and phytoplankton, but prevented movement of zooplankton and excluded fish. The phytoplankton community in the pens and open water were similar throughout the summer but densities of *Daphnia* were much higher in the enclosures. They also had higher

Fig. 4.13 Aerial view of experimental enclosures in Hoveton Great Broad, eastern England. A barrier that excludes fish but allows water movement can be seen, together with a series of circular experimental enclosures (photograph courtesy of Geoff Phillips, Environment Agency).

egg-ratios and mean brood sizes because of the greater proportion of larger animals in the pens. This again suggested that zooplankton populations in the lake were controlled by size-selective predation by fish and not by algal production. It was suggested that the removal of fish would extend a brief clear-water phase, caused by zooplankton grazing throughout the summer, allowing macrophytes to establish.

Mesocosm studies have also examined the role of different species of young fish in structuring the plankton community (Kurmayer and Wanzenböck, 1996). Enclosures with bleak (*Alburnus alburnus*), roach (*Rutilus rutilus*) and perch (*Perca fluviatilis*) all showed reduced *Daphnia* populations, transparency and phosphorus concentrations, and increased chlorophyll *a* levels, compared with mesocosms without fish. The strongest effects, however, were in enclosures containing perch, emphasizing this species as a key predator in structuring plankton communities in some lakes.

Experimental ponds (area 500 m^2, depth 1 m) in subtropical Mexico have been used to examine the effect of carp (*Cyprinus carpio*) on turbidity and macrophyte growth (Zambrano and Hinojosa, 1999). Carp caused a marked increase in turbidity of the water but the effect was non-linear, occurring at carp densities above 0.8 fish m^{-2}. Only the macrophyte *Sagittaria mexicana* was reduced in the presence of carp, because it was eaten by the fish; at a depth of only 1 m, the ponds were considered to be too shallow for the increased turbidity to affect plant growth.

Whole lake manipulations

Scientists at the Freshwater Institute of the Fisheries Research Board of Canada have observed the effects of nutrient additions to whole lakes. Some 46 small lakes, within an area containing several hundred in Ontario, have been set aside for experimental research (Experimental Lakes Area). Lake 227 has a surface area of 5 ha and a maximum depth of 10 m. Phosphorus and nitrogen were added weekly from 1969 to 1972, the annual addition amounting to 0.48 g P m^{-2} yr^{-1} and 6.29 g N m^{-2} yr^{-1}, these loadings increasing the natural inputs by about five times.

The transparency of water became on average less in each year (Fig. 4.14). In 1969 the maximum biomass of phytoplankton was 16 160 mg m^{-3}, compared with 5000 mg m^{-3} in 1968, before fertilization. In 1970 the peak standing crop had more than doubled to 35 000 mg m^{-3}. Poor weather conditions in 1971 prevented further increase in the maximum biomass, but the peak biomass in 1972 increased to 63 000 mg m^{-3}.

Before fertilization the phytoplankton of Lake 227 was dominated by Chrysophyceae (golden-brown algae). In 1969 Chlorophyta (green algae) were predominant in the summer. In 1970 cyanobacteria, which had never occurred before fertilization, became dominant in the late summer. In 1971, cyanobacteria (especially *Oscillatoria*) again showed periods of dominance, although the situation was complicated by unusually cold weather in July and early August. In 1972 cyanobacteria and Chlorophyta were again abundant, with the latter remaining dominant. The phytoplankton production in Lake 227 was some ten times higher than that occurring in natural lakes in the area (Schindler *et al.*, 1973; Schindler and Fee, 1973).

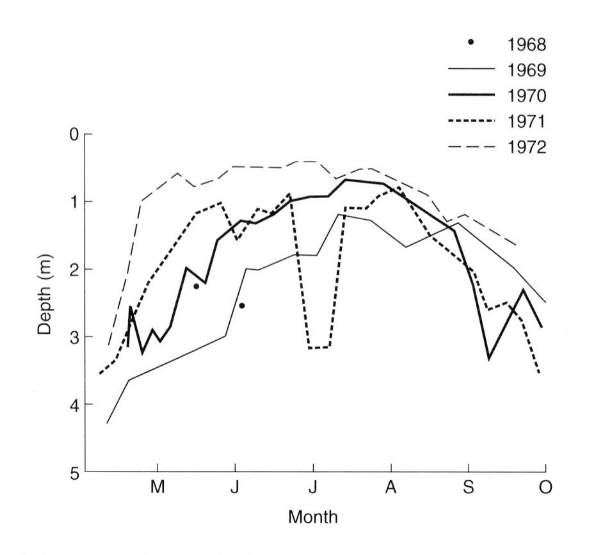

Fig. 4.14 Seasonal changes in the Secchi disc visibility in Lake 227, Ontario, Canada, 1968–72 (from Schindler *et al.*, 1973).

In Lake 227, algal blooms consisted primarily of chlorophytes, with cyanobacteria only occasionally dominant. By contrast a lower rate of fertilization to another lake, 226, consistently resulted in blooms of nitrogen-fixing cyanobacteria. As well as the lower level of nutrient addition, Lake 226 was receiving a different ratio of N : P (5 : 1 by mass) compared with Lake 227 (14 : 1). It therefore appeared that the onset of a cyanobacterial bloom was triggered by a scarcity of nitrogen relative to phosphorus, rather than merely a phosphorus input. This was tested in Lake 227 in 1975 when nitrogen additions were reduced while phosphorus loading was kept constant (Schindler, 1988a; Findlay *et al.*, 1994). Cyanobacteria dominated the lake in summer, fixing atmospheric nitrogen to compensate for deficiencies in nitrogen supply.

These observations on whole-lake fertilization were supplemented by experiments in replicated enclosures (large polyethylene tubes) in which nutrients were added singly or in various combinations (Schindler *et al.*, 1971). The addition of either organic or inorganic carbon resulted in no increase in algal standing crop. The addition of nitrogen and phosphorus together always produced large standing crops of algae. When added singly, phosphorus always caused some increase in standing crop, whereas the addition of nitrogen alone never elicited a response. So, both whole-lake fertilization and manipulations in enclosures demonstrate that it is the concentration of phosphorus that controls the standing crop of phytoplankton. If nitrogen becomes limiting, cyanobacteria will fix atmospheric nitrogen and will dominate the phytoplankton, often to the exclusion of other groups. Limiting the supply of phosphorus to lakes would therefore appear to be the key to overcoming problems caused by eutrophication.

The whole-lake manipulations in Canada have concentrated on the role of nutrients in fuelling eutrophication. The earlier section on mesocosms looked especially at the role of predators in promoting algal blooms in eutrophic lakes. Experimental lakes in Wisconsin have been used to explore how predators structure plankton communities – the trophic cascade. The study is described in detail by Carpenter and Kitchell (1993). The role of nutrients and predators are the basis, respectively, of bottom-up and top-down control of eutrophication to be described below.

MODELLING EUTROPHICATION

Many lakes are exhibiting eutrophication as a result of human activities and it is clearly impossible to study all of them in detail. It would obviously be extremely valuable to managers of freshwater ecosystems to be able to predict the changes caused by rates of enrichment, but with the measurement of as few parameters as possible.

Modelling approaches to eutrophication are described in Jørgensen (1980) and Henderson-Sellers and Markland (1987). They vary widely in their complexity. Vollenweider (1975) has emphasized that satisfactory models should meet three essential criteria. They should be general, realistic and precise. In practice one of these criteria usually has to be sacrificed in order to maximize the others.

Models of eutrophication have concentrated especially on the relationships of nitrogen and phosphorus to algal production, with particular emphasis on phosphorus because this nutrient can explain much of the variance in chlorophyll levels in lakes (Heyman and Lundgren, 1988).

The best known conceptual model is that of Vollenweider (1969, 1975):

$$\bar{P} = \frac{L}{\bar{z}(r_s + r_f)}$$

Table 4.5 Vollenweider's permissible loading levels for total nitrogen and total phosphorus (biochemically active) (g m^{-1} yr^{-1})

	Permissible loading		**Dangerous loading**	
Mean depth (m)	**N**	**P**	**N**	**P**
5	1.0	0.07	2.0	0.13
10	1.5	0.10	3.0	0.20
50	4.0	0.25	8.0	0.50
100	6.0	0.40	12.0	0.80
150	7.5	0.50	15.0	1.00
200	9.0	0.60	18.0	1.20

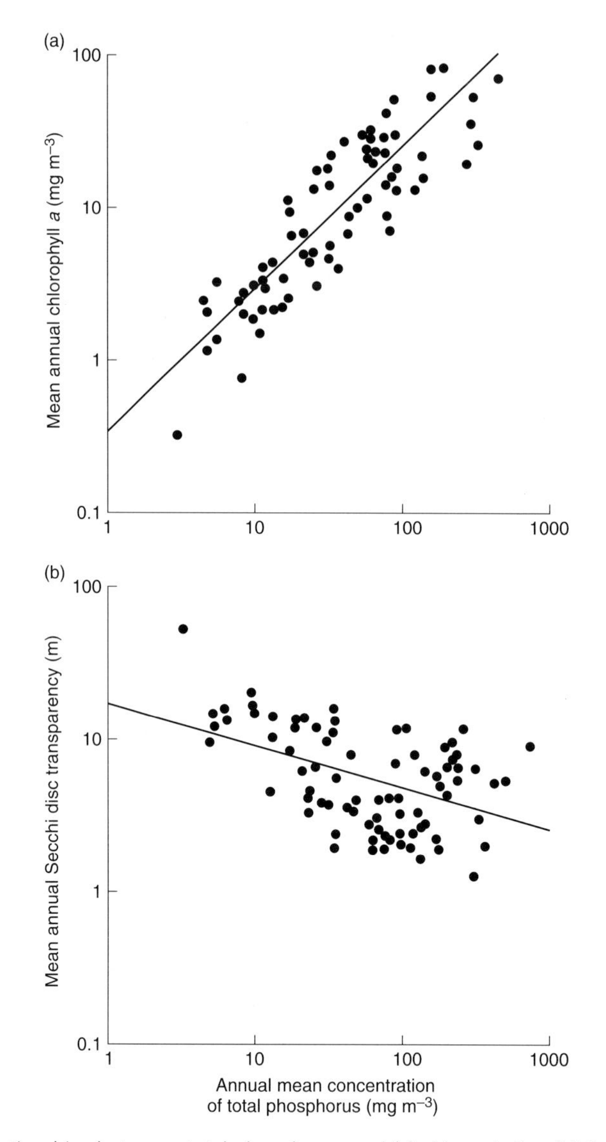

Fig. 4.15 Relationships between total phosphorus and (a) chlorophyll *a*; (b) Secchi depth from an international study of lake eutrophication (adapted from OECD, 1982).

where $\bar{P}$ is the concentration of total phosphorus ($g\,m^{-3}$), often measured at the time of lake overturn in spring, L is the phosphorus loading ($g\,m^{-2}$ lake area per year), $\bar{z}$ is the mean depth of the lake (m), r_s is the sedimentation rate coefficient (the fraction of P lost per year to the sediments) and r_f is the hydraulic flushing rate (the number of times the water in the lake is replaced each year).

This input–output model assumes that the lake is in a steady state, uniformly mixed and with loading and flushing rates constant over time. Most of the parameters can be measured using straightforward techniques, only r_s presenting special difficulties. The model has been used to produce loading values of nitrogen and phosphorus which would be permissible for lakes of different depths, and loading values which would cause deterioration of the lake (Table 4.5). The ratio of nitrogen to phosphorus in the water can also provide useful information. If the N : P ratio exceeds 16 : 1, phosphorus is likely to be limiting to algal growth; if the ratio is less than 16 : 1, nitrogen may be limiting.

The simple relationship between loading and depth has been further developed by including such parameters as the internal loading (such as from the sediments) and the length of the shoreline. From a study of many lakes, worldwide empirical relationships between phosphorus loading, chlorophyll *a* concentration and transparency (measured using a Secchi disc) have been calculated (OECD, 1982), as shown in Fig. 4.15.

It should be noted that the relationships illustrated in Fig. 4.15 are logarithmic and there is considerable variation around the regression line. The model illustrates general relationships across lakes of widely differing nutrient status but it may be less good at predicting the impact of nutrients in individual waterbodies. There is now considerable evidence from some lakes that it is nitrogen, not phosphorus, that is limiting productivity. Biological productivity may not always be adequately predicted by simple phosphorus (or indeed nitrogen) load approaches because the links between nutrient fluxes and biological production are subject to complex biotic and abiotic interactions (Güde and Gries, 1998).

Despite these caveats, there are instances where simple phosphorus loading models have proved very useful in managing eutrophication. In Ontario, Canada,

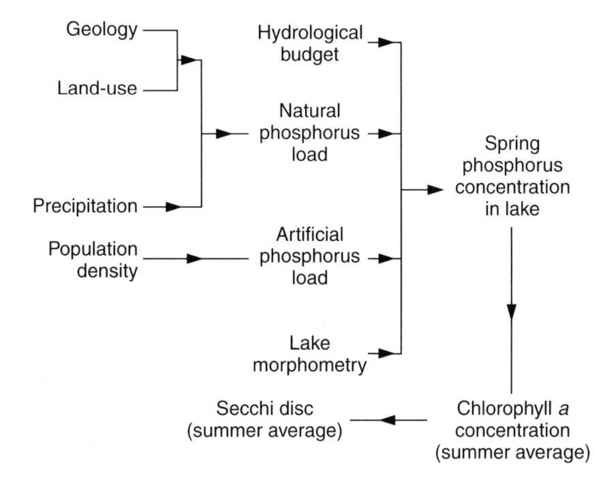

Fig. 4.16 Scheme of empirical models used to assess the effects of development on the trophic status of lakes (from Dillon and Rigler, 1975).

there is a high demand for summer cottages and other recreational developments by lakes, and their often rudimentary sanitation is a potential source of nutrients to the water. A model was developed by Dillon and Rigler (1975) to predict the effect of such developments on lakes. The model consists of a sequence of equations relating nutrient loadings to watershed characteristics, lake nutrient concentrations to loading, and biological responses to nutrient concentrations. The approach is illustrated in Fig. 4.16. The *total phosphorus* supplied to a lake from natural sources (J_N) derives from the catchment (J_E) and direct precipitation (J_{PR}). An estimate of the export of phosphorus from a catchment can be made by examining the land-use (e.g. the proportions of forestry, arable and pasture) and the geology (e.g. whether igneous or sedimentary). The export coefficient is multiplied by the area of the catchment and these are summed for all tributaries to the lake to obtain J_N. J_{PR} can be determined from long-term average rainfall values and the average concentration of phosphorus in rainwater.

The total natural phosphorus supplied to a lake in a year is:

$$J_N = J_E + J_{PR} \text{ mg yr}^{-1}$$

and the total natural loading (L_N) is J_N divided by the surface area of the lake (A_O) in $\text{mg m}^{-2}\text{ yr}^{-1}$.

The artificial supply (J_A) to the lake will depend on a number of factors, such as the number of cottages, caravans etc., their methods of sewage disposal, the number of inhabitants, and the time they spend in the cottage each year. Dillon and Rigler assumed that the average North American supplies 0.8 kg P to the environment each year. An estimate of the number and usage of cottages requires a field survey.

The total supply of phosphorus to the lake (J_T) is:

$$J_T = J_N + J_A \text{ mg yr}^{-1}$$

and the total loading (L_T) is:

$$L_T = J_T/A_O \text{ mg m}^{-2}\text{ yr}^{-1}$$

To predict the spring phosphorus concentration ($\bar{P}$) in the lake, Dillon and Rigler used the Vollenweider model described above. The sedimentation rate (r_s) was obtained indirectly through the retention coefficient, which is highly correlated with the areal loading and hence easily predicted.

Knowing the spring phosphorus concentration ($\bar{P}$), the summer average chlorophyll *a* concentration can be predicted:

$$\log_{10} (\text{chl } a) = 1.45 \log_{10} (\bar{P}) - 1.14 \text{ mg m}^{-3}$$

Chlorophyll *a* can then be related to transparency (the Secchi disc reading, Fig. 4.17).

Changes in the loading of phosphorus to a lake result in changes in the phosphorus concentration and hence water quality, but the response is gradual, rather than immediate, and follows an exponential relationship. The response time can be

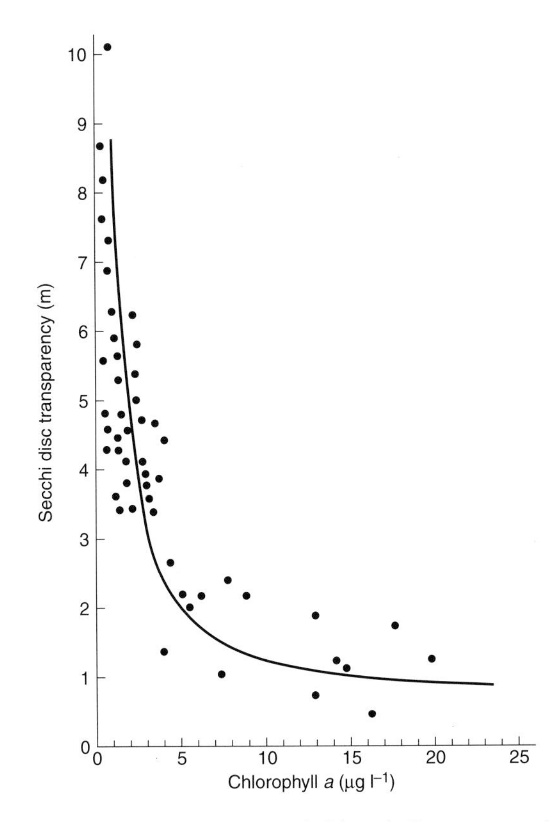

Fig. 4.17 Relationship between transparency and chlorophyll *a* concentration for lakes in southern Ontario. The values for each lake are based on means collected over the period of stratification (June–September) (from Dillon and Rigler, 1975).

described by the half-life of the change in concentration ($t_{1/2}$) which depends only on the rate coefficients and

$$t_{1/2} = \ln 2/(r_s + r_f)$$

Lakes with rapid flushing times will have short half-lives and hence response times, while lakes with slow flushing rates will respond only slowly to changes in loading.

The Dillon–Rigler model is mostly used by managers in the reverse direction. A minimum summer lake transparency or maximum chlorophyll *a* concentration which is considered acceptable is decided. The maximum permissible loading for phosphorus from artificial sources (L_A) is then determined using the model and this is translated into the maximum allowable development in terms of cottages etc.

It would obviously not be feasible in terms of resources or time to scientifically evaluate every lake in Ontario where development is anticipated, so the model allows reasonable predictions of the effects of development to be made with little or no field work. While some of the parameters used in this model apply only to Ontario, models in general have wide application in the management of water resources. Refinements of the Dillon–Rigler model allow lake managers to predict the effect of lakeside developments throughout an entire watershed, including lakes situated downstream (Hutchinson *et al.*, 1991). Schindler (1987a) considered that the development of loading models to control phosphorus inputs has been the greatest single success of ecological knowledge over environmental problems.

CONTROLLING EUTROPHICATION

Table 4.4 lists the guidelines for classifying the trophic status of a waterbody following a survey. In Ardleigh Reservoir, the plankton of which was described on p. 136, peak phosphorus concentrations were some 250 times the minimum concentration for assigning a waterbody to the eutrophic category, while peak concentrations of nitrogen were 10 times the minimum concentration. We have seen that the productivity of freshwaters is generally limited by phosphorus. Eutrophication may therefore be most effectively controlled by reducing the load of phosphorus. Even in those cases where eutrophication is fuelled by nitrogen, there are a number of reasons why phosphorus control is likely to be more effective than nitrogen control. Phosphate is present in only trace amounts in oligotrophic lakes, while their inflow streams are low in phosphate when they are not influenced by human activities. In contrast, inflow streams may contain large quantities of nitrates. Nitrates are leached readily from agricultural lands whereas phosphates tend to be more tightly bound. Some cyanobacteria and bacteria can fix gaseous nitrogen and so will not be limited by nitrogen provided other essential nutrients are available. It is also relatively easier and cheaper to remove phosphorus than nitrogen during the sewage treatment process. Nevertheless, where water is used for potable supply it may be necessary to reduce the level of nitrogen for public health or legal reasons (p. 146), while nitrogen is the most important cause of eutrophication in coastal waters.

Rast and Holland (1988) have provided a basic, practical framework for the development of a eutrophication management strategy. It consists of the following steps:

1. Identify eutrophication problem and establish management goals.
2. Assess the extent of available information about the lake or reservoir.
3. Identify available, feasible eutrophication control methods.
4. Analyse all costs and expected benefits of alternative management strategies.
5. Analyse adequacy of existing institutional and regulatory framework for implementing alternative management strategies.
6. Select desired control strategy and disseminate plan to interested parties.
7. Use institutional mechanisms to minimize future eutrophication problems.

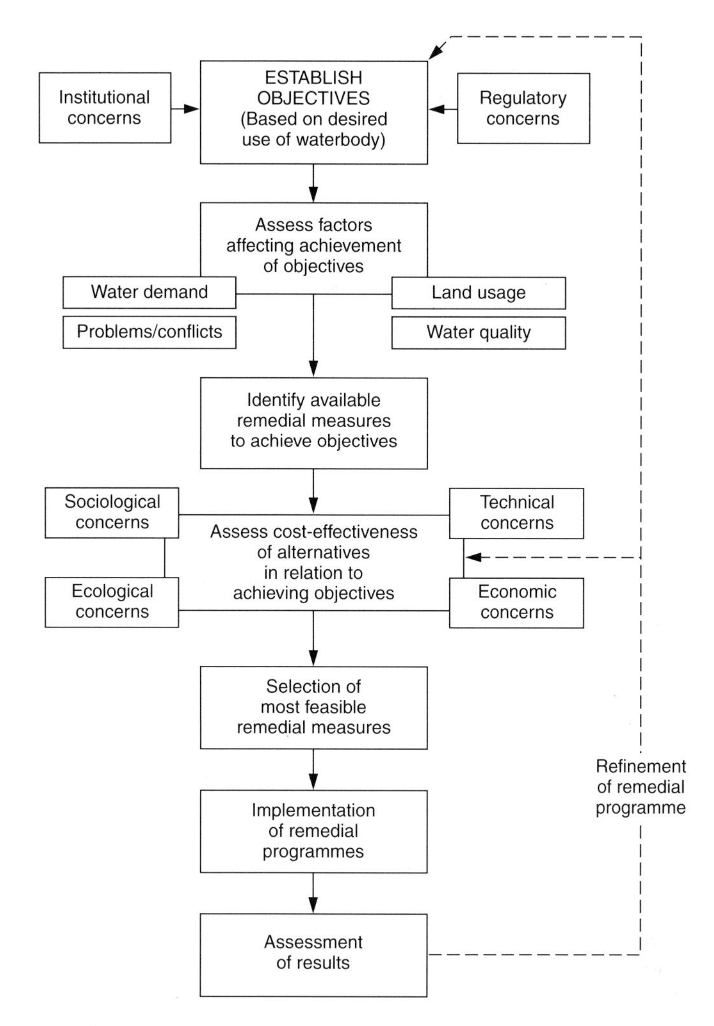

Fig. 4.18 Sequence of decisions to be made in the development and implementation of eutrophication control programmes (from Rast and Holland, 1988).

A schematic representation of these steps is illustrated in Fig. 4.18.

Restoration methods may be divided into two classes:

- Reduction of external loading of nutrients.
- Manipulations within the lake ecosystem.

In practice it is often necessary to apply both techniques simultaneously to ensure some measure of success. Ryding and Rast (1989) and Cooke (1993) describe in detail the steps needed for eutrophication control.

Reduction in external loading

The discharge of nutrient-rich sewage and animal wastes into watercourses and lakes is often the major source of enrichment problems, while detergents are often the most important source of phosphorus in wastewaters. In the United States 40–70 per cent of phosphorus within sewage came from phosphate detergents and the output from municipal sewage plants may contain up to 70 per cent of the total input of phosphorus to a lake. The elimination of phosphorus as an essential component of detergents could therefore remove about 50 per cent of the total phosphorus entering some lakes. Effective phosphate-free detergents have been developed, for example based on zeolites, and some states and cities in the United States have banned or restricted the phosphorus content of detergents. Such a ban has been resisted in the European Union.

Primary sewage treatment removes only about 5–15 per cent of nutrients and secondary treatment only 30–50 per cent, so that a tertiary treatment process is necessary to eliminate the majority of phosphorus contained within sewage. Treatment may involve chemical, physical or biological removal.

Phosphates are precipitated, a process known as phosphate stripping, using coagulants of lime or compounds of aluminium or iron. The precipitate is then separated in a sedimentation unit and combined with the other sludges generated in the treatment of sewage. The process can be more than 95 per cent efficient at removing phosphorus and in a number of European countries (e.g. Denmark, the Netherlands, Sweden, Switzerland) it is a routine part of the treatment process. This has led, for example, to an 80 per cent reduction in the concentration of phosphorus in discharges from sewage treatment works in Denmark over the period 1984–95.

Biological removal uses the ability of some microorganisms to take up phosphorus in excess of their immediate nutritional requirements and store it within the cells in the form of polyphosphates. These can then be separated from the system with the sludge. Biological removal of phosphorus can be achieved by modifying existing processes for nitrogen removal.

Phosphorus may also be removed by passing water into treatment ponds, where much of it is absorbed on to particulate matter which will settle out in the pond. If the water is retained for three or more days nutrients will be taken up by algal blooms, which die and transfer phosphorus to the sediment. It may also be possible to use simple floating plants, such as duckweeds (*Lemna*) or water fern (*Azolla*) which can be regularly harvested. Details are provided by Viessman and Hammer (1993).

The European Union Urban Waste Water Treatment Directive is an important tool in controlling some of the major sources of phosphate pollution, though its action is limited to treatment works serving conurbations of more than 10 000 people. Morse *et al.* (1998) point out that some 250 000 t of phosphorus annually are potentially removable from wastewater in Western Europe alone, comparable to the requirements of the industrial phosphate industry, which is currently based almost entirely on mining phosphate rock, a non-renewable resource. Phosphorus recovery should therefore be viewed not as a problem but as an opportunity, leading towards a sustainable industry in phosphorus.

The operating costs of tertiary treatment plants removing phosphate may almost double the overall costs of sewage treatment. It is therefore essential to know accurately the proportion of the total phosphorus loading due to sewage effluents and the efficiency of the proposed tertiary treatment plant before it is installed. A nutrient budget for Barton Broad, eastern England, showed that 80 per cent of a total phosphorus loading of 18–22 mg P $m^{-2} yr^{-1}$ was derived from sewage sources and it was considered that an 80 per cent reduction in the phosphorus content of sewage effluents was required to permit the re-establishment of aquatic macrophytes (Osborne, 1981). Subsequently phosphorus removal was initiated by adding ferric (iron(III)) sulphate to final effluents from sewage treatment works upstream of Barton Broad and this eventually reduced the annual phosphorus load to the lake by 90 per cent. Total phosphorus concentrations within the lake, however, remained high, summer values exceeding 100 μg P l^{-1}, while the annual mean chlorophyll *a* concentration exceeded 100 $\mu g\, l^{-1}$ and the Secchi depth transparency was only 30 cm in summer (Phillips and Kerrison, 1991). Internal loading (see below) was found to be as high as 130 mg P $m^{-2} d^{-1}$, compared with an external load of only 12 mg P $m^{-2} d^{-1}$ (Phillips *et al.*, 1994). Macrophytes have not returned. Sediments from this lake have been removed as part of the restoration programme (Moss *et al.*, 1996).

An alternative to tertiary sewage treatment is to divert nutrient-rich wastes away from vulnerable lakes, which can only be done if there is somewhere, usually the sea, to put the diverted water. Lake Washington (p. 132) was one example where this procedure was applied successfully. Effluent diversion from Little Mere, northwest England, reduced the concentration of nitrogen and phosphorus by 91 per cent and 92 per cent respectively (Beklioglu *et al.*, 1999).

Where the majority of external loading is derived from one source, as it usually is in a pumped-storage reservoir scheme, it is possible to treat the inflow water in a holding lagoon before it enters the main waterbody. In Ardleigh Reservoir, 90 per cent of the load of soluble reactive phosphorus was derived from water pumped from the River Colne, which has a number of sewage treatment works upstream (Redshaw *et al.*, 1988). This water is now treated with ferric sulphate to precipitate out phosphorus before it enters the reservoir.

Some 50 per cent of phosphorus originates from agriculture, two-thirds of which derives from livestock. The development of methods for dealing with animal wastes is an important but somewhat intractable problem. Livestock also damage river banks, causing erosion, sediment contamination of river gravels and the release of phosphorus. Providing animals with alternative sources of water, while fencing river banks, is a solution.

The erosion of arable soils is a major problem. Some 80 per cent of erosion events studied in England and Wales were on land cropped to winter cereals, resulting in some 18 per cent of overall phosphorus losses from agriculture

(Chambers *et al.*, 2000). Poor crop cover, soil compacted by farm vehicles, and natural features that concentrated surface run-off were associated with the major erosion events. The drilling of seeds into stubbles (minimum tillage) and early drilling to ensure good crop cover during winter were among techniques suggested to reduce erosion. Changes in the timing and a reduction in the frequency of fertilizer applications would also reduce nutrient losses to water. Putting land down to permanent pasture may be undertaken. In Europe, Nitrate Vulnerable Zones have been established in areas where groundwater is heavily contaminated with nitrate. These cover some 600 000 ha in the United Kingdom and there are restrictions on the amount and timing of manure applications, while farmers must keep records of crops grown and the addition of fertilizers.

The planting of buffer strips between farmland and watercourse may also be effective (Haycock *et al.*, 1993). A buffer strip is a vegetated strip of land some 5–50 m wide adjacent to a watercourse. It may be planted and managed as grassland, or allowed to develop as marshland or riparian woodland if hydrological conditions allow. The immediate environs of the river at least should ideally be allowed to develop a natural riparian vegetation. Buffer strips filter out particulate material and hence much of the phosphorus associated with them, while nitrate will be taken up by the plants or, if the soil is wet and anaerobic, converted to nitrogen gas. A strip of 2 m may remove all of the nitrogen. A wetland buffer strip will prevent some 10 t N $ha^{-1} yr^{-1}$ from reaching an adjacent watercourse, resulting in savings of US\$2000 $ha^{-1} yr^{-1}$ in reduced water treatment costs. Buffer strips also act as habitats and corridors for wildlife, while adding organic matter to fuel aquatic food webs. Ryding and Rast (1989) provide a detailed assessment of methods to control external sources of nutrients.

Cullen and Forsberg (1988) recognized three main types of response of lakes to a reduction in the external load of phosphorus:

1. Reduction in lake phosphorus and chlorophyll concentration sufficient to change the trophic category of the water, e.g. from hypertrophic to eutrophic or eutrophic to mesotrophic.
2 Reduction in lake phosphorus and chlorophyll concentration that is insufficient to change the trophic status, though lakes become less eutrophic.
3 Reduction in lake phosphorus results in little or no reduction in chlorophyll or algal biomass, though nuisance species may be reduced.

In a survey of 43 lakes where point sources of phosphorus had been reduced, Cullen and Forsberg consider that 15 fell into type 1, 9 into type 2 and 19 into type 3, that is, in 44 per cent of cases there had been no significant improvement.

In the Netherlands, for example, the external phosphorus load to the Loosdrecht lakes ecosystem and to the Naardermeer was substantially reduced in the mid-1980s but there was no immediate improvement in water quality (Liere *et al.*, 1992; Bootsma *et al.*, 1999). In Esthwaite Water in the English Lake District, the removal, from 1986, of the phosphorus from the sewage effluent, contributing

47–67 per cent of the total P loading, resulted in no marked changes in water quality and no reduction in the cyanobacterial dominance of the phytoplankton (Heaney *et al.*, 1992). Reduction of phosphorus loadings, usually in excess of 70 per cent, to 27 Danish lakes resulted in very little improvement in conditions (Jeppeson *et al.*, 1991). Phosphorus controls on sewage treatment works discharging to the Vaal River in South Africa led to a 37 per cent reduction in phosphate concentration at an abstraction point to a reservoir but no reduction in chlorophyll levels, while there was a ten-fold increase in cyanobacteria populations, with an increased incidence of taste and odour problems in the drinking water (Heath *et al.*, 1998). The example of Barton Broad was mentioned above. Therefore, external phosphorus control alone is often not sufficient to achieve the desired objectives of lake restoration. Within-lake measures are also required.

Manipulations within the lake ecosystem

In stratified eutrophic lakes oxygen depletion occurs in the hypolimnion, which may result in the death of fish. Anaerobic conditions at the sediment surface result in the release of phosphate which then becomes available to algae at overturn, resulting in more photosynthesis and more organic matter to deplete oxygen further in the hypolimnion. This cycle can be broken by aeration. Destratification increases the turbulence in the water which may affect the ability of the phytoplankton to grow, especially the nuisance cyanobacteria. The techniques of destratification and re-aeration are described in detail by Henderson-Sellers and Markland (1987) and Harper (1992).

In the Biesbosch Reservoirs in the Netherlands, fed by polluted and highly eutrophic water from the River Meuse, the injection of air at the bottom of the reservoirs prevents thermal stratification and hence a serious deterioration in water quality is avoided. It also mixes algae over a sufficient depth so that light, rather than nutrients, limits growth. This, combined with grazing by zooplankton, maintains the biomass at acceptable levels (Oskam and Breeman, 1992). Artificial mixing in Lake Nieuwe Meer, the Netherlands, prevents nuisance blooms of the cyanobacterium *Microcystis* developing, the plankton switching to one dominated by flagellates, green algae and diatoms (Visser *et al.*, 1996).

One of the major problems following a reduction in external phosphorus is an internal loading of phosphorus from the sediments. There is normally a net flow of phosphorus to the sediments but release occurs under conditions of low oxygen. Aerobic release may also occur, probably dependent on desorption of phosphorus from inorganic complexes and the mineralization of phosphorus associated with organic material, especially where sediments are in contact with overlying water low in orthophosphate. In Ardleigh Reservoir, the net sediment release of soluble reactive phosphorus between July and September was equal to 23 mg P m^{-2} d^{-1}, equivalent to 33 per cent of the mean annual loading of the nutrient (Redshaw *et al.*, 1988). At other times of the year the sediments acted as a sink, but release coincided with the development of large algal blooms. In the Norfolk Broads internal loading has been recorded as high as 278 mg P m^{-2} d^{-1} (Phillips *et al.*, 1994). When external loading has been reduced, the internal loading is responsible for variations in lake phosphorus concentrations (Molen and Boers, 1994). Phosphorus release may take place at depths down to 25 cm in sediments and internal loading may persist for at least 30 years after a reduction of external loading (Søndergaard *et al.*, 1999).

Phosphorus cycling between sediments and water is complex and is still poorly understood. It is clear, however, that it is not only a chemical process, dependent on redox and pH, but that microbial activity is also involved. Microorganisms may release or bind phosphorus through metabolic reactions, extracellular release and cell lysis. Brunberg and Boström (1992) have shown that colonies of *Microcystis* in lake sediments stimulate bacterial activity, eventually resulting in a release of phosphorus to the water column. As *Microcystis* colonies may last for several years in the sediments, they may be important for delaying the recovery of lakes.

One approach to preventing the release of phosphorus from the sediment is chemically to 'seal' the nutrient in. Foxcote Reservoir, in the English Midlands, was highly eutrophic and periodically the water was untreatable for up to six months of the year. Following laboratory tests, pumping of river water to the reservoir was stopped in early April 1981 and the waterbody was dosed, from a boat, with 9000 l of ferric sulphate liquor (3.5 mg Fe l^{-1}). Within five minutes of dosing, the soluble reactive phosphorus in the water dropped from 0.025 mg l^{-1} to 0.001 mg l^{-1}. A layer of ferric hydroxide floc, about 1 cm thick, formed over the reservoir sediment, acting as a barrier for phosphorus release (Hayes *et al.*, 1984). In June pumping of river water recommenced but it was treated with ferric sulphate before it was released into the reservoir and this treatment has continued ever since. By the summer following the treatment the algal populations had declined sharply (Fig.

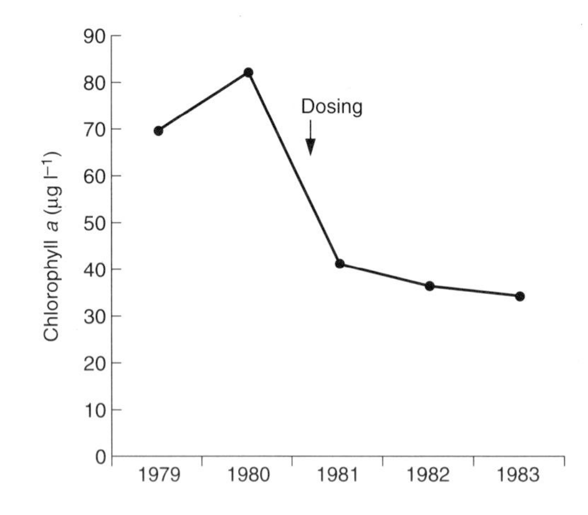

Fig. 4.19 Peak algal standing crops (µg chlorophyll *a* l^{-1}) in Foxcote Reservoir before and after treatment with ferric sulphate (from data in Hayes *et al.*, 1984).

4.19) and in the next year a dense growth of macrophytes had developed in the reservoir. The macrophyte community showed marked changes over the following decade, with an increase in *Elodia* and an initial increase, followed by a decline, in *Chara* (Daldorph, 1999). The sudden reduction in the phosphorus loading in Foxcote Reservoir switched the waterbody from phytoplankton dominated to macrophyte dominated. Two other reservoirs in the region have their inflow water treated but the reservoirs themselves have not been dosed. The improvement in water quality has been less spectacular than at Foxcote Reservoir because, without sediment sealing, internal loading continues (Daldorph and Price, 1994).

Jaeger (1994) reported a 90 per cent decline in maximum chlorophyll levels in a small glacial lake following treatment with ferric chloride. Boers *et al.* (1994) observed a similar rapid decline in phosphorus and chlorophyll following the treatment of a shallow Dutch lake with ferric chloride. Their success was only transitory, however, owing mainly to the high external phosphorus loading and the short residency time of the water in the lake (35 days). This emphasizes the need to control external P sources before undertaking within-lake treatment. The ferric chloride may also cause a depression of feeding rate and mortality in zooplankton (Randall *et al.*, 1999). Aluminium salts have been used with some success to coagulate and seal phosphorus into sediments (Cooke, 1993); the treatment effectiveness in six Washington State (USA) lakes ranged from 50 to 80 per cent and lasted for at least five years (Welch and Shrieve, 1994). The metal is potentially toxic, however, and is best avoided in any water likely to be used for public supply.

The Riplox technique attempts to oxidize the sediments, favouring the formation of ferric phosphate (Ripl, 1976). The oxidizing agent, a concentrated solution of nitrate, is injected into the sediment. Iron may be added in conjunction with nitrate to increase the phosphate binding capacity of the sediment. Good results have been obtained in some Swedish lakes, but it is expensive and requires skilled operators.

The alternative, generally more expensive technique is to remove the sediment entirely, thus both taking away the internal source of phosphorus and deepening the lake. Moss *et al.* (1986) described the restoration of two small lakes in the Norfolk Broads system. In Alderfen Broad, diversion of an inflow stream in 1979 led to effective isolation from the main source of phosphorus. At Cockshoot Broad a dam was constructed to isolate the lake from the nutrient-rich river (Fig. 4.20) and the sediment was pumped out.

In the first two years after isolation the phytoplankton in Alderfen Broad was greatly reduced, the water became clear and a dense growth of the macrophyte *Ceratophyllum demersum* developed. Over the following three years, however, the plants gradually disappeared, to be replaced by a large phytoplankton bloom, including cyanobacteria, as phosphorus was released from the sediments. It is not known why phosphorus release did not occur in the early years after diversion but it may have been the result of decreased input of organic matter from reduced spring plankton populations. Once plants re-established, the organic matter produced by them may have led to the increase in phosphorus release from the

Fig. 4.20 The dam across the channel linking Cockshoot Broad to the River Bure. Nutrient-rich water is excluded and phosphorus-rich sediment was pumped out of the broad (photograph by the author).

sediments (Phillips, 1992), the plants being at least partly responsible for their own demise. Poor fish recruitment and adult fish mortality in the late 1990s led to an increase in large-bodied cladocerans and a return of *Ceratophyllum*, which produced a substantial biomass over the period 1988–91, despite high phosphorus concentrations (Perrow *et al.*, 1994). The community in Alderfen Broad remains unstable and the sediment has been removed to reduce internal loading and deepen the lake, though a successful outcome is not yet guaranteed. At Cockshoot, after isolation and sediment removal, phytoplankton growth declined, the water became clear and a diverse assemblage of aquatic plants developed in part of the broad, possibly helped by a scarcity of planktivorous fish (many of which had been removed by netting). Most of the open water, however, was not colonized by plants despite relatively low phosphorus levels and the absence of release from the sediments, and phytoplankton populations began to increase. It was considered that a developing fish population had preyed on the *Daphnia*, which were previously keeping the lake clear. There were insufficient refuges because macrophytes were thought to be grazed by a water-bird, the coot (*Fulica atra*) (Moss *et al.*, 1996), though subsequent studies have demonstrated that coots have a negligible impact on macrophyte growth (Perrow *et al.*, 1997). Further biomanipulation was required (see below).

Other techniques to prevent the recycling of nutrients or to accelerate the outflow of nutrients have included sealing lake bottoms with polyethylene sheeting, selectively discharging hypolimnetic water in water supply reservoirs, or diluting and/or flushing with water from an oligotrophic source. The removal of nutrient-rich macrophytes, algae or fish (biotic harvesting) has also been attempted.

Barley straw has been used successfully to reduce the growth of filamentous and planktonic algae and cyanobacteria in canals, ponds and reservoirs. Reductions of more than 90 per cent in algae growing on microscope slides downstream of the straw, and in laboratory experiments, have been achieved with no change in nutrient levels (Welch *et al.*, 1990; Ridge and Barrett, 1992; Newman and Barrett, 1993). Suppression of algae in a water-supply reservoir has been successful since 1993 using this technique (Barrett *et al.*, 1999). It appears that phenolic compounds released during decomposition are toxic to algae (Everall and Lees, 1996, 1997). The method has considerable potential, and barley straw is in great surplus in many cereal-growing areas, though success is not guaranteed (Kelly and Smith, 1996).

Biomanipulation

The restoration techniques so far discussed have been largely concerned with controlling the sources of phosphorus that fuel algal growth, so-called *bottom-up techniques*. On p. 141 the complex interactions were described that occur between the members of the aquatic community, and how a breakdown in one of these interactions could lead to the rapid collapse of the system, terminating in a stable, phytoplankton-dominated community. Once phosphorus reduction has been achieved it may be possible to recreate the complex community and control excessive algal growth by *top-down* methods or *biomanipulation*. Biomanipulation can be defined as the *manipulation of the food web of aquatic ecosystems to increase the numbers of grazers on algae* (Moss, 1992).

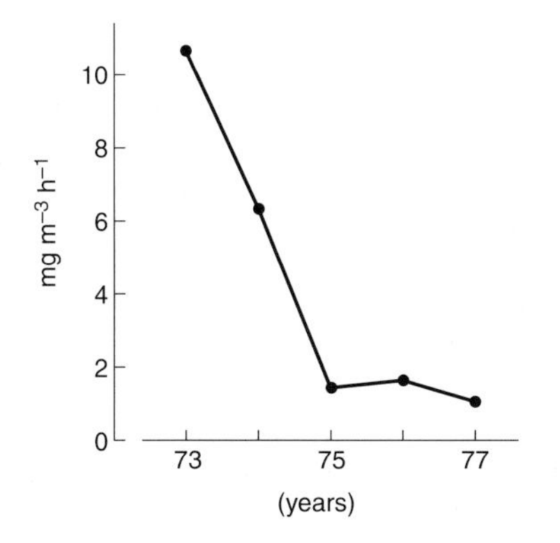

Fig. 4.21 Algal primary production (mean midday values) in Lake Lilla Stockelidsvatten following removal of fish by rotenone in November 1973 (from de Bernardi, 1989, after Henrikson *et al.*, 1980).

Fig. 4.22 Two small connected lakes on the campus of the University of Essex. The upper lake (a) developed a dense algal bloom in August 1994 as it did in previous years. The lower lake (b) had a large population (30 000) of fish (rudd and bream) removed in autumn 1993 after several years of study. In contrast to previous years no algal bloom developed in the lower lake in 1994, the water remained clear and macrophytes began to establish. Swards of macrophytes, mainly *Potamogeton crispus*, dominated the lake in summer 1995 (photographs by the author).

Fish may be significant predators of larger zooplankton, especially in the absence of cover (p. 152) and such predation can result in intense algal blooms. The addition of increasing numbers of blue-gill sunfish (*Lepomis macrochirus*) to enclosures in a US lake led to an increase in chlorophyll *a* concentration and a decrease in transparency in proportion to fish numbers (Lynch and Shapiro, 1980). In another experiment, Round Lake, Minnesota, was treated with rotenone, a fish poison, in autumn 1980. Most of the fish, mainly planktivores, were killed and the lake was then stocked with piscivorous species. Over the following two years there were decreases in chlorophyll *a* and total phosphorus in the water and increases in transparency, in the body size of zooplankton, and in numbers of *Daphnia* (Shapiro and Wright, 1984). Figure 4.21 illustrates the reduction in algal primary production in a Swedish lake, Lilla Stockelidsvatten, following treatment with rotenone in November 1973. The improvement in conditions was dramatic and Secchi disc transparency more than doubled. Successful results in reducing algal growth following fish removal have also been achieved in a water supply reservoir (Benndorf, 1987). Figure 4.22 dramatically illustrates the effects of fish removal on water quality in the University of Essex lakes.

Manipulations to control eutrophication in Cockshoot Broad and their lack of success were described above. An attempt to remove all of the fish was made in the winter of 1988/99 and a 20-fold reduction in stocks was achieved. There was an immediate return to clear water in the summer of 1989 and populations of *Daphnia hyalina* recovered. Macrophytes were more slow to establish but a diverse community had developed by 1995 (Moss *et al.*, 1996). Biomanipulation following sediment removal is now considered to be the most reliable method to restore the Broads, though the mechanisms that enable a stable macrophyte community to be maintained are not clearly understood (Phillips *et al.*, 1999). Phillips *et al.* (1996) provide an overview of fish manipulations in the Norfolk Broads.

Dutch lakes are shallow and very similar to the Norfolk Broads. They have suffered a similar loss of macrophyte vegetation with eutrophication. Once macrophytes have gone, populations of the predatory pike (*Esox lucius*) decline. Pike need cover for hunting and refuges to reduce the risk of intraspecific predation. In waters with dense vegetation, populations of young pike are high and they effectively control the abundance of young bream. Pike reduce the use of macrophyte beds by zooplanktivorous fish, so enhancing the value of the refuge (Jacobsen and Perrow, 1998). In the absence of pike, bream come to dominate the fish community, eating zooplankton, stirring up the bottom sediments and accelerating eutrophication (Hosper, 1989). Lake Bleiswijkse Zoom was divided into two compartments. In April 1987 the majority of planktivorous and bottom-feeding fish (at a density of 650 kg ha^{-1}) were removed from one compartment and young individuals of the predatory zander were introduced. The other compartment was used as a reference. Removal of the fish resulted in low concentrations of chlorophyll *a*, total phosphorus, total nitrogen and suspended solids, while Secchi disc transparency increased from 20 cm to 110 cm (Meijer *et al.*, 1989). By June charophytes (the Phase 1 community, p. 134) had become abundant, creating a habitat for the re-establishment of the main predator, the pike. The success in biomanipulation was due not only to increased grazing pressures on algae but also to a decreased availability of phosphorus. This resulted from the sedimentation of detritus as the increased growth of macrophytes reduced turbulence in the water (Boers *et al.*, 1991).

Biomanipulation has been attempted on 18 shallow lakes in the Netherlands, the largest, Lake Wolderwijd, covering 2700 ha but with a depth of only 1.5 m. In all cases the fish stock was drastically reduced. In eight of the lakes there was a rapid increase in transparency and a strong growth of aquatic macrophytes, while eight more showed some increase in transparency. More than 75 per cent of the fish population had to be removed to ensure significant improvements (Meijer *et al.*, 1999). A review of Dutch experiences with biomanipulation is provided by Hosper (1998).

As well as food web effects, fish may also have direct effects on macrophytes. The removal of fish, predominantly bream, from a gravel pit resulted in a 90-fold increase in plant cover and an increase in plant biomass from 1 $g\,m^{-2}$ to 47 $g\,m^{-2}$. Zooplankton numbers remained high and unchanged. The re-introduction of fish into an enclosure resulted in a decline of plant biomass from 47 $g°m^{-2}$ to 5 $g°m^{-2}$ (Wright and Phillips, 1992). Fish both inhibited the initial establishment of seedlings and reduced the growth of existing stands. Bottom-feeding fish such as bream also resuspend sediments, reducing the transparency of shallow lakes, while the excretory products of large populations of fish add significant amounts of nutrients to the water (Breukelaar *et al.*, 1994).

Fish population manipulation as a technique to control eutrophication is a recent approach and fish stocks are easier to manipulate than nutrients or the plankton (Lammens, 1999). However, the long-term success of fish population reduction is by no means guaranteed. Manipulated lakes are often dominated by single, unstable populations of macrophytes. For example, following manipulation in Lake Vaeng, Denmark, in 1986–88, *Potamogeton crispus* became dominant as Secchi depth doubled. By 1990, however, *Elodea canadensis*, which is not native to Europe, became exclusively dominant in the lake (Lauridsen *et al.*, 1994). There is also evidence that the clear-water phase resulting from fish removal may only be temporary. For example Lake Bleiswijkse Zoom (see above) gradually increased in turbidity from the second year of biomanipulation, while the conditions in some other Dutch lakes deteriorated after five years (Meijer *et al.*, 1994, 1999). Planktivorous fish populations build up despite the presence of predators, because of high recruitment rates following a reduction in competition, and they prey heavily on zooplankton (Hansson *et al.*, 1998). In Lake Ringsjön, Sweden, it was suggested that the removal of fish, in addition to stimulating fish recruitment, led to an increase in benthic invertebrates, which encouraged the number of waterfowl, which in turn had a negative impact on macrophytes, resulting in no improvement in overall water quality and continued cyanobacterial blooms (Bergman *et al.*, 1999; Cronberg *et al.*, 1999). The removal of all fish may also lead to the development of substantial populations of predatory populations of invertebrates, such as

Neomysis (see above p. 152) or *Chaoborus*, which may lead to poorer water quality than occurs in the presence of planktivorous fish (Wissel and Benndorf, 1998). For biomanipulation to be successful it seems clear, at least, that ongoing control of planktivorous fish populations may be necessary. Biomanipulation is best seen as one tool in a number of techniques to be used to combat eutrophication in any one lake (Annadotter *et al.*, 1999).

Economically it is the eutrophication of water supply reservoirs that is of most concern. Very often these reservoirs also support recreational fisheries and are intensively stocked. In Ardleigh Reservoir, referred to several times above, the zooplankton community is dominated by small-bodied forms, while *Daphnia hyalina* has a small adult size relative to populations in similar waters (Mason and Abdul-Hussein, 1991). The reservoir is managed as a put-and-take fishery for rainbow trout (*Oncorhynchus mykiss*) which are unable to breed in such eutrophic waters. Trout stomachs contain many zooplankton. From the discussion above, a reduction in fish populations would seem to be a sensible measure as part of the restoration process, which at present consists of treating the inflow water with ferric salts, with only limited success. However, the costs of water treatment problems have never been weighed against the benefits of the revenue received from angling and even the countenancing of such an analysis seems politically unacceptable. In many such reservoir fisheries, pike are destroyed whenever possible and even fish-eating birds, especially the cormorant (*Phalacrocorax carbo*), are tolerated only because they are legally protected, but then with deep resentment by the angling fraternity.

Lakes can be restored successfully, especially when a reduction in phosphorus loading is followed by biomanipulation. Biomanipulation is likely to be most effective in shallow waters where macrophytes are a major component of the ecosystem, or in artificial waterbodies where fish removal is a feasible and acceptable option. Each lake must be considered as an individual however, and will require its own unique management prescription, so that a detailed limnological study is always necessary. Adequate funding and political will are also essential. That so many attempts at restoration have been only marginally successful is due in part to a lack of imagination by lake managers, and to the often grudging way such schemes are sanctioned by the authorities following prolonged public campaigns.

5 CHAPTER

ACIDIFICATION

Acid rain and the acidification of freshwaters have received considerable attention over the past 30 years, but the problem is not new. There were observations of lakes in Scandinavia losing their fish populations as early as the 1920s, while studies of diatom remains in sediment cores from lakes in southwest Scotland have indicated that acidification began around the middle of the nineteenth century. There has certainly been an acceleration of the process in the last three decades. For example, of 87 lakes in southern Norway surveyed in both the periods 1923–49 and 1970–80, 24 per cent had a pH below 5.5 in the earlier period, compared with 47 per cent in the latter.

SOURCES OF ACIDITY

The main pollutants responsible for acid rain are sulphur dioxide (SO_2) and the oxides of nitrogen (NO_x). Figure 5.1 illustrates the rate of emission of sulphur dioxide in the United Kingdom since 1970. Around 60 per cent of the SO_2 is derived from power stations and 30 per cent from industrial plants, the amount of emission having been reduced since 1970 with improved air pollution control legislation. By contrast the emission of nitrogen has continued to rise, with 45 per cent derived from power stations and 30 per cent from the exhausts of vehicles, the burgeoning number of the latter causing increases in nitrogen. Ammonium compounds, mainly from the excreta of intensively farmed livestock, are also involved in acidification.

When SO_2 and NO_x reach the atmosphere they react with moisture and undergo oxidation, resulting in the formation of sulphuric and nitric acids, which exist mainly in the clouds and fall to earth in rain or snow (*wet deposition*). Conversion rates are very rapid, being approximately 100 per cent per hour in summer and 20 per cent per hour in winter. Alternatively, in a dry atmosphere complex photochemical reactions, involving highly reactive oxidizing agents such as ozone, result in the production of sulphuric and nitric acids, conversion rates being approximately 16 per cent per day in summer and

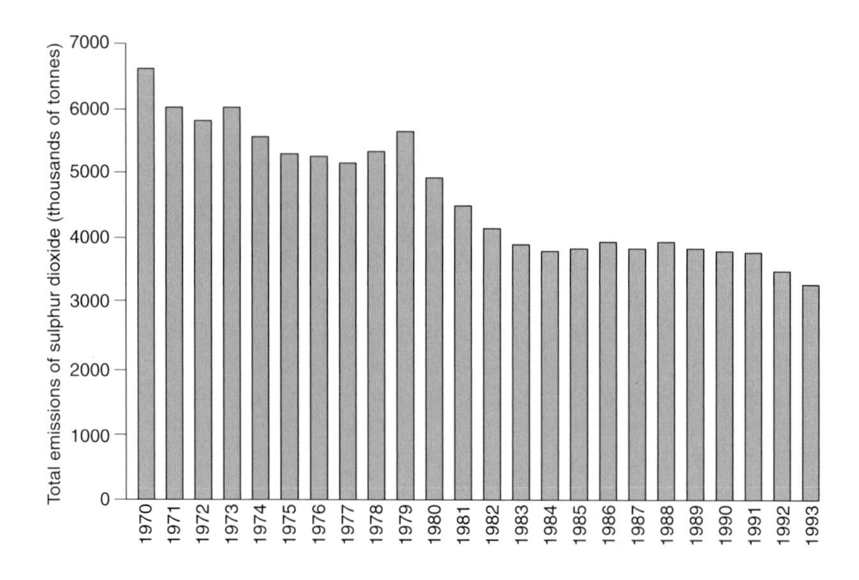

Fig. 5.1 National emissions of sulphur dioxide in the United Kingdom, 1970–93.

3 per cent per day in winter, much slower than the reactions in a moist atmosphere. The acids from these reactions reach the ground in gaseous or particulate form (*dry deposition*). The production of acids from precursor gases seems to be limited by the availability of oxidizing agents, hydrocarbons and ultraviolet radiation from the sun. There is evidence that the whole atmosphere in the northern hemisphere is more reactive than it was a few decades ago and this accelerates the process of acidification. A review of chemical reactions in the atmosphere is provided by Harrison (2001).

Uncontaminated rainwater, in equilibrium with atmospheric carbon dioxide, has a pH of 5.6. Almost everywhere in the world the pH of rain and snow is lower than this. In Europe, the highest annual fall-out of sulphur, is in central England, central Europe and the Alps. The fall-out of nitrogen follows a similar pattern. This pattern is reflected in the average pH of rainfall, with eastern Britain, parts of Scandinavia and central Europe averaging less than pH 4.3, more than ten times more acid than 'natural' rainfall, pH being measured on a logarithmic scale. North America is similarly affected, with large areas of the northeast suffering acidification, as well as the southeast, mid-West and far West. Some 2500 lakes and 36 000 km of streams in the United States have been identified as either acidic or sensitive to acidification from acid rain (Downey *et al.*, 1994). A rainfall sample from Pennsylvania in 1978 was pH 2.32!

Sulphuric acid contributes roughly 70 per cent to the mean total annual acidity of precipitation in northwest Europe and about 60 per cent in eastern North

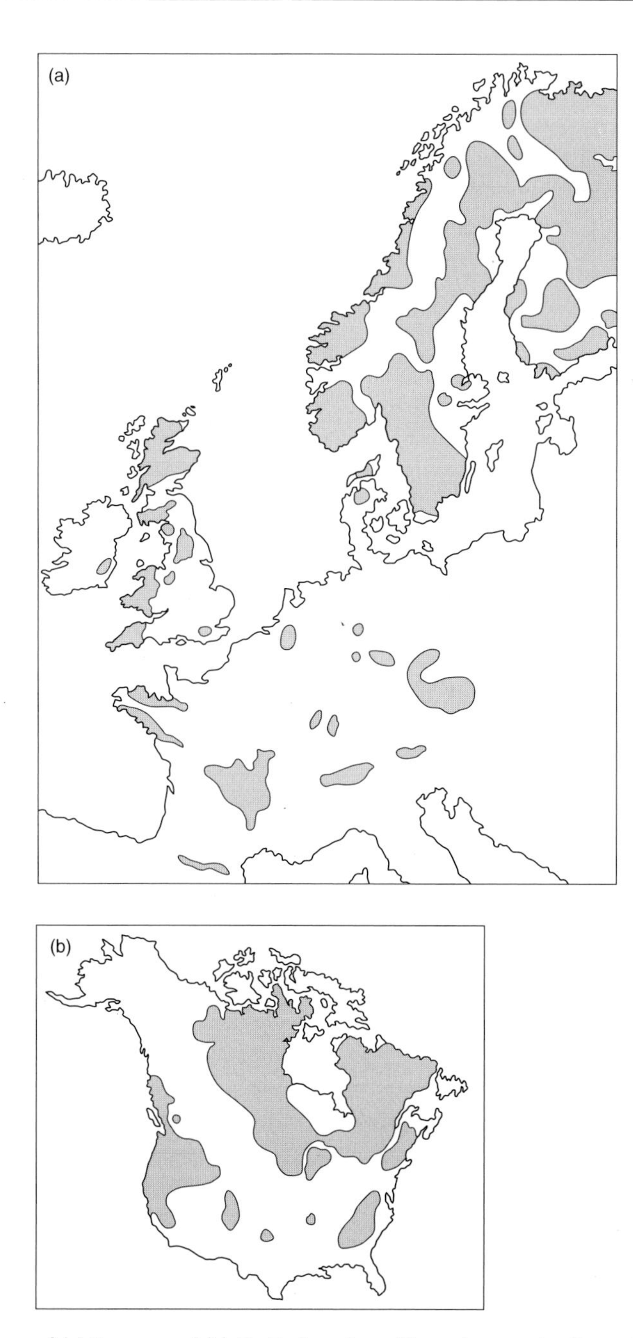

Fig. 5.2 Areas of (a) Europe and (b) North America with geology and soils rendering freshwaters liable to acidification.

America, most of the remainder being due to nitric acid. Although much of the pollution falls fairly locally in the area of production, some of it will be transported by winds for several thousand kilometres before it eventually falls in rain. It is calculated, for example, that about 17 per cent of the acid falling on Norway derives from Britain and about 20 per cent of that falling on Sweden is from eastern Europe.

The input of acidic materials to freshwaters may be from three sources: directly from the atmosphere, indirectly from the atmosphere via run-off in the catchment and from the generation of acidity within the catchment (e.g. by soil acidification). The impact of acid precipitation on freshwaters is dependent on the surrounding geology and soils, which determines the capacity of the water to neutralize acids. Regions with a calcareous geology are not sensitive to acidification and even small amounts of limestone in a drainage basin exert considerable influence in areas that would otherwise be very vulnerable. Acidification is most likely to occur where the bedrock is granite or gneiss, with thin soils that have insufficient base cations freely available to neutralize the acidity which is deposited. Areas in Europe and North America with sensitive geology and soils are shown in Fig. 5.2.

The acids naturally present in soil, organic and carbonic acids are so-called weak acids. Although they contribute significantly to the overall acidity of the soil, they do not dissociate into their respective anions and cations to the same extent as the strong acids (H^+, SO_4^{2-} and NO_3^-), which are derived from acid precipitation. The transfer of acidity from soil to surface waters requires mobile negative ions to bind to the acid hydrogen ions. As the rate of supply of dissociated hydrogen ions increases in soil water from acid precipitation, the rate of cation exchange is increased: that is, cations such as Na^+, K^+ and Mg^{2+} are displaced from exchange sites on soil particles. These can then move through the soil, provided that mobile ions are present to transport them and acid precipitation provides these in the form of SO_4^{2-}. The nitrate ions could serve the same purpose, but most are taken up by vegetation as soils are generally nitrate deficient. Nevertheless, total organic nitrogen contributes up to 60 per cent of total acid anion concentrations in water in autumn and winter when the vegetation ceases growing (Jenkins *et al.*, 1996). Strong acids also mobilize the aluminium ion (Al^{3+}), which is of particular significance because of its toxicity. Heavy metals may also be mobilized (Vesely, 1994). The sulphate ion thus efficiently transfers acidity from soils to surface waters.

Land-use influences the rate of acidification of freshwaters, forestry being especially important. Figure 5.3 illustrates the headwaters of the River Severn in central Wales, which winds through the conifer plantation forest of Hafren. A minimum pH of 4.4 was recorded in the forested stretch of the river, while there were pH values below 5.5 for several kilometres downstream of the forest. By contrast in a tributary of the Severn, the River Dulas, which drains open moorland, a minimum pH of 5.0 was recorded near the source, but most of the river had a mini-

Fig. 5.3 Upper reaches of the River Severn in mid-Wales showing extensive afforestation with conifers, resulting in marked acidification of the water (photograph by the author).

mum pH above 5.5 (Mason and Macdonald, 1987). Vegetation scavenges both dry deposition and pollutants held in mists and fogs (*occult deposition*) very efficiently and the rainfall reaching the soil beneath vegetation is more acid than incoming rain because it contains both these scavenged pollutants, washed from the surfaces of the vegetation, and materials leached out of the plants. Conifers are especially efficient at scavenging pollutants. Data from the Llyn Brianne catchment in mid-Wales are shown in Table 5.1. Both acidity and sulphate deposition increase beneath spruce. The higher rate of evapotranspiration from trees compared with grassland results in a reduced amount of run-off of water through soils with a greater concentration of pollutants. Trees also take up nutrients for growth, including ions that could potentially neutralize acids. The accumulation of litter as forests develop also leads to the natural acidification of soils. For example, significant acidification of soils over only 15 years took place in broad-leaved plantations on

Table 5.1 Acidity and sulphate reaching soils beneath different types of vegetation at Llyn Brianne, Wales (adapted from Gee and Stoner, 1988)

	Mean pH	***Mean SO_4 (μequiv. l^{-1})***
Ambient precipitation	4.60	54
Moor grass	4.70	124
Oak	4.70	115
12-year-old spruce	4.27	144
25-year-old spruce	4.32	296

former arable soil, despite their buffering capacity (Hughes-Clarke and Mason, 1992).

The drainage channels cut through plantation forests expose sulphur and nitrogen to the atmosphere, acids being produced by oxidation. The relationship between forest cover and stream chemistry is illustrated in Figure 5.4. Comparing an unafforested catchment and one with 30 per cent forest cover, mainly of Sitka spruce, a sharp decline in pH is apparent while both mean sulphate and mean aluminium increase markedly. In an extensive survey in Wales, pH was found to decrease and aluminium concentration of stream water increased with increasing percentage of plantation forest cover within catchments (Ormerod *et al.*, 1989).

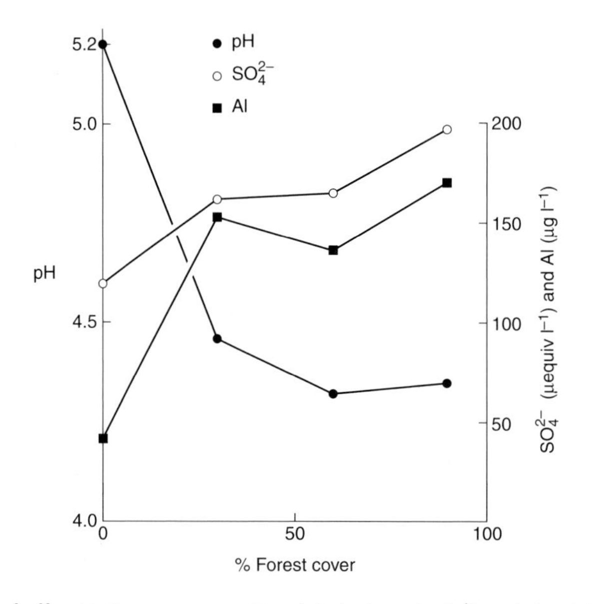

Fig. 5.4 Effect of afforestation on mean pH, sulphate (μequiv. l^{-1}) and aluminium (μg l^{-1}) in streams in adjacent catchments in Scotland (adapted from Harriman and Wells, 1985).

The felling of blocks of conifers may also lead to a temporary increase in acidification, as the mobilization of nitrates results in increases in nitrification (Reynolds *et al.*, 1995). However, if the amount of nitrate released is small, then there is a reversal of acidification (Neal *et al.*, 1998): the impact of felling on stream acidification appears to depend on local soil conditions.

The practice of liming upland grasslands to raise productivity increased in the United Kingdom from a total usage of less than 20 thousand tonnes per annum in the 1940s to more than 60 thousand tonnes in the late 1950s. With a change in the liming subsidy, it has since gradually declined to around 30 thousand tonnes per annum. Ormerod and Edwards (1985) have suggested that liming could have had a substantial effect on the mineral concentrations of soft water streams in the past, for increased acidification in some Welsh streams since the mid-1960s parallels a decline in the application of lime to agricultural land.

LONG- AND SHORT-TERM CHANGES IN ACIDITY

It is extremely difficult to determine long-term trends in acidification by examining series of water quality records. Historically these have been taken for other purposes and usually in lowland, buffered reaches of rivers. Furthermore, analytical techniques have been refined in recent years. In particular, accurate pH measurement in waters of low conductivity is difficult and earlier analysts would have been unaware of this. In an analysis of 75 data sets from the United Kingdom, Ellis and Hunt (1986) could find evidence of a downward trend in only six. The most

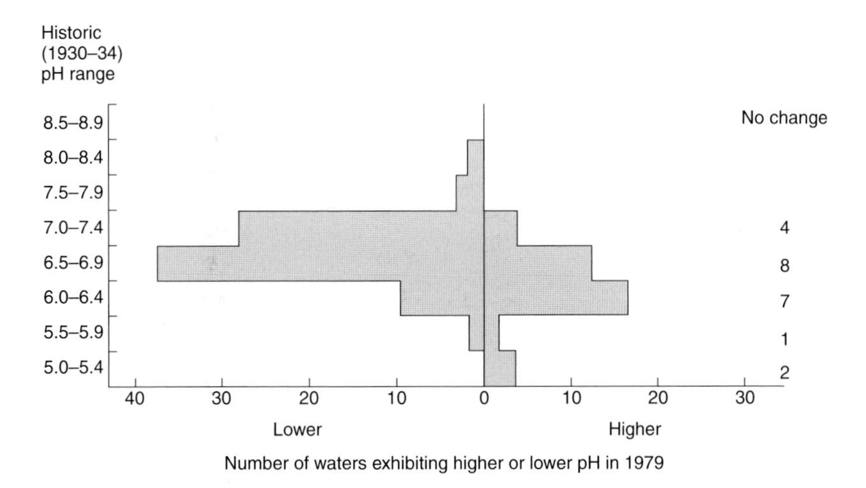

Fig. 5.5 Comparison of historical (1930–34) and more recent (1979) pH values for 138 lakes in the Adirondack Mountains, United States (adapted from Havas *et al.*, 1984).

convincing was the River Ystwyth in west Wales, where mean pH fell from 6.5 to 5.5 over the period 1970–84. Figure 5.5 shows a comparison of historical (1930–34) and more recent (1979) pH values, measured colorimetrically for a series of lakes in the Adirondack Mountains in the United States. There were fewer lakes with pH above 7.0 and more with pH below 6.0 in the later survey, acidification being greater in those lakes that had a pH originally above 6.5. In southern Norway, of 87 lakes surveyed in both the periods 1923–49 and 1970–80, 24 per cent had a pH below 5.5 in the earlier period compared with 47 per cent in the second period.

In Chapter 4 the analysis of diatom remains in sediment cores in order to track progressive eutrophication was described (p. 138). The same technique has been used to follow the course of acidification in lakes. In Galloway, southwest Scotland, acidification of lakes began around 1850. Species of diatoms typical of acid waters, such as *Tabellaria binalis* and *T. quadriseptata*, began to increase, while species indicating less acid conditions, such as *Anomoeoneis vitrea* and *Fragilaria virescens*, gradually declined in abundance. Species indicative of strongly acid lakes ($pH < 5$) dominated in various sites between 1930 and 1950 (Flower *et al.*, 1987). Figure 5.6 illustrates the inferred pH for the Round Loch of Glenhead, southwest Scotland, and Lake Gårdsjön, Sweden, reconstructed from diatom communities in sediment

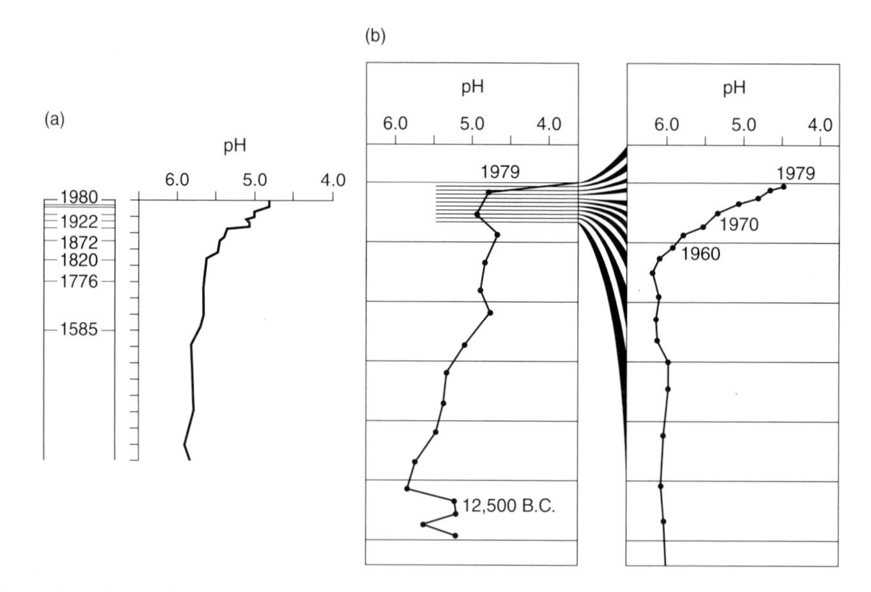

Fig. 5.6 Historical reconstruction of pH using sediment diatom assemblages of (a) Round Loch of Glenhead, southwest Scotland, and (b) Lake Gårdsön, Sweden. In the latter, the species composition dates back to the last Ice Age, with changes since 1900 shown in more detail (adapted from Henriksen, 1989, after Flower and Battarbee, 1983, and Renberg and Hedberg, 1982).

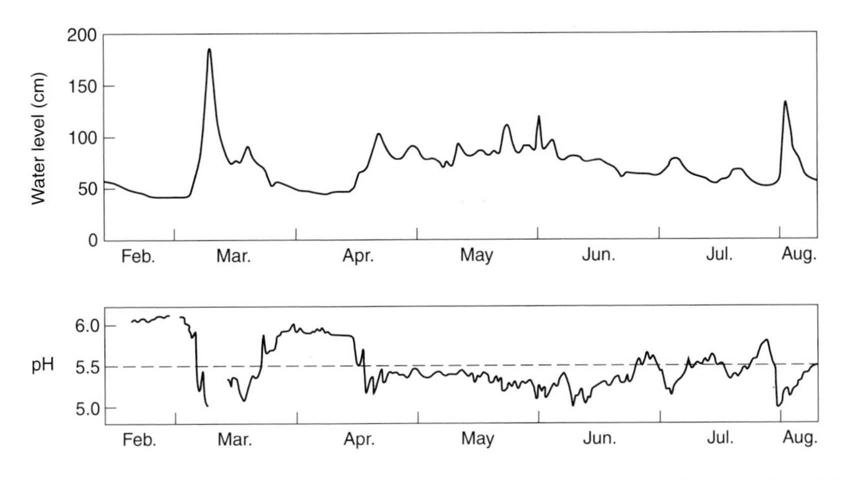

Fig. 5.7 Variations in pH and water level in the Vikedal River, Norway, February–July 1983 (after Henriksen *et al.*, 1984).

cores. The pH in the Scottish loch had dropped from pH 5.5 to 4.5 during the past 130 years, with the most dramatic change in recent decades. In Lake Gårdsjön changes became detectable in the early 1950s, but accelerated during the 1960s and 1970s. From a review of a number of reconstructions using diatoms it has been concluded that acidification will occur only when the ratio of calcium in water (μequiv. l^{-1}) to sulphur deposition ($g\,m^{-2}\,y^{-1}$) is less than 60 : 1 (Battarbee, 1994a,b; Battarbee and Charles, 1994).

There are also marked short-term changes in the acidity of freshwaters. Seasonal changes occur, the summer composition of streamwater being closer to that of groundwater, having a higher pH. Lower pH values will be more common in winter, as the composition of surface run-off will more closely reflect that of precipitation. In addition, brief pulses, or episodes, of high acidity may occur during heavy rainfall or snow-melt. During dry periods, acids will be deposited and accumulate on vegetation, to be washed off during heavy rain, increasing the input of acid to surface waters. Similarly acids accumulate in lying snow, to be rapidly mobilized during a thaw. An example of these acid episodes is illustrated for the Vikedal River, southwestern Norway, in Fig. 5.7. A continuous monitor was installed in the river and large fluctuations in pH were related to changes in water flow. During heavy rainfalls, pH fell by more than half a unit within three hours. A survey of episodic acid events in Canada caused by snow-melt produced maximum pH depressions of between 0.4 and 2.6 units, the minimum pH value recorded being 3.2 in one lake. Alkalinity was depressed by 100–200 μequiv. l^{-1} (Tranter *et al.*, 1994). The major cause of these episodic events was sulphate ions, deriving ultimately from anthropogenic sources. These episodes can be highly damaging to the river ecosystem, fish kills being one regular consequence

(Muniz, 1991), but routine monitoring may be insufficiently frequent to detect them.

EFFECTS OF ACIDIFICATION ON THE AQUATIC ENVIRONMENT

Water chemistry

The generalized changes occurring during the acidification of lakes are shown in Fig. 5.8. The buffering capacity of the water will be determined by the concentration of bicarbonate ions and the process of acidification can be seen as taking place in three stages (Henriksen, 1989). In the first stage, the dissolved bicarbonate buffers the inputs of strong acids:

$$H^+ + HCO_3^- \rightarrow H_2O + CO_2$$

The pH generally remains above 6 and the plant and animal communities remain stable as the lake loses alkalinity. In the second stage, transition lakes, the bicarbonate buffer may be lost during long periods of acid inputs, resulting in large fluctuations in pH and periodic fish kills. In the final stage the loss of alkalinity is complete and the lake retains a low but stable pH, usually below 5, while metal levels, especially aluminium, may be elevated, resulting in the extermination of fish populations. Those lakes that have a high buffering capacity will never reach the permanently acid state.

Primary producers

Acidification results in a decrease in diversity of the phytoplankton. Acidic lakes

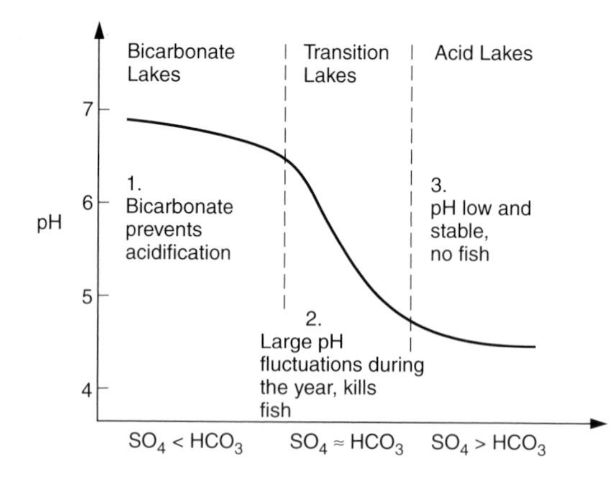

Fig. 5.8 The acidification process in lakes.

have 10–20 species, compared with 30–80 species in circumneutral oligotrophic lakes (Muniz, 1991). During acidification, the percentage of green algae decreases and dinoflagellates and cryptomonads increase (Brettum, 1996). Dinoflagellates tend to dominate many acid lakes, but despite the decline in species richness there is little evidence of a decrease in biomass or photosynthesis if the supply of phosphorus is maintained (Olaveson and Nalewajko, 1994). Some species survive acidic conditions because they have evolved acid and aluminium tolerance in phosphate metabolism (Smith, 1990).

The attached algae (*periphyton*) show a somewhat different response. Although diversity generally declines, there is a proliferation of growth owing to a reduction in grazing and an increase in the clarity of acid waters. Periphyton growth begins to increase as pH drops to below 6 and is extensive below pH 5.5 (France and Wellbourn, 1992). Filamentous algae, such as *Mougeotia* and *Zygnema*, often dominate, while mats of cyanobacteria may form in conditions too acid to support phytoplanktonic cyanobacteria. In Welsh streams, the green algae *Mougeotia*, *Ulothrix* and *Stigeoclonium* were most frequent at low pH (Ormerod and Wade, 1990).

The species richness of macrophytes is less in lakes of lower pH (Jackson and Charles, 1988). Changes in the macrophytes of lakes have included a reduction in the dominant *Lobelia*, especially at pH 4 or less, and an increase in the dominance of *Sphagnum* (Grahn, 1986). It has been suggested that increases in *Sphagnum* may accelerate the acidification process because the moss has a high ion-exchange capacity in its cell walls (Hendrey and Vertucci, 1980). Ormerod *et al.* (1987a) have studied the macro-flora of acid streams in Wales and believe that the presence or absence of indicator species can be used in the assessment of stream acidity. The presence of the liverworts *Scapania undulata* and/or *Nardia compressa*, and the absence of the macroscopic alga *Lemanea*, indicate waters in the pH range 4.9–5.2. With the moss *Fontinalis squammosa* present, but *Lemanea* absent, the pH range is 5.6–5.8, and with both present the pH is likely to be in the range 5.8–6.2. Streamwater is likely to be above pH 6.2 where *Lemanea* is present, but *F. squammosa* absent.

Decomposers

Some studies have indicated that in acid waters there is a reduced rate of decomposition and, in laboratory experiments, the rate of oxygen consumption by micro-organisms using birch leaves as a substrate was halved when the pH dropped from 7.0 to 5.2. A marked reduction in the decay rate of rice grains added to acid waters compared with circumneutral waters has been reported (Ormerod and Wade, 1990). There is a shift in dominance from bacteria to fungi (Haines, 1981; Perry *et al.*, 1987), though some studies have shown a decline in the diversity and abundance of hyphomycetes (Hall *et al.*, 1980; Chamier, 1987).

It has been suggested that, with increasing acidification, there will tend to be a build-up of organic matter such as leaves and twigs, which could lead to a reduction in mineralization. The consequent decrease in limiting nutrients such as phosphorus could then lead to a reduction in primary production. This premise is not universally supported, however, and several studies have failed to show a decrease in decomposition with acidification.

Invertebrates

Not only is there a change in the phytoplankton of lakes, but also changes in the zooplankton, a topic examined in detail by Brett (1989). Figure 5.9 illustrates the relationship between species richness of the crustacean fauna and pH in over 70 acidic waterbodies in upland areas of Great Britain. There is a marked decrease in the number of species with increased acidity, but no such relationship occurred with calcium. The greatly reduced scatter of points below pH 5 indicates that pH is the controlling factor on the community in acidic waters, whereas above pH 5 other factors are involved. Common dominant species in acidic lakes in North America include *Diaptomus minutus*, *Bosmina longirostris* and the rotifer *Keratella taurocephala*, these species often representing the entire zooplankton community (Mierle *et al.*, 1986; Brett, 1989). *Eudiaptomus gracilis*, *Eubosmina longispina* and *Keratella serrulata* are typical species of acidic lakes in Scandinavia. Species of

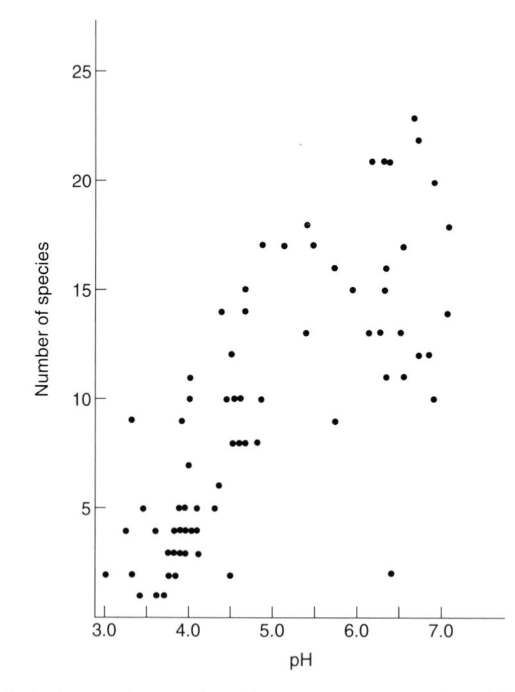

Fig. 5.9 Relationship between the number of crustacean species in acidic waterbodies and pH (adapted from Fryer, 1980).

Daphnia and *Cyclops* are mostly scarce or absent in highly acid lakes, which tend to have an increased representation of generalist species.

A number of factors may explain the changes in zooplankton community structure in lakes after acidification. High concentrations of H^+ and metals impose physiological stresses. For example, Havas and Likens (1985) studied the combined effects of acidity and aluminium on mortality and sodium balance in *Daphnia magna*. Both H^+ and Al interfered with sodium balance. At pH 6.5, raised Al (1.02 $mg\,l^{-1}$) led to a decreased rate of Na^+ influx of 46 per cent and an increased rate of outflux of 25 per cent, there being a net loss of sodium. At pH 4.5, Na^+ influx was inhibited by 73 per cent, compared with treatments at pH 6.5, whether or not Al was present. However, Al decreased Na^+ outflux by 31 per cent at pH 4.5, so reducing the net loss of sodium and temporarily prolonging the survival of *Daphnia*. Physiological stress may be most important in highly acid environments where the community has been reduced to just a few species, a shift in competitive fitness favouring those species, frequently generalists, which are most tolerant of acidity.

Other important factors are shifts in the structure of phytoplankton populations, for example towards less edible species, changes in the availability of bacterioplankton and detritus, and changes in the predator community, including the elimination of fish (Brett, 1989). In an artificially acidified lake, the population of the dominant crustacean *Holopedium* collapsed, the main cause being the disappearance of phytoplankton, notably *Oocystis* (Hessen and Lydersen, 1996). Havens and DeCosta (1985) described how, in acidified enclosures, both the abundance and mean body

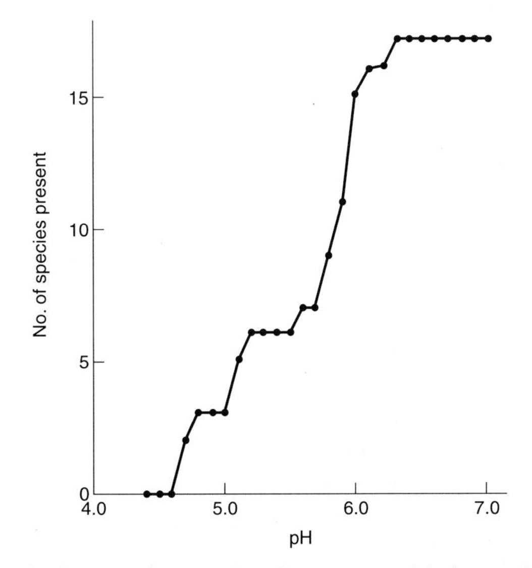

Fig. 5.10 Lower limits of tolerance of 17 species of crustaceans, bivalves and snails, based on a survey of 1500 Norwegian lakes. All species were present in waters greater than pH 6.3 (adapted from data in Økland and Økland, 1980).

size of *Bosmina longirostris* increased relative to the controls, which they attributed to a decline in its predator, *Mesocyclops edax*.

Marked changes also occur in the benthos of lakes during acidification. Figure 5.10 illustrates the tolerance limits of 17 common species of crustaceans, bivalves and snails to pH, based on a survey of 1500 Norwegian lakes. Snails and bivalves, with calcareous shells, largely disappeared below pH 6, though some small mussels survived down to pH 4.7 and some snails to pH 5.2, albeit at low densities. The crustaceans *Lepidurus arcticus* and *Gammarus lacustris*, important food items for fish, were sensitive to acidity, whereas *Asellus aquaticus* was found down to pH 5.2 and less frequently down to pH 4.8.

In North America the most abundant amphipod of lakes, *Hyalella azteca*, is acid sensitive, being found, for example, in 69 of 70 lakes in Ontario with positive alkalinity, but in none of 9 acidified lakes (Stephenson and Mackie, 1986). The biomass and production of *Hyalella azteca* in two lakes that suffered pulses of acidity below pH 5 in spring were found to be lower than for populations in lakes where acid episodes did not occur (France, 1996). Three of five species of crayfish of soft waters in North America are sensitive to acidification, with delayed hardening of the exoskeleton, parasitism, egg mortality and recruitment failures being reported (Mierle *et al.*, 1986). The physiology of crayfish in acid waters is reviewed by McMahon and Stuart (1989).

Studying headwater streams of the River Tywi in west Wales, Stoner *et al.* (1984) found that, where pH was greater than 5.5 and hardness greater than $8\,mg\,l^{-1}$, the invertebrate community consisted of 60–78 taxa, whereas in streams with a mean pH less than 5.5 and hardness less than $10\,mg\,l^{-1}$, only 23–37 invertebrate taxa were present. In a study of 104 upland streams in Wales, pH and aluminium concentration strongly influenced species richness and the structure of the macroinvertebrate community (Wade *et al.*, 1989). Species such as *Gammarus pulex*, *Ancylus fluviatilis* and *Hydropsyche* spp., as well as several herbivorous mayflies, were absent from acidic streams, even though potential food sources were present (Ormerod *et al.*, 1987a). Filter feeders are scarcer in acidic streams and shredders more abundant, especially stoneflies such as *Leuctra* (Friberg *et al.*, 1998a, b; Ventura and Harper, 1996). In the Adirondack Mountains of New York State streams with pH in the range 5.8–7.2 had relatively diverse invertebrate assemblages, the mayfly *Ephemerella funeralis* and the beetle *Oulimnius latiusculus* dominating. In the more acidic streams (range 4.4–5.0) these two species were absent and the community had less than half as many taxa. Stoneflies dominated (Simpson *et al.*, 1985). This impoverishment of the invertebrate community with acidification of streams is a general phenomenon (Sutcliffe and Hildrew, 1989). In contrast, the relationships between acidification and population density and biomass of macroinvertebrates are weak and inconsistent.

As with the plankton, the decline in richness of the stream fauna could be a combination of physiological stress, a change in food supply and a reduction in predators. Aquatic invertebrates need to actively take up sodium, chloride, potassium and calcium ions for survival and uptake is dependent on external concentrations. In acid waters ion concentrations may be too low, while hydrogen and aluminium ions may become dominant in the water. Being small and mobile, these may be transported inwards instead of essential ions, disturbing the normal internal equilibrium and possibly leading to a fatal loss of vital ions from blood and tissue (Sutcliffe and Hildrew, 1989). A net loss of sodium ions at low pH, upsetting osmoregulation, has been reported for stoneflies and mayflies (Twitchen, 1990; Frick and Herrmann, 1990), while increases in aluminium have been shown to decrease sodium ions in body fluids.

Differences in susceptibility to aluminium in closely related species has been observed. For example, the caddis *Arctopsyche ladogensis* and *Hydropsyche siltalai* showed damage to their ion-regulatory organs with increasing aluminium concentrations, whereas *Hydropsyche angustipennis* was much more tolerant; relatively low levels of aluminium during acid episodes will limit the distribution of the two sensitive species in streams (Vuori, 1996). In artificial streams, the mortality of a number of species was enhanced at pH 4 in the presence of elevated aluminium, but the toxic effect of aluminium was less in the presence of organic matter (Burton and Allan, 1986). Dissolved organic carbon probably complexes with aluminium, reducing its toxicity, and its presence is extremely important in influencing the survival of stream biota under acidic conditions. Crustaceans and gastropods appear especially susceptible to low ion concentrations in the presence of acidity, probably because they are more permeable, while all arthropods are vulnerable when they moult, a time when permeability is greatly increased.

There is currently little evidence to support the view that food availability and a reduction in predation pressure may influence the fauna of acid streams. A decrease in the decomposition of leaf packs and a substantial decline in the invertebrate populations within them was reported at pH 4 (Burton *et al.*, 1985), but the decline appeared to be due to invertebrate mortality rather than a change in food quality. Sutcliffe and Hildrew (1989) review the limited evidence suggesting that food quality for grazing invertebrates may be reduced in acid streams.

Metals are generally more soluble at low pH and laboratory experiments have often demonstrated increased metal uptake by invertebrates under acid conditions. However, to date, field observations lend little support to laboratory findings (Wren and Stephenson, 1991; Gerhardt, 1993).

Fish

The first reports of acidification came from fisheries inspectors in Norway in the early 1900s, who reported kills of Atlantic salmon (*Salmo salar*). Salmon showed a sharp decline in numbers, beginning in the 1970s, in seven southern Norwegian rivers receiving acid precipitation, whereas there has been no overall change in rivers not receiving acid precipitation (Fig. 5.11). Brown trout (*Salmo trutta*) began disappearing from mountain lakes in Norway in the 1920s and 1930s and by the 1970s lakes devoid of fish were reported from many regions in southern Scandinavia (Fig. 5.12). A survey in 1986 found that the number of lakes without

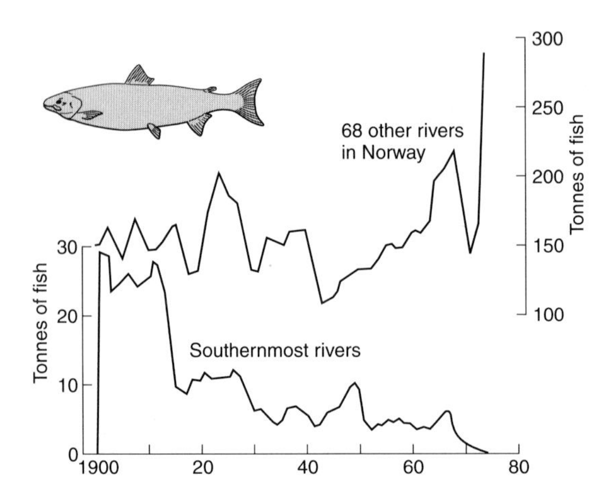

Fig. 5.11 Average catch size of salmon in seven rivers in southern Norway receiving acid precipitation and 68 other rivers that do not receive acid precipitation (from Henriksen, 1989).

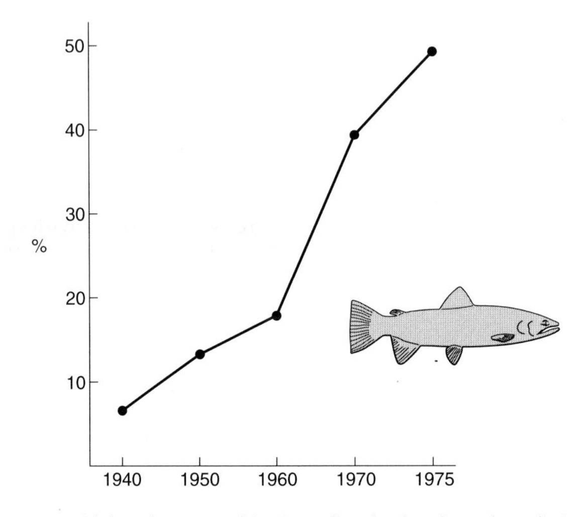

Fig. 5.12 Percentage of lakes (n = 2850) in Scandinavia that have lost their populations of brown trout.

fish in southern and southwestern Norway had doubled since a previous survey in 1971–75 (Henriksen *et al.*, 1989). Fish populations are directly affected by acidification in some 68 000 km^2 of Norway.

The evidence that fish populations were being reduced by acidification came much later in other countries. Haines and Baker (1986) estimated that 200–400 lakes in the Adirondack Mountains may have lost their fish populations due to acidification. Acidification has affected fish in several east coast states and eastern

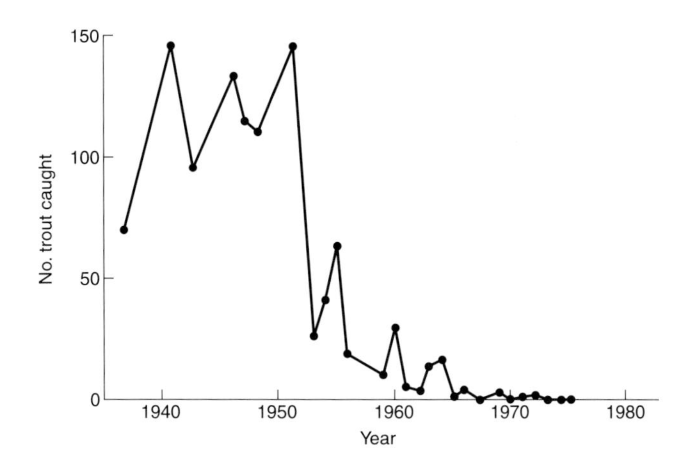

Fig. 5.13 Number of trout caught by rod and line in Loch Fleet, southwest Scotland (from Turnpenny, 1989).

provinces of Canada. Historical data on changes in fish populations are largely lacking in Britain. In Loch Fleet, southwest Scotland, catches of trout by rod and line have been recorded since the 1920s, where a sharp decrease in catches occurred in the early 1950s (Fig. 5.13) and no fish were caught after 1975 (Turnpenny *et al.*, 1988). Declines in fisheries due to acidification are widespread in this region of Scotland (Harriman *et al.*, 1987), though this appears to be the only area of the country seriously affected, despite the susceptibility of Scottish waters because of geology and soil type. There is, however, a widespread decline in stream fisheries in Scotland associated with extensive conifer plantations (Egglishaw *et al.*, 1986), which increase the acidity of the drainage basin. Fisheries in upland areas of central and west Wales have also declined, especially in afforested catchments (Stoner and Gee, 1985).

Turnpenny (1989) has examined the occurrence of fish species in streams of various pH ranges in Great Britain. Brown trout were found in only 28 per cent of streams with pH less than 5, but in 95 per cent of streams with pH greater than 6.5. No streams with pH less than 5.5 contained Atlantic salmon. Population densities of salmonids in streams with pH less than 5 was 16.1 per cent that of streams with pH greater than 6.5, while biomass was greatly reduced. Eels (*Anguilla anguilla*) were widely distributed with respect to acidity, but biomass was three to four times lower in streams with pH less than 5.5 than those with pH greater than 6. Minnows (*Phoxinus phoxinus*) and bullheads (*Cottus gobio*) were scarce in streams with pH less than 6.5. In lakes in Finland, roach (*Rutilus rutilus*) were found to be the most sensitive species, disappearing from a number of study sites in the 1980s. Even the least sensitive species, whitefish (*Coregonus peled*) and perch (*Perca fluviatilis*) showed signs of stress and reproductive impairment in both males and females. Laboratory experiments with newly hatched fry showed sensitivity to pH and aluminium in the order roach > zander (*Stizostedion lucioperca*) > whitefish, perch > pike (*Esox lucius*), the last species

Table 5.2 Generalized short-term effects of acidity upon fish

pH range	Effect
6.5–9.0	No effect
6.0–6.4	Unlikely to be harmful except when carbon dioxide levels are very high (> 1000 mg l^{-1})
5.0–5.9	Not especially harmful except when carbon dioxide levels are high (> 20 mg l^{-1}) or ferric ions are present
4.5–4.9	Harmful to the eggs of salmonids and to adult fish when levels of Ca^{2+}, Na^{+} and Cl^{-} are low
4.0–4.4	Harmful to adult fish of many species that have not been progressively acclimated to low pH
3.5–3.9	Lethal to salmonids, although acclimated roach can survive for longer
3.0–3.4	Most fish are quickly killed

tolerating one pH unit lower than roach (Rask, 1992; Vuorinen and Vuorinen, 1992). Similar experiments exposing fish species to an acute aluminium challenge found sensitivities in the order Atlantic salmon, roach, minnow, perch, grayling (*Thymallus thymallus*), brown trout and Arctic char (*Salvelinus alpinus*). Minnow, roach and brown trout exposed to the acid medium with low aluminium also showed some mortality. Arctic char showed a high resistance to aluminium (Poleo *et al.*, 1997).

Acid waters are associated with two types of fishery problem, fish kills during episodes of acidity and the gradual decline of fish populations in waters known to be acid. The generalized short-term effects of acidity on fish are given in Table 5.2, though this does not take into account the influence of associated events, such as increases in aluminium.

The effect of acidity on fish is mediated by way of the gills, five major functions being affected: ion regulation, osmoregulation, acid–base balance, nitrogen excretion and respiration (Brakke *et al.*, 1994). High levels of sodium and chloride ions are found in the blood plasma of fish, and ions that are lost from the urine or from the gills must be replaced by active transport across the gills against a large concentration gradient. When calcium is present in the water it reduces the egress of sodium and chloride ions and the ingress of hydrogen ions. The excessive loss of ions such as sodium, which cannot be replaced quickly enough by active transport, is the main cause of mortality in acid waters. When the concentrations of sodium and chloride ions in the blood plasma fall by about a third, the body cells swell and extracellular fluids become more concentrated. Potassium may be lost from the cells to compensate for this but, if it is not eliminated from the body quickly, depolarization of nerve and muscle cells occurs, resulting in uncontrolled twitching of the fish before it dies.

Aluminium is toxic to fish in the pH range 5.0–5.5. Aluminium ions precipitate on to the gill surfaces, clogging them with mucus and interfering with respiration.

At lower pH, aluminium ions bind to the gill surface and interfere with the regulation by calcium of gill permeability, therefore enhancing the loss of sodium in the critical pH range. Respiratory stress caused by aluminium is still severe in the presence of calcium, though ion loss is reduced. Electrolyte loss and gill dysfunction affect especially the younger life stages of the fish (Appelberg *et al.*, 1992). In experiments young brown trout have been shown to avoid aluminium (Åtland, 1998).

High concentrations of other metals, such as copper, zinc and mercury, may also sometimes prove toxic in acid waters. In one study a small lake was divided into two, one half being acidified. Yellow perch (*Perca flavescens*) accumulated 16 per cent more mercury over two years in the acidified section. Processes within the acidified half seemed to result in greater methylation of mercury (Wiener *et al.*, 1990). In one small acidified lake (pH 5.2–5.6) in Canada mercury accumulated in fish to much higher levels than in fish from a nearby circumneutral (pH 6.3–6.9) lake, and were considered to be hazardous to fish-eating birds (Scheuhammer and Graham, 1999).

Eggs may be especially sensitive to acidity. The relationship between pH levels and calcium levels to the survival of freshly fertilized trout eggs is shown in Figure 5.14. Survival after eight days was 100 per cent at pH 5.1, irrespective of calcium concentration, and at the highest concentration of calcium (400 μequiv. l^{-1}) irrespective of pH. At low pH (4.2) and low calcium (less than 50 μequiv. l^{-1}) all eggs died. In experiments with brook trout (*Salvelinus fontinalis*) eggs, Hunn *et al.* (1987) found that mortality exceeded 80 per cent at pH 4.5, was 15–18 per cent at pH 5.5 and was less than 2 per cent at pH 7.5. Embryo mortality was reduced at the lowest pH in the presence of aluminium. All larvae died within 30 days at pH 4.5, none surviving for more than ten days. After fertilization, which itself may be influenced by acidity, fish

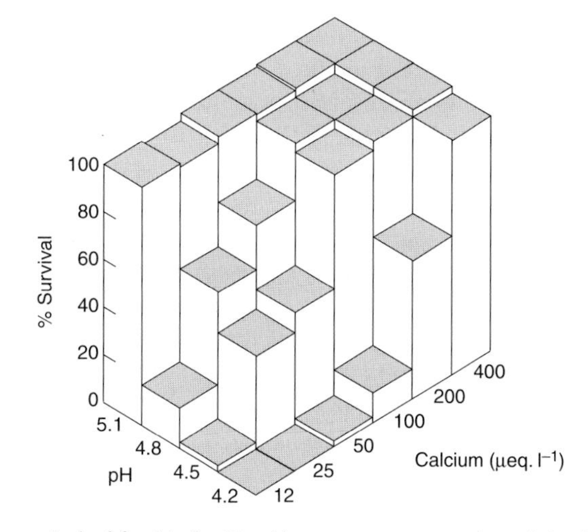

Fig. 5.14 Percentage survival of freshly fertilized brown trout eggs after eight days in a range of pH and calcium concentrations (from Brown and Sadler, 1989).

embryos appear to be susceptible through their entire developmental period, though the times immediately following fertilization and shortly before and during hatching appear to be especially critical (Rosseland, 1986). Low pH reduces the activity of the hatching enzyme chlorionase, which dissolves the inner wall of the egg, so reducing hatching success. The period of metamorphosis, when the gills become exposed to the environment and fully functional, is also critical (Stallsmith *et al.*, 1996).

The effects of acidification on fish growth are somewhat confused, some workers reporting reduced growth rates, others finding no reduction. In Atlantic salmon, the period of smoltification, when the fish change physiologically and behaviourally for their downstream migration, appears to be a supersensitive time to acidification (Rosseland, 1986). Skeletal deformities have also been found in fish from acid waters (e.g. Campbell *et al.*, 1986), owing probably to a malfunction in calcium metabolism. There is, however, some evidence, at least for perch, that fish populations can adapt to relatively rapid acidification and this could be genetically determined (Vourinen *et al.*, 1994).

In conclusion, fish populations may disappear from acidified waters because of periodic mortality caused by acid episodes, or because mortality in the susceptible early stages of development and growth may mean that the population fails to maintain itself or grow. This will result in a gradual population decline and extinction even in waters where conditions are not overtly toxic to adult fish (Rosseland, 1986).

Higher vertebrates

As with fish, reproductive failure is the major effect of acid water on amphibians. In general they appear more tolerant than fish. Aston *et al.* (1987) reported high densities of spawn of the common frog (*Rana temporaria*) in waters down to pH 4.2. Experiments showed that the survival of embryos to normal free-swimming larvae was not affected by pH in the absence of aluminium but the survival of embryos decreased with increasing aluminium concentrations (Tyler-Jones *et al.*, 1989). For 26 species of amphibian the lethal pH, causing 100 per cent mortality of embryos, ranged from 3.4 to 4.5 (Freda, 1986). Andrén *et al.* (1988) found that egg mortality and the embryonic development time of three species of frog (*Rana arvalis*, *R. temporaria* and *R. dalmatina*) increased when pH declined. Raised levels of aluminium had no effect on the eggs, but pH and aluminium both influenced larval mortality, deformities being recorded. *Rana arvalis* was the most tolerant. Salamanders, which are much more aquatic than frogs and toads, are more sensitive to reduced pH (Mierle *et al.*, 1986). Changes in the food supply are also likely to be significant and Freda (1986) concluded that pond acidity is having a major effect on the local distribution and abundance of amphibians in North America.

Dippers (*Cinclus cinclus*) are song birds that live along swiftly flowing streams where they feed mainly on aquatic invertebrates, such as mayfly and caddis larvae, which they collect by 'flying' underwater or foraging on the stream bottom. Densities of dippers have been found to be reduced in both Scotland and Wales

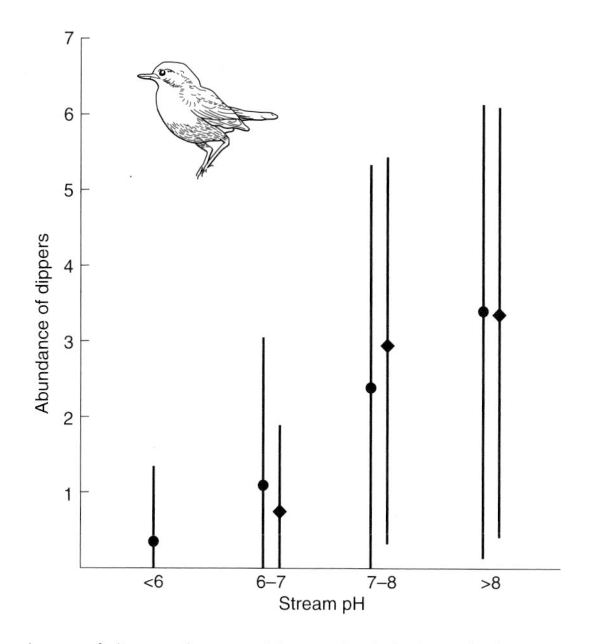

Fig. 5.15 Abundance of dippers (mean with standard deviation) along Welsh streams of different pH during autumn 1986 (dots) and 1987 (diamonds). Abundance is measured as catch per unit effort of ringing activity (adapted from Ormerod and Tyler, 1993).

along acid streams that otherwise provide suitable habitat (Ormerod *et al.*, 1986; Vickery, 1991; Logie, 1995; Buckton *et al.*, 1998). There is a direct relationship between overall dipper abundance and stream pH in Wales (Fig. 5.15) and their breeding density (number of pairs per 10 km) was also highly correlated with the abundance of mayfly and caddis. Dippers were scarcest along those streams with low pH (less than 5.7–6.0) and elevated concentrations of aluminium (greater than 0.08–0.10 $g\,m^{-3}$). On the River Irfon, historical data indicate a decline of 1.7 pH units between the 1960s and 1984, while the dipper population fell by 70–80 per cent. Dippers breeding along acidic streams showed a delayed start to the breeding season, smaller clutch size, reduced egg mass, thinner egg shells, reduced brood size, slower nestling growth and poor adult condition. Mayfly and caddisfly larvae are scarce in these acid streams and dippers spend longer foraging and deliver food at a slower rate to nestlings. Fledgling survival is also reduced (Tyler and Ormerod, 1992; Ormerod and Tyler, 1993). Dipper populations outside of the breeding season are also influenced by the effects of acidification on their food supply (Logie, 1995).

Fish-eating great northern divers or common loons (*Gavia immer*) in Canada suffer a high level of brood mortalities on acidic lakes, which is thought to be the result of a shortage of suitable food (Alvo *et al.*, 1988). In contrast lake birds that feed on benthic insects may benefit from a reduction in competition with fish. For example, in Finland the densities of both pairs and broods of the goldeneye duck

(*Bucephala clangula*) were found to be highest in the most acid lakes with the lowest density of perch and the greatest density of invertebrates. Brood density of goldeneyes increased after a sudden mortality of perch in an experimental lake (Poysa *et al.*, 1994).

Those areas of Great Britain liable to suffer acidification of freshwaters are also the regions which hold thriving populations of otters (*Lutra lutra*) (p. 6), so they are of especial importance for the conservation of this species. The River Severn, in Wales, holds a good population, but otters did not much use the top 15 km of the river, which flows through forestry plantations (Fig. 5.3). They frequented a tributary, however, the Dulas which rises in open moorland, almost to the source. The minimum pH of the Severn was less than 5.5 for the top 15 km or so of its length, whereas only the top 2 km of the Dulas had a minimum pH as low as this (p. 178). Fish were almost entirely absent from the top of the Severn, whereas good populations existed in the Dulas. This lack of fish, because of acidification, in the headwaters of the Severn prevented otters living there (Mason and Macdonald, 1987).

In a broader study of three regions of Scotland, 72 one-kilometre stretches of river bank were surveyed and signs of otters were found at all of them. There were, however, highly significant correlations between the use made by otters of stretches and both pH and conductivity of the water. The effect was greatest in southwest Scotland, the region with the most acidified waters. Fewer signs of the presence of otters were found at those sites where minimum pH was low enough to adversely affect fish populations (Mason and Macdonald, 1989).

Humans

As mentioned previously, acid waters bring toxic metals into solution. These may then accumulate in fish, to be eaten by people; mercury being of special concern (p. 63). Alternatively heavy metals may enter the drinking water supply, while acid water also dissolves metals from the pipes of the water distribution system. Particular concern has been expressed over lead and aluminium. The drinking water of a number of cities is distributed through old lead pipes and increased lead concentrations may result in blood and nervous disorders, a reduced intellectual development in children and behavioural abnormalities (p. 68). High concentrations of aluminium have been linked to osteomalacia, a bone-wasting disease, and to Alzheimer's disease, premature senile dementia. The effects on the respiratory system of gaseous emissions is a more direct problem.

DEMONSTRATING THE EFFECTS OF ACIDIFICATION BY ECOSYSTEM MANIPULATION

The previous chapter (p. 153) described how an entire lake was fertilized to chart the progress of eutrophication. In the same Experimental Lakes Area of Canada, Lake 223 was acidified over an eight year period, causing dramatic changes in ecosystem function (Schindler *et al.*, 1985; Schindler, 1988a). After a two year preliminary study, sulphuric acid was added from 1976 to decrease the pH gradually from an initial 6.8 to an eventual 5.0 in 1981. The pH was then maintained at 5.0 to 5.1 from 1981 to 1983. Over the period there was no overall change in either production or decomposition processes in the lake, but marked changes took place in the species composition. In the phytoplankton, *Dinobryon* decreased and *Gymnodinium* and *Merismopedia* increased below pH 5.5, whereas at pH 5.6 a filamentous alga, *Mougeotia*, not previously recorded in the lake, formed thick mats in the littoral (shallow water) areas. The zooplankton community shifted from one dominated by copepods to one dominated by cladocerans. The cladoceran *Daphnia galeata mendotae* was largely replaced by *Daphnia catawba* and *Holopedium gibberum*, most probably because of a reduction in fish predation. The overall biomass of the zooplankton, however, remained relatively constant.

Among the first animals to disappear from Lake 223 were opossum shrimps (*Mysis relicta*) and fathead minnows (*Pimephales promelas*). The estimated population of shrimps before the experiment was almost seven million, but at pH 5.9 they had all but disappeared. Fathead minnows failed to reproduce and died out a year after the shrimps. Both white sucker (*Catostomus commersoni*) and lake trout (*Salvelinus namaycush*) produced young in abundance at pH 6.4, though recruitment in both species failed at pH 5 and no first year fish of any species were observed in the following year. The condition of adult fish was poor. Cannibalism was noted in trout as many of their food items were killed. At pH 5.6 the exoskeleton of crayfish (*Oronectes virilis*) began to lose its calcium and animals became infected with a microsporozoan parasite. They were absent by the end of the experiment, as were leeches and the mayfly *Hexagenia*. Some taxa, such as chironomid midges, did well as they were released from the pressure of predation in the simplified ecosystem. As no species of fish reproduced at values of pH below 5.4, it was predicted that the lake would become fishless within ten years, based on knowledge of the natural mortalities of long-lived species. In addition to species replacement and an overall loss of diversity, acidification resulted in modifications in animal behaviour, alterations in predator–prey and parasite–host relationships, and changes in physiology and reproduction. Ecosystem stress resulted in the decline of many species before they were affected by direct toxicity of hydrogen ions (Schindler, 1988b).

The changes in Lake 223 were caused solely by increases in hydrogen ion concentration and not by any secondary effects such as aluminium toxicity. This might explain some unexpected results, such as the lack of decline in rates of primary production, decomposition or nutrient cycling. Furthermore, the minimum pH was 5.0 in Lake 223, whereas many acidified lakes have a mean pH much lower than this.

The effects of adding acid and aluminium to a small headwater stream in central Wales were studied by Ormerod *et al.* (1987b). Sulphuric acid was added over a 24 h period to reduce pH from 7.0 to 4.2–4.5, the upstream stretch acting as a reference reach. Some 200 m downstream aluminium sulphate was added continuously to increase the concentration from background ($0.05\,g\,m^{-3}$) to 0.3–$0.4\,g\,m^{-3}$. Toxicity tests were carried out *in situ* using six species of macroinvertebrates, brown trout and salmon. Three species of invertebrates showed no increase in mortality, while three others showed a 25 per cent increase in mortality in both treatment zones. Brown trout and salmon showed a 7–10 per cent mortality in the acid zone and a 50–87 per cent mortality in the aluminium zone, salmon succumbing more rapidly than trout. The feeding rate of *Gammarus pulex* declined during the dosing period (McCahon *et al.*, 1989b). Downstream drift of blackfly larvae (Simulidae) increased in both acid and aluminium zones, while several species increased in the drift in the aluminium zone, the most pronounced response being that of the mayfly *Baetis rhodani*, which was the only benthic species showing a significant decline in density at the end of the experiment. In an experimental study in California, where *Baetis* was also susceptible to acid, a high proportion of drifting invertebrates (45–100 per cent depending on species) was found to be dead (Kratz *et al.*, 1994). In a further experiment in Wales citrate was added at a downstream site to bind aluminium. The drift density of *Baetis rhodani* increased six-fold above that in the control stretch in both acid and labile aluminium treatments, but was not affected in the zone of organically bound aluminium (Weatherley *et al.*, 1988). Greater rates of invertebrate drift during aluminium additions than during acid additions have also been reported from an experimentally dosed stream in New Hampshire, in the United States, by Hall *et al.* (1987), who considered that fluctuations in aluminium concentration at a pH not toxic to invertebrates may be an important factor regulating their distribution and abundance in poorly buffered streams. The Welsh experiments are reviewed by Weatherley *et al.* (1990).

Long-term experiments in acidifying whole catchments have been conducted in Norway (Wright *et al.*, 1994). Eight years of acid additions resulted in marked changes in the chemistry of run-off. Sulphate and base cations (sodium, potassium, calcium and magnesium) increased while the acid-neutralizing capacity decreased. The run-off was acidic, rich in aluminium ions and toxic to fish.

CRITICAL LOADS

The above discussion has shown how acid deposition has had severe effects on many habitats. To restore these we need an estimate of the acid input that can be tolerated by the flora and fauna of sensitive sites. The critical load is *an estimate of exposure, below which significant harmful effects on particular parts of the environment do not occur, according to present knowledge* (Bull, 1991; Lükewille, 1994). Using this approach the numerical loads at which adverse effects are likely to occur can be estimated. There are two approaches to determining critical loads, a chemical method and one based on changes in diatom communities.

The Steady-State Water Chemistry Method involves determining the acid neutralizing capacity (ANC) of the water in the catchment. The ANC is defined as the ability of a solution to neutralize inputs of strong acid to a pre-selected equivalence (Henriksen *et al.*, 1992). It is effectively the alkalinity of the water, the sum of bicarbonate and dissolved organic salts. It usually has a small positive value. A threshold

ANC is defined for each target species (e.g. mayflies, trout) depending on their presence or absence. The critical load is set to allow all, or part, of the ANC to be used up, preserving a critical acid-neutralizing capacity for the target organism that is to be protected. In Scandinavia the critical ANC for fish is taken as 20 μequiv. l^{-1}, values above that being the critical load exceedance.

The diatom method is based on the observation that, in a sediment core from a lake, there is a distinct horizon that characterizes the onset of acidification (see p. 182). This is taken as the critical load exceedance (Battarbee *et al.*, 1993). There is a strong correlation between the changes in diatom species and the relationship between the base cation concentration of lake water and the loading of acid due to the deposition of sulphur. By studying the water chemistry and sulphur deposition of a range of lakes, a ratio of calcium (μequiv. l^{-1}) to sulphur (kequiv. $ha^{-1}\,yr^{-1}$) of 94 : 1 is taken as the critical load point (Allott *et al.*, 1995).

The advantages and disadvantages of the critical load approach are discussed by Cresser (2000). Maps can be drawn of critical loads and critical load exceedances. By modelling, it is possible to calculate the reduction in emissions required to reduce deposition to acceptable levels in particular sensitive areas, while the costs relative to the benefits can also be considered. During the 1980s it was predicted that, in southern Norway, a 30 per cent reduction in sulphur emission could reduce the number of acid lakes by 20 per cent, while a 50 per cent reduction may reduce the number by 35 per cent (Henriksen *et al.*, 1989). For the River Tywy catchment in Wales, modelling has predicted that a 50 per cent reduction in atmospheric deposition would prevent further deterioration in soft water streams draining moorland, but in streams draining conifer plantations the rate of deterioration would only be slowed down, mainly because of the continued leaching of toxic aluminium from forest soils (Edwards, 1989; Ormerod *et al.*, 1990). Similarly for southwest Scotland predictions from a mathematical model have suggested that a 50 per cent reduction in acid emissions is necessary to prevent further increases in stream acidity on moorlands, while afforested catchments will require greater reductions (Whitehead *et al.*, 1987).

Acidification of freshwaters will remain a problem long after emissions have been reduced because of the large amount of sulphur deposited in soils. A small catchment in Norway was provided with a roof to protect it from acid rain. It led to significant reductions in acidity and aluminium in run-off water but after five years water quality was still not suitable for fish and decades will be required for complete recovery of the catchment (Hultberg and Skeffington, 1998).

REVERSING ACIDIFICATION

In Lake 223, described above, sulphuric acid additions were decreased in 1984, allowing the pH to recover from 5.0 to 5.4, while in 1985 it increased further to 5.5–5.6 owing to unusually wet weather (Schindler, 1987b). White suckers and pearl dace (*Semotilus margarita*), which had survived without breeding, spawned when the pH reached 5.4 and the zooplankton species *Holopedium gibberum* and *Daphnia catawba* declined, presumably because of heavy predation by larval fish. Lake trout recovered their overall condition, but evidence of breeding was not obtained in the year of recovery. The crayfish, opossum shrimp and fathead minnows had not returned by 1985 and the pH was not considered sufficiently high to support them. Nevertheless, signs of recovery in the lake fauna were seen rapidly after the rise in pH.

Another lake in the Experimental Lakes Area (Lake 302) was acidified to pH 4.5 and then allowed to recover to pH 5.8 over a six year period, by the end of which there had been no consistent signs of recovery in the zooplankton associated with the littoral zone of the lake (Hann and Turner, 2000). It appears that the rate of recovery from acidification is dependent, not surprisingly, on the severity of the acid stress.

There are two ways of reversing the acidification of freshwaters – reducing emissions and adding lime. Almost all of the countries in Europe, together with Canada and the United States, committed themselves in 1983 to reduce sulphur emissions by 30 per cent within a decade. Flue gas desulphurization, using scrubbers, is the favoured technique of cutting emissions at power stations. One method is to inject limestone slurry and air to convert sulphur dioxide to gypsum ($CaSO_4$). Scrubbers are technically simple to install and operate, and are highly efficient. A major problem is that sources of limestone are often situated in areas of outstanding landscape and the movement of limestone and gypsum on the scale required involves many thousands of lorry journeys each year.

There have been marked reductions in acidifying emissions over the last two decades. Sulphur dioxide concentrations in air decreased by 63 per cent in northern and central Europe between 1985 and 1996, by 32 per cent in Great Britain between 1979 and 1993, and by 28 per cent in the United States and Canada between 1980 and 1995. This has led to reversal of acidification of waters in some regions (Stoddard *et al.*, 1999; Tipping *et al.*, 2000). Despite this, it has been predicted that by 2010, the critical load for sulphur in Sweden, for example, will still be exceeded in 1.6 million ha, concentrated in the south and southwest of the country. In eastern Canada, 95 000 lakes, an area the size of Texas, are still being harmed by high levels of acidity. It appears that, while there has been a marked decline in sulphuric acid in precipitation, there has also been a decline in the base cations in dust and dirt, which would neutralize acids. More sulphate is also being released from wetlands as their soils have dried out in recent droughts, to release acid when they are re-wetted (Pelley, 1999).

The liming of waters has been practised for some years. As early as 1926, in Norway, intake water to fish hatcheries was successfully limed to prevent mortality. In Sweden thousands of lakes have been limed since 1976, at an annual expenditure of millions of US dollars, with similar activities on a smaller scale in Norway (Henriksen, 1989). Pulverized limestone ($CaCO_3$), hydrated lime ($Ca(OH)_2$) and quicklime (CaO) may be used, but the first is preferred because, though less reactive than the other compounds, it is less caustic and is cheaper.

The effectiveness of liming depends on the retention time of water in lakes. In those with short retention times liming must be carried out annually or biannually, whereas with retention times of 2–5 years, re-acidification following liming may take 5–10 years (Wright, 1985). Lake Hovvatn in southern Norway was limed in 1981 and over 11 000 brown trout were stocked over the next four years. During this period the lake re-acidified to levels considered critical to fish and severe mortality occurred a year later. Growth rates reduced as acidification progressed (Barlaup *et al.*, 1994). Liming does restore the diversity of the lake zooplankton, though its composition is strongly influenced by predation (Stenson and Svensson, 1995; Nyberg, 1998). The addition of lime to rivers and streams reduces the toxic effect of acidification to salmonids (Rosseland *et al.*, 1986; Weatherley *et al.*, 1989), though there is the possibility of precipitating out aluminium salts, which may then prove toxic at higher pH levels. Fish kills have occurred in lakes following liming when aluminium levels remained high (Leivestad *et al.*, 1987) and, as reported above, aluminium is especially toxic to fish in the pH range 5.0–5.5, so that fish mortality could be increased during the liming process. Evidence suggests that salmon and brown trout can avoid the toxic mixing zones when rivers are limed (Åtland and Barlaup, 1995). The concentration of mercury in fish is markedly reduced following the liming of lakes (Andersson *et al.*, 1995). In the Appalachian Mountains of Virginia a low-cost direct liming has been practised in trout streams to offset trends in acidification (Downey *et al.*, 1994). Single treatments with limestone (costing up to US$1500 per treatment) were sufficient to maintain native trout populations.

Good results following liming are by no means guaranteed, however. Rundle *et al.* (1995) limed acidic headwaters and restored the water chemistry to levels similar to circumneutral streams. Over a five year period there was an increase in the diversity and abundance of acid-sensitive macroinvertebrate species but species richness remained substantially less than in natural circumneutral streams and no species became permanently established.

For long-term results, the liming of land in catchments rather than water might prove more effective. Sub-catchments of Loch Fleet, in southwest Scotland, were subjected to experimental liming and, from data collected over 3–5 years, it was predicted that limestone, applied as a slurry or as dry powder, at rates of 20 t ha^{-1} could maintain acceptable water quality for 15 years (Howells *et al.*, 1992), though subsequent measurements have shown the prediction to be rather optimistic (Howells and Dalziel, 1995). Fish were re-stocked, they bred and maintained good condition (Turnpenny *et al.*, 1995). However concern has been expressed over damage to terrestrial vegetation, especially *Sphagnum*, while the efficacy elsewhere has been questioned (Weatherley and Ormerod, 1992), particularly in view of the expense.

Because of the widespread nature of the acidification problem, and the expense of treating acidified waters (many operations, for example, use helicopters to apply doses), liming is likely to offer only a local solution to protect fisheries resources of particular commercial or recreational value. The causes, not the symptoms, must be tackled, requiring international commitments to reducing sulphur emissions. It is hoped that the declines in sulphur emissions described above will lead to a recovery in the ecology of acidified lakes and rivers in the not too distant future.

ACID MINE DRAINAGE

Drainage water from mines causes very considerable pollution problems over wide areas. In the United States, some 19 300 km of watercourses and 73 000 ha of lakes are adversely affected by mining pollution. In the Clark Fork River complex in Montana, waste material (tailings) covers 35 km^2 and is estimated to contain 9000 t arsenic, 200 t cadmium, 90 000 t copper, 20 000 t lead, 200 t silver and 50 000 t of zinc (Sell, 1992). As well as severe adverse environmental effects, still detectable in rivers 200 km downstream, the surrounding area has one of the highest mortality rates in people aged between 35 and 74 in the United States.

Substantial lengths of river in many industrial countries are affected by acid water from both coal and metal mines. There are also severe problems in developing countries. In the Amazon, for example, prospecting for gold has resulted in rivers and fish being severely contaminated with mercury used in the refining process (Cleary and Thornton, 1994). In Chapter 1 the ecological damage caused by the collapse of dams holding back mine waste was described (p. 4). A similar accident poisoned some 300 km of river in the Andes of Brazil in 1996, while the Fly and Ok Tedi rivers in Papua New Guinea are virtually biologically dead because of mine waste (Swales, 1998). Environmental concerns seem very low in the list of priorities of many mine owners, often multinational companies, even though the impacts on the ecosystem and human communities can be catastrophic.

Abandoned mines are more of a problem than working mines and this is causing great concern in the UK with the closure of many works during the early 1990s. An example is the Wheal Jane tin mine in Cornwall, southwest England. When in operation, pumping depressed the water table by some 400 m but the mine closed at the end of 1991 and pumping ceased. In January 1992 an orange plume of some 45 million litres of highly acid water, with dissolved metal concentrations as high as 5000 mg l^{-1} flooded into the nearby river and down into the estuary.

When coal is mined, substantial quantities of the mineral pyrite, a crystal composed of reduced iron and sulphur (FeS_2), are often exposed to the oxidizing action of air, water and chemosynthetic bacteria and these utilize the energy obtained from the conversion of the inorganic sulphur to sulphate and sulphuric acid. The bacteria involved are *Thiobacillus thiooxidans*, *T. ferrooxidans* and *Sulfolobus acididocaldarius*. The reactions are as follows:

$$2FeS_2 + 2H_2O + 7O_2 \rightarrow 2FeSO_4 + 2H_2SO_4$$

$$4FeSO_4 + O_2 + 2H_2SO_4 \rightarrow 2Fe_2(SO_4)_3 + 2H_2O$$

The ferric ions formed under acidic conditions are soluble and can oxidize more pyrite to ferrous and sulphate ions:

$$FeS_2 + 14Fe^{3+} + 8H_2O \rightarrow 15Fe^{2+} + 2SO_4^{2-} + 16H^+$$

The ferrous ions produced are again oxidized to ferric ions by bacteria, which again react with pyrite, so that the oxidation of pyrite accelerates. The highly acid conditions that result bring other metals into solution. When living in streams, *Thiobacillus* are frequently embedded in filamentous streamers, white or cream in colour, which are composed of a network of polymer fibres secreted by the bacteria (Wakao *et al.*, 1985). Harrison (1984) provides a review of the biology of *Thiobacillus* and its associates.

The types of water issuing from mines depends on the kind of mine and the surrounding geology. Where drainage is in the pH range 2–4.5, aluminium may be as high as $2000\,mg\,l^{-1}$, though at higher pH it is usually never more than $20\,mg\,l^{-1}$. At this low pH range, ferrous iron may be as high as $10\,000\,mg\,l^{-1}$ and there will be no ferric iron. As stated above the high acidity of mine water also brings heavy metals into solution, adding to the pollution problems.

Those organisms in the receiving stream responsible for breaking down organic matter may be destroyed by the high acidity so that the self-purification process is inhibited and oxygen-depleted water, often with suspended solids, extends much further downstream than would otherwise be the case. Washings from active mines also result in a heavy load of suspended matter. As the pH increases downstream, owing to the dilution of mine water with streamwater and run-off, the iron is oxidized by the dissolved oxygen in the stream which turns a bright orange colour. Ferric hydroxide precipitates out on to the stream bottom, covering the substratum with a brown slime and smothering benthic algae and macrophytes.

Hargreaves *et al.* (1975) have described the photosynthetic flora from 15 sites in England, 13 associated with coal mines, having a pH of less than 3.0. All of the sites also had high levels of one or more heavy metals and of silica, while most had high concentrations of phosphate, ammonia and nitrate. Twenty-four species of photosynthetic plants were found in flowing waters, with an additional four species restricted to pool sites. The alga *Euglena mutabilis* proved to be the most widespread species and was also the most abundant, sometimes forming 80 per cent cover. Some species, although widespread, could not survive the lowest pH and species richness was greatest in the range pH 2.5–3.0. Four species were found to have been recorded both in the English survey and in studies of acid streams in the United States, namely *Euglena mutabilis*, *Lepocinclis ovum*, *Eunotia exigua* and *Ulothrix zonata*. In acid-polluted tundra pools in northern Canada, *E. mutabilis* has been recorded at pH 1.8 (Sheath *et al.*, 1982). The presence of heavy metals may influence the composition of the species. For example *Eunotia exigua* is more likely to be found where copper and zinc concentrations are increased (Whitton and Diaz, 1981).

Acid mine drainage has severe impacts on invertebrate communities. In a mining complex in Canada, in a pool with water at pH 3.2 and contaminated with metals, the fauna was made up of 99.5% Chironomidae (Wickham *et al.*, 1987). This domination of acid mine waters by chironomids appears characteristic, while cranefly larvae (Tipulidae) and alderfly larvae (Megaloptera) are also frequently common. Taxon richness is generally reduced, with mayflies and stoneflies sensitive (Soucek *et al.*, 2000a; Winterbourn *et al.*, 2000). In Sweden, species common at undisturbed sites have been found to be absent at neighbouring sites impacted by metals from mine drainage, namely the mayflies *Ameletus inopinatus*, *Ephemerella aurivilli* and *Heptagenia dalecarlica*, the stonefly *Protonemura meyeri*, and the caddis *Apatania* sp. (Malmqvist and Hoffsten, 1999). Crustaceans are usually absent from acid mine waters. Experiments using *Daphnia magna* as an *in situ* test organism showed that water chemistry had a greater impact on toxicity than sediment chemistry in acid-impacted streams in Virginia, USA (Soucek *et al*., 2000b).

In a study of 46 metal-contaminated sites in 12 streams in Cornwall, copper was found to have most influence on the invertebrate community, followed by aluminium. Interactions between toxic metals and pH, alkalinity, hardness and dissolved organic matter added to the complexity of the response of the invertebrate community. At the most severely contaminated sites, with copper exceeding 500 $\mu g\ l^{-1}$, the community was reduced to just four species – a flatworm (*Phagocata vitta*), two chironomids (*Chaetocladius melaleucus* and *Eukiefferiella claripennis*) and a net-spinning caddis (*Plectrocnemia conspersa*) (Gower *et al*., 1994). In New Zealand streams contaminated by acid mine drainage (pH range 2.6–6.2) there was no correlation between the concentrations of aluminium and iron in algae, bryophytes, and invertebrates and pH, conductivity and metal burdens in the stream water (Winterbourn *et al*., 2000), though high levels of these metals in bryophytes have been reported in other studies (e.g. Englemann and McDiffet, 1996).

Typically acid waters below a discharge result in a low diversity of species, but with large populations, released from competition and from predation by fish. Further downstream, in the zone of neutralization, the precipitation of iron, which alters the characteristics of the substratum and clogs the gills of invertebrates, results in both low diversity and low population density.

Fish are very severely affected by acid mine drainage and low pH. They may be eliminated for many kilometres below a discharge in severe conditions; in less severe conditions a reduced number of species may survive at very low densities.

If the source of pollution is water percolating through mine waste heaps, it can often be treated successfully by covering the tip with clay and top soil. Water coming from springs or underground water sources tends to be more acidic, with higher loads of metals, and is much more difficult to control. It can be neutralized with lime and metals flocculated out but this is very expensive and frequently ineffective. Artificial wetlands can be constructed to purify mine wastewater, generally by bacterial action, and there are very many of these in the United States (Fennessy and Mitsch, 1989; Cairns and Atkinson, 1994). At the Wheal Jane mine mentioned above a pilot scheme has been developed to treat the water. An aerobic cell containing a water-filled reedbed removes iron and arsenic, and produces a pH suitable for bacterial growth. The water then passes to an anaerobic cell, containing a mixture of sawdust and cattle manure, where metals are removed by forming insoluble sulphides. A final cell contains rocks on which algae grow, generating a high pH to remove manganese. Three pre-treatment methods have been examined before water flows into the aerobic cell. One, consisting of a layer of limestone operated under anoxic conditions, has been shown to remove 95 per cent of incoming metals.

CHAPTER 6

ENERGY AND POLLUTION

Most electricity generating plants, whether they use fossil or nuclear fuels, operate through the thermodynamic process known as the Rankine cycle, in which high-pressure steam is produced in boilers and then expanded through turbines, which convert thermal energy into mechanical energy. The basic plan is illustrated in Fig. 6.1. In a fossil fuel plant, water from the condenser is pumped into feedwater heaters and pressurized, from where it passes to the boiler to be converted to saturated steam. This superheated steam is expanded through the turbine, creating mechanical energy to drive the turbine and hence the generator. The resulting expanded low-pressure steam is condensed, the heat being removed by cooling water circulating through the condenser.

The condensers require large amounts of cooling water, which is removed from the environment and returned at a higher temperature. About one-third of all the water abstracted from groundwaters and surface waters in England and Wales is used by the electricity supply industry. Most of this water, however, is returned to the environment and most of it is free of contamination other than heat, though there may be reduced levels of ammonia and organic nitrogen and increased levels of nitrate, chlorine and suspended solids compared with water in the power station intake. These chemical contaminants, though small in relation to the volume of cooling water, may in fact have a greater impact on the ecology of the receiving stream (Langford, 1983).

The overall efficiency of power stations is less than 40 per cent. Nuclear power plants generate at lower temperatures and pressures, so that less energy is added to the power cycle, but the Rankine cycle is less efficient, so that they still discharge a large amount of waste heat.

In 1998 in the United Kingdom, coal made up 33 per cent of the mix of fuels used to generate electricity, substantially less than a decade previously, while the use of natural gas had increased to a third of the total, having been negligible in 1990. Oil makes up 1.5 per cent and nuclear sources 26 per cent. Renewables still account for only 2.5 per cent of the total (Department of Trade and Industry, 1999). As well as thermal discharges, fossil fuels, especially coal and oil, add acidifying gases and carbon dioxide to the atmosphere, adding substantially to global

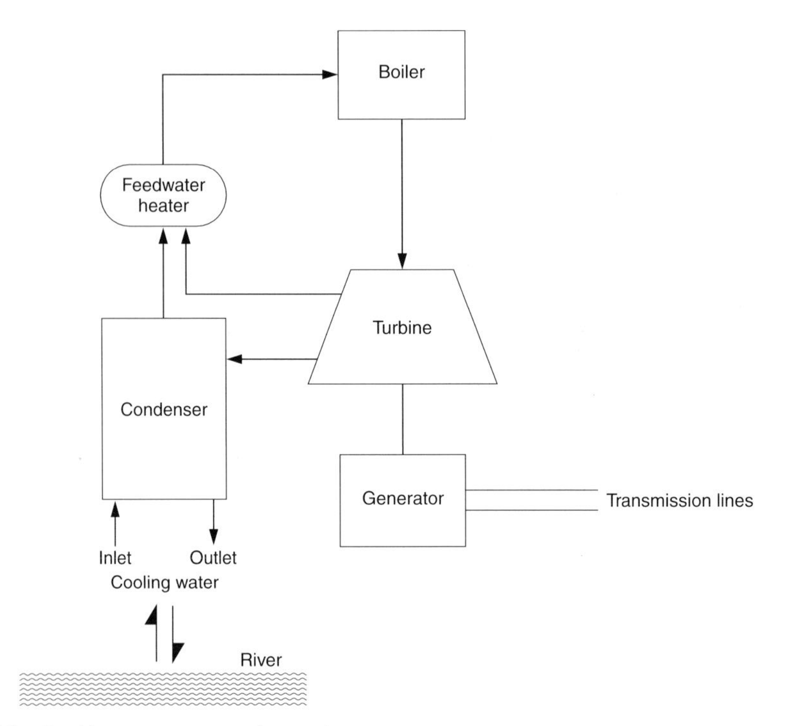

Fig. 6.1 Rankine power conversion cycle.

warming. Oil is a hazard to the aquatic environment at its point of abstraction and in accidents during transport and storage, and adds greenhouse gases to the atmosphere when burnt, the exhausts of motor vehicles being an important source. Electricity production from nuclear reactors adds low levels of radioactivity to waters but it is the threat of a nuclear accident that causes particular concern.

THERMAL POLLUTION

The effects of a power station at Ironbridge (Fig. 6.2), in the English West Midlands, on the temperature of the River Severn have been studied in detail. The natural annual range of temperature was increased by up to 6 °C (e.g. the maximum temperature in 1970 400 m above the power station intake was 22 °C, while 2000 m below the outfall it was 28 °C). The daily increments downstream ranged from 0.5 °C during spates to 7.2 °C in low flow conditions and the diurnal variation in summer was increased by more than 100 per cent. In spring the rising mean temperature was advanced by three to four weeks, while the fall in temperature in autumn was delayed by one to three weeks (Langford, 1970).

An increase in temperature alters the physical environment in terms of both a reduction in the density of water and its oxygen concentration, which varies

Fig. 6.2 A typical coal-fired electricity generating station with cooling towers, at Ironbridge on the River Severn (photograph by the author).

inversely with temperature. Nevertheless the oxygen concentration below a cooling water discharge may be substantially above that of the intake because of turbulence and agitation produced within the cooling tower (Langford, 1983).

Effects on the biota

Possibly the most damaging environmental effect of a power station is that many organisms may be sucked in through the water intake. Larger creatures, such as fish, are killed on the intake screens while smaller species pass through the plant. Even algae may be damaged, with permanent impairment of the photosynthetic mechanism (Nalewajko and Dunstall, 1994). At the Bay Shore power station, Lake Erie, over the period 1976–77, 284 million larval fish and 426 million fish eggs were destroyed by the intake, which was situated in shallow water where young fish congregate (Laws, 1993).

There are various effects of heated effluents on the biology of receiving waters. Those species intolerant of warm conditions may disappear, while others, rare in unheated water, may thrive so that the structure of the community changes. Declines in species richness have been recorded in bacteria, benthic invertebrates and zooplankton living in thermal effluents. The standing crop and productivity of attached algae usually increases in heated effluents whereas species richness declines. Cyanobacteria are most tolerant and become dominant if temperatures remain above 32 °C for any length of time. Exotic species, such as the tubificid worm *Branchiura sowerbyi*, are confined to heated waters, in which they build up large populations. Respiration and growth rates may be changed and these may alter the feeding rates of organisms. The reproductive period may be brought forward and development may be speeded up. Parasites and diseases may also be affected. Increased temperatures may render organisms more vulnerable to the effects of toxic pollutants present in the water. Any reduction in the oxygen concentration of the water, particularly when organic pollution is also present, may result in the loss of sensitive species.

As temperature increases, the respiration and heart rate of a fish will increase in order to obtain oxygen for an increased metabolic rate but at the same time the oxygen concentration of the water is decreased. At 1 °C, for example, a carp (*Cyprinus carpio*) can survive in an oxygen concentration as low as $0.5\,mg\,l^{-1}$, whereas at 35 °C the water must contain $1.5\,mg\,l^{-1}$.

Many species are able to acclimate to the normal range of temperatures occurring below thermal discharges, so that when exposed to increasing water temperatures, their upper lethal limit is raised. Temperature polygons can be constructed, enclosing the limiting temperatures which vary according to the history of previous exposure (Fig. 6.3). In this example for the roach (*Rutilus rutilus*) an increase in the acclimation temperature from A to B raises the upper tolerance from D to E. A further increase in the acclimation temperature, however, does not increase tolerance, so that the upper limit is reached (E to F). For fish acclimated to high temperatures (P), a sudden exposure to a low ambient temperature (4 °C at Q in Fig. 6.3) may put the fish outside its tolerance polygon, so that it dies. Fish must, however, do more than survive and different activities have different temperature tolerances. Figure 6.4 illustrates the temperature tolerance polygon for different activities of the brown trout (*Salmo trutta*). Whereas the fish can tolerate temperatures greater than 20 °C, even after acclimation it will not feed or grow. At 19 °C the swimming speed of trout declines, making them less efficient predators, and prolonged exposure to such temperatures may lead to death by starvation.

Dheer (1988) maintained the fish *Channa punctatus* at temperatures of 23 °C, 30 °C and 35 °C, with a control group at 14 °C. At 30 °C and 35 °C fish lost weight and there was an increased mortality compared with the control group. Fish maintained at 14 °C and 23 °C showed normal growth. Within a week at 35 °C there were significant falls in blood glucose level and depletion of glycogen reserves in liver and muscle, such stress symptoms becoming apparent after four weeks in those *Channa* kept at 30 °C. Mosquitofish (*Gambusia holbrooki*) from a population

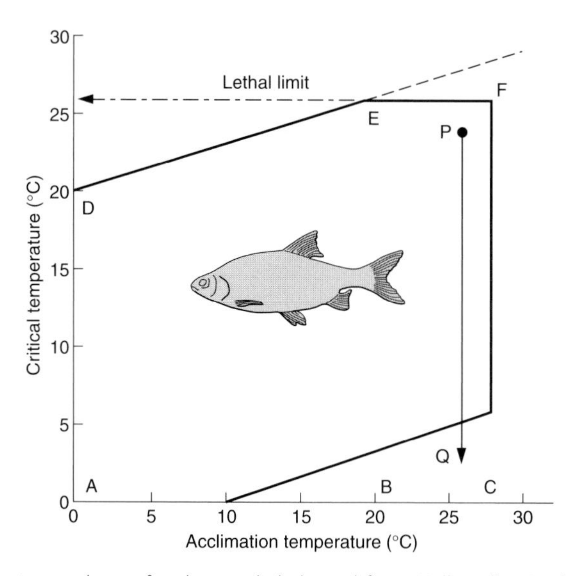

Fig. 6.3 Temperature polygon for the roach (adapted from Hellawell, 1986).

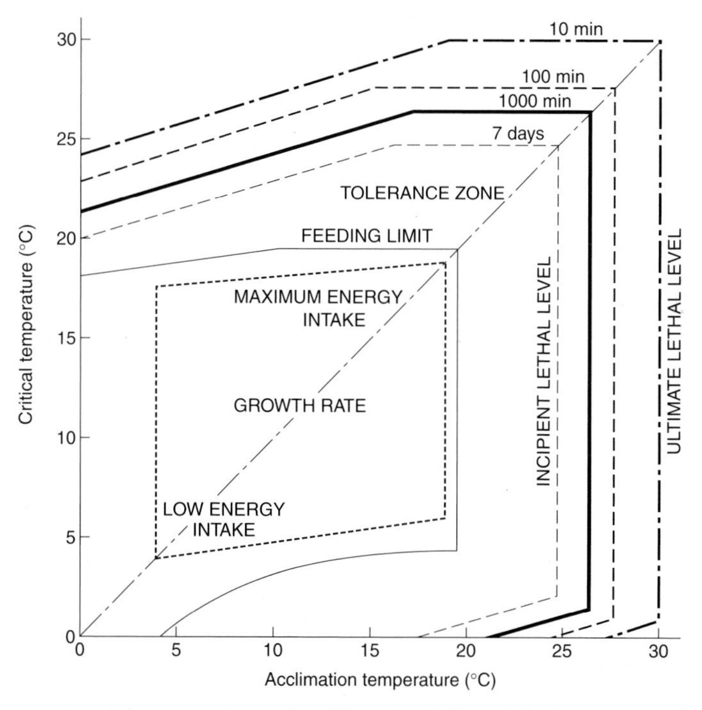

Fig. 6.4 Temperature tolerance polygon for different activities of the brown trout (from Hellawell, 1986, after Elliott, 1981).

exposed over many years to a heated, but fluctuating, discharge from a nuclear power station, bred throughout the year, whereas those from natural habitats ceased breeding for six months in the winter. The exposed population had larger clutch sizes and reproductive biomass, but produced smaller offspring than those from natural habitats (Meffe, 1991).

The general ability to acclimate means that large-scale mortalities of fish are infrequent, occurring mainly when populations are trapped in an effluent channel or when there are sudden discharges of hot water. Some fish populations are unaffected by rapid temperature fluctuations of 12–15 °C but these temperature changes may be too great for cold water fish. Discharges of thermal effluents into water that is already polluted may also tip the balance for survival.

With temperatures up to 26 °C below thermal discharges fish tend to be attracted, whereas at temperatures above 30 °C they tend to move away. Anglers certainly make use of the attraction of heated discharges to fish.

The effects of a thermal effluent on the growth of an organism will depend on the relative apportionment of energy into respiration and the scope for growth. If the increase in temperature results in increased metabolism, with no increase in feeding rate, then less energy may be diverted to growth. However, the increased temperature may extend the period over which growth can occur, bringing forward the onset of growth in spring and extending the growing period into the autumn. Mobile species may avoid periods when discharges result in higher temperatures than the species' optimum, so that fishes living in thermally polluted areas will not show significantly different rates of growth from fish in neighbouring unpolluted waters. Other species of fish have shown an increase both in growth rate and in the maximum size attained. Faster growth rates and larger maximum sizes have also been described for invertebrates (Langford, 1983).

The life-histories of organisms may be altered by thermal pollution, for instance by accelerating development time, and some fish spawn earlier in power station effluents than they do in neighbouring unheated waters, though others, such as brown trout, may not spawn successfully (Langford, 1983). In an experimental stream maintained 10 °C above a control stream at natural temperatures, peak macroinvertebrate density was three to four weeks earlier, while some species began reproduction two to three months earlier (Arthur *et al.*, 1982). The development of eggs was more rapid, while life cycles were shortened in some invertebrates living in a river below a heated discharge (Howells and Gammon, 1984). The emergence of insects from the River Severn, however, whose temperature regime was discussed above, was not noticeably altered by heated water and natural variability due to other environmental factors masked any effect due to temperature (Langford, 1975).

Temperature changes may increase the vulnerability of a species to predation and parasitism. Coutant *et al.* (1976) observed that rapid temperature decreases of about 6 °C, which may occur when the quantity of a thermal effluent is suddenly reduced, increased the susceptibility of juvenile channel catfish (*Ictalurus punctatus*) to predation by large-mouth bass (*Micropterus salmoides*). With reference to parasites Aho *et al.* (1976) found that, with the mosquitofish (*Gambusia affinis*), the metacercariae of the brain parasite *Ornithodiplostomum ptychocheilus* occurred at higher densities in fish from thermally polluted waters, but, in contrast, the metacercariae of the body-cavity digenean *Diplostomum scheuringi* were most numerous in fish from waters not receiving thermal pollution. These differences may have been due to the effects of temperature on the life cycles of the parasites or to effects on either the intermediate or definitive hosts.

Temperature changes are a feature of natural ecosystems so that organisms have the ability to adapt to the altered conditions provided by thermal effluents. Except for unusually severe thermally polluted conditions it appears that macroinvertebrate communities of rivers are relatively little affected by thermal discharges. The same is generally true of fish, which have the ability to vacate water that is temporarily inimical to them. The overall conclusion from an extensive body of research is that thermal pollution has not been as damaging to aquatic ecosystems as was originally feared and, within the usual range of temperature increases caused by power stations, the homeostatic mechanisms at work within the community minimize the damage.

Beneficial uses of thermal discharges

The production of electricity is an inefficient process and there have been a number of attempts to put the waste heat in cooling water to use. Power stations can, for example, be linked to sewage treatment works. Mixing warm power station effluents with raw sewage accelerates the treatment process, while the circulation of partially treated sewage effluent around the cooling tower accelerates the conversion of organic nitrogenous compounds and ammonia to nitrites and nitrates. Cooling water is also being used to heat greenhouses for the production of high-value crops such as tomatoes.

Most effort has been put into aquaculture, growing marine fish such as sole (*Solea solea*) and turbot (*Scophthalamus maximus*), and freshwater fish such as eel (*Anguilla anguilla*), carp and channel catfish in cooling water, resulting in greatly increased growth rates. Figure 6.5 illustrates the growth of eels in power station condensers, tower ponds and tanks containing water pumped from the River Trent. The mean weight increased from 1 g to 25 g over the year in the condenser water, a markedly greater increase than in the pond and river water, though individual eels varied widely in size. A marketable eel takes 2–2.5 years to produce in cooling water, compared with 10–14 years in the wild. Carp have shown similar good growth in condenser water, averaging 958 g in weight after ten months, from a starting weight of 13 g, compared with a mean weight of 172 g in river water. Eels are highly prized in continental Europe and many rivers have been overfished. Britain exports large numbers of eels to the continent and beyond so that there is a potentially lucrative market in growing at least this species in cooling waters. Aquaculture units do, however, have to be sited close to, and preferably within, power station complexes, because the temperature advantage is quickly lost if water

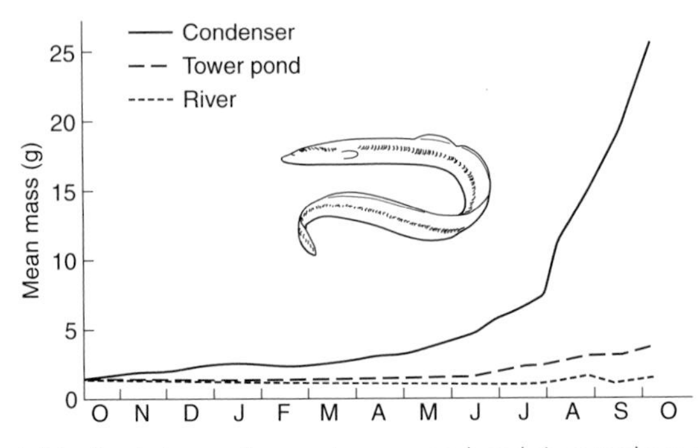

Fig. 6.5 Mean weight of eels in condenser, tower pond and river tanks over a 12 month period (adapted from Langford, 1983).

has to be transported. Rapid fluctuations in temperature and the frequent presence of chlorine, used to prevent fouling within the power station pipework, are also problems that have to be overcome.

RADIOACTIVITY

The nucleus of an atom contains positively charged protons and electrically neutral neutrons and is surrounded by orbiting electrons, each of which carries a negative charge equal to the positive charge on a proton. Normally they are balanced. Atoms of the same chemical element may vary in the number of neutrons they have and are known as isotopes. Some isotopes (*radioisotopes* or *radionuclides*) are unstable and they seek stability by giving off particles or electromagnetic rays. There are four main types of radiation, with differing powers of penetration (Fig. 6.6):

1. *Alpha particles*, consisting of two protons and two neutrons, which have very little penetrating power and lose their energy in a short distance. If, however, they do penetrate living tissue, by inhalation or ingestion, they can do considerable damage because they are strongly ionizing.
2. *Beta particles* have greater penetrating power than alpha particles, but they can be stopped by layers of water, glass or metal, though like alpha particles they can be hazardous if taken into the body. Beta particles may be negatively charged electrons or positively charged positrons, depending on whether a neutron spontaneously changes into a proton, or vice versa, in an unstable nucleus.
3. *Gamma rays*, like X-rays, are short-wave electromagnetic radiation, having neither mass nor charge, but they are capable of penetrating thick material.
4. *Neutron radiation* also occurs, but it is generally found only in nuclear reactors. Some heavy, unstable elements have nuclei with a large excess of neutrons and the nucleus breaks into two fragments (spontaneous fission), with the production of free neutrons which are highly penetrating.

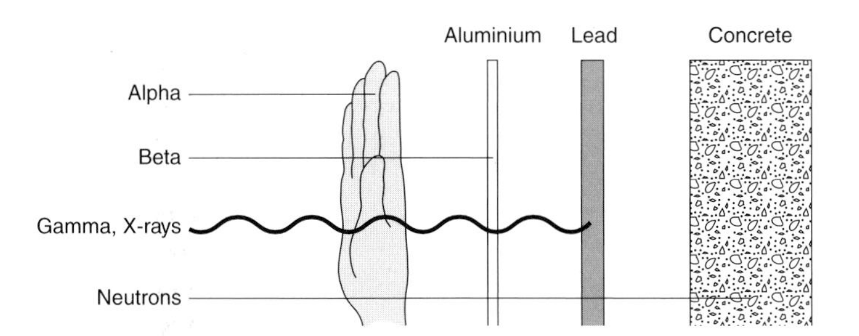

Fig. 6.6 The penetrating power of different types of radiation (source, United Kingdom Atomic Energy Authority).

Apart from their radioactivity, radionuclides behave chemically like the stable form of the element because the pattern of orbital electrons is the same. An important feature of radioactive material is that radioactivity steadily declines with time. Each material has a fixed time for the radioactivity of a given quantity of the material to fall by 50 per cent, known as the half-life. In two half-lives the radioactivity falls to a quarter of its original level, in ten half-lives to about a thousandth. Radioactivity is inversely related to the half-life of a substance, so that substances with short half-lives are intensely radioactive, while those with long half-lives emit very little radiation, have generally low penetrating power and are only hazardous if taken into the body.

There are several units used in the measurement of radioactivity:

- **Becquerel** (Bq). Radioactivity is measured by the frequency with which radioactive disintegrations of the nuclei take place in a substance. The becquerel is one nuclear disintegration per second. The becquerel supersedes an older unit, the curie (Ci) and 1 Bq = 27.03 pCi.
- **Gray** (Gy), the measure of the amount of radioactivity absorbed by a tissue or organism. One gray is the amount of radiation causing 1 kg of tissue to absorb 1 joule of energy.
- **Sievert** (Sv), an arbitrary unit that accounts for the fact that different kinds of radiation do different amounts of damage to living tissue for the same energy. Neutrons and alpha particles have about ten times the effect of beta or gamma particles for the same number of grays.

Sources of radioactivity

The environment is naturally radioactive, with radiation coming from outer space, from the earth, from the atoms within the body and from normal activities such as burning fuel and cultivating soil. The average person in Britain receives an annual dose of 2500 μSv of radiation. In general some 88 per cent of radiation received is of natural origin, over which we have no control. Natural sources include cosmic rays from outer space (10 per cent of total), terrestrial gamma rays from rocks and soil (14 per cent, but very variable depending on geology), radon and thoron gas inside buildings (52 per cent but variable according to local geological sources) and radiation accumulated in tissues from food and drink (12 per cent). Levels of radionuclides in the human food chain in the United Kingdom are discussed in Ministry of Agriculture, Fisheries and Food (1994). Despite our constant exposure to natural sources of radiation there is no threshold limit for radiation damage and the amount of damage caused is roughly proportional to the total amount of radiation absorbed, so that it is essential to restrict man-made emissions of radiation to the environment.

The major controllable sources of radiation to the general environment are those from nuclear weapons testing and from the nuclear industry. Nuclear weapons testing began in the 1940s and fallout from this source reached a peak in the 1950s. With the signing of a test ban treaty in 1963 between the USSR, the United States and the United Kingdom inputs from this source have been greatly reduced, but unfortunately not entirely eliminated. Of course, while many nuclear weapons are retained for defensive purposes, there remains the possibility of nuclear war, but in this event, the release of radioactivity would be only one of many catastrophes facing the world, including the possibility of nuclear winter and major global climatic change.

A nuclear power station is in many ways similar to a conventional power station in that steam is used to power generators which produce electricity. The major difference is that the source of heat is the splitting of atoms rather than coal, oil or gas. The steam is produced in the reactor building, in which the basic fuel is uranium, usually enriched so that the more reactive ^{235}U is concentrated relative to the more abundant ^{238}U, making up 3.1 per cent of total uranium in the reactor compared with 0.7 per cent in natural uranium. The fuel is cast into ceramic pellets which are built into a stack and encased in a zirconium alloy to prevent direct contact with cooling water.

The uranium nucleus can be split into two roughly equal pieces (nuclear fission) if hit by a neutron. The total mass of the fragments is slightly less than that of the original nucleus and this reduction in mass (m) appears as energy (E) according to Einstein's equation:

$$E = mc^2$$

where c is the velocity of light. Even a small loss of mass releases a considerable amount of energy. The fission process releases two or three neutrons which can split other atoms, releasing more neutrons in a chain reaction. The energy of the fission reaction appears as the velocity of the fission fragments and is converted into heat, to produce steam for power generation, as the fragments are slowed down by collision with surrounding material. To increase the deceleration of neutrons, the fuel is embedded in a material known as a moderator and the neutrons bounce off atoms in this moderator, losing energy with each collision. In the Pressurized Water Reactor (PWR), the most common type of reactor in the world, the moderator is water, which is also used as a coolant to carry heat away. The power of the reactor is controlled by adjusting the number of available neutrons by using neutron absorbers, either cadmium or boron, which can be raised or lowered into the reactor core.

To carry away the heat, water is pumped through the fuel assembly and passed to a heat exchanger or steam generator, where it boils water that is in a separate circuit connected to the steam turbines. The water in the circuit passing through the fuel assembly must not boil, so it is kept under pressure. The steam emerging from the turbine must be condensed using cooling water drawn in from the environment. In Great Britain most of the nuclear power stations are situated on the coast and use seawater for cooling. In continental land masses nuclear power stations are sited by major rivers. Inevitably the cooling waters acquire some radioactivity and the discharge also contains liquid wastes, with low radioactivity, which come from processes involved with the handling of spent fuel rods from the reactor. The fuel rods in the reactor eventually lose their efficiency, but they still contain large amounts of ^{235}U which has not undergone fission. They can be reprocessed and the facilities that undertake this release large amounts of wastewater with a low content of radioactivity, but in overall quantities much greater than is released by nuclear power stations. They also produce small quantities of high-level waste that has to be stored for several decades for the radioactivity to decay.

The nuclear industry is strictly controlled and it is generally considered that the amounts of radioactivity normally released are not harmful either to the environment or to people. However, it is the potential problems with the nuclear industry that cause so much concern. The use of radioactive materials produces wastes which must be disposed of. High-level waste is produced only by the nuclear industry, but there is much intermediate and low-level waste from nuclear power stations, defence establishments, the radioisotope industry, laboratories and hospitals. In Britain, some 2500 m^3 of intermediate-level waste and 25 000 m^3 of low-level wastes are produced each year, compared with only 1100 m^3 of high-level waste over 30 years. Some low-level liquid and gaseous wastes from laboratories and hospitals are disposed of directly to the environment, but solid wastes, such as glassware, clothing, sludges and used reactor components and many radioactive liquid wastes need very careful management. Disposal presents a political if not a scientific problem. Much low-level waste has previously been dumped at sea, after being packed in steel drums and encased in concrete, but land disposal seems to be the current politically preferred option. The concern is that such disposal sites, whether deep or shallow repositories, may eventually leak, leading to the contamination of surface and groundwaters. The potential pollution arising from the decommissioning of nuclear power plants and the disposal of their wastes is a current problem as some power stations are reaching the end of their useful life. Openshaw (1992) and Hewitt (2001) discuss the sources and environmental impacts of radioactivity.

Other major concerns are those of sabotage or accidents at nuclear installations.

The nuclear power industry necessarily has a good safety record. Of several accidents, only that at Chernobyl (discussed below) has resulted in widespread and long-term environmental contamination. Nevertheless, that any accidents occur explains the unease of many people over the nuclear industry. However well regulated and safety conscious it may be, accidents can happen, and eventually will, with dire consequences for ecosystems and human populations over large areas. Therefore, from the standpoint of public safety, the nuclear power industry should probably be considered unsafe.

Biological effects

The main concern over radionuclides is obviously the contamination of those food chains that lead to humans. A small number of radionuclides tend to be of special importance. These are the ones that are incorporated into the food chain and biomagnify. They are either radioisotopes of essential nutrients or show chemical behaviour similar to essential nutrients. Some radionuclides, for example uranium and plutonium, are of concern because they may be inhaled with dust and concentrated in the lungs. Table 6.1 lists radionuclides of particular biological concern.

At high concentrations radionuclides cause acute toxicity. The radiosensitivity of a tissue is directly proportional to the mitotic activity of its cells (i.e. whether they are actively dividing), chromosome damage being the major cause of cell death. Different organs therefore cease to function at different times after exposure or at different levels of exposure. After a whole-body X-irradiation of 2–5 Gy, for example, a deficiency in lymphocytes becomes evident within two days, while a reduction in erythrocytes is delayed for two to three weeks. In mammals a dose of 6–10 Gy leads to failure of the tissues that generate the blood cells, 10–50 Gy causes intestinal failure, including vomiting and diarrhoea, and more than 50 Gy causes central nervous system syndrome, including tremor, convulsions and lethargy. These effects are only likely in the event of nuclear energy accidents or warfare.

Table 6.1 Radionuclides of especial biological concern

Radionuclide	Half-life	
^{3}H	12.4 yr	Assimilated by body in water
^{14}C	5730 yr	Passed up food chain
^{32}P	14.3 days	Concentrated in bones
^{40}K	1.3×10^{9} yr	Found throughout body
^{90}Sr	28.9 yr	Concentrated in bones
^{131}I	8.1 days	Concentrated in thyroid
^{137}Cs	30.2 yr	Found throughout body
^{226}Ra	1622 yr	Concentrated in bones
^{238}U	4.5×10^{9} yr	Concentrated in lungs and kidneys

Particular concern is expressed over levels of exposure that may cause chronic effects, either somatic (affecting the body cells) or genetic (affecting the germ cells). Somatic effects are those that result in various types of cancer. Genetic effects are defects and abnormalities that occur in the next generation as a result of irradiation of the parent's reproductive organs. As such conditions normally appear long after the period of exposure to radiation they are very difficult to relate to particular exposure events.

Those aquatic organisms living below discharges from nuclear power stations or research installations have been most exposed to the long-term chronic effects of radiation. Chironomid midge larvae in White Oak Lake, receiving low-level radioactive waste from Oak Ridge National Laboratory in the United States, were subjected to 0.6–3.6 mGy per day of radiation from sediments. They had an increased frequency of chromosomal aberrations, though there was no effect on the abundance of the larvae (Blaylock, 1965). Mosquitofish from the same area showed increased damage to liver DNA and produced smaller broods with a higher incidence of dead and abnormal embryos (Theodorakis *et al.*, 1997). High-level damage to liver DNA was also recorded in two species of pond turtles (*Trachemys scripta* and *Chelydra serpentina*) from the same site (Meyers-Schöne *et al.*, 1993). Salmon spawning below a nuclear installation on the Columbia River in the United States, have apparently been unaffected by radiation doses of 1–2 mGy per week (Laws, 1993). In British waters no evidence of radiation damage, either acute or chronic, to aquatic organisms has ever been recorded. This does not mean that environmental effects have not occurred, but that investigations have not been sufficiently sophisticated to detect them.

Because of the concern over accumulation, the distribution of radionuclides in waters receiving effluents from nuclear installations is carefully monitored. It is, however, impracticable to monitor the whole system and usually only one or two pathways are important to humans, so-called *critical pathways*. If radionuclide contamination is controlled along critical pathways, then adequate control can usually be ensured along other pathways. A generalized critical pathway model is shown in Fig. 6.7. The critical pathways and critical materials (those materials receiving or leading to the greatest degree of exposure) are identified after carrying out a study, usually a desk study, of the various pathways. The limiting environmental capacity is then set by comparison with primary exposure standards and on the basis of concentration of critical materials per unit rate of introduction. The critical material may be the receiving medium or some living component of the ecosystem that is affected by the radionuclide or is important in the human diet. Jackson (1992) describes the approach in more detail.

Below a nuclear power station on the Columbia River, it was considered that ^{32}P was the critical material, the human population being exposed through eating fish. The average concentration in river water below the plant was about 5.6 mBq ml^{-1} and the average concentration in the flesh of whitefish was 9 Bq g^{-1}. Concentration factors from water to fish were estimated at around 5000 in summer but less than 1 in winter. In those fish species eaten in greatest numbers, the average concentration

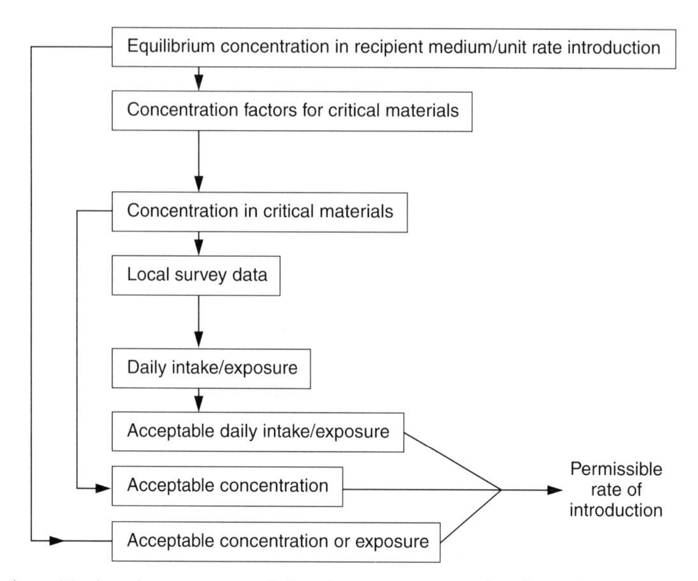

Fig. 6.7 The critical pathway approach for the assessment of radionuclides in the environment (adapted from Preston, 1974).

of radiation was 1.5 Bq g^{-1}. A person who ate 200 fish meals a year (around 40 kg of fish) would receive a dose to the bone, the critical organ, of 3 mSv yr^{-1}, which is 20 per cent of the annual limit (Laws, 1993).

From the evidence available, the discharge of radionuclides from the normal operation of the nuclear power industry appears to cause no measurable effects on the biota *in situ* and despite potentially large accumulation factors does not result in levels of radioactivity in food organisms that approach dose limits set for people by the International Commission for Radiological Protection.

The Chernobyl accident

On 26 April 1986 there was an accident at the Chernobyl nuclear power station, some 100 km north of Kiev in the Ukraine. A sequence of events occurred, involving a sudden surge of power in the reactor while it was at low power, followed by an explosion of steam and hydrogen and a fire. Initial reports of an accident came the following day from Sweden, where higher than normal background levels of radiation were recorded. The accident was reported in Moscow on 28 April. It was estimated that 3.5 per cent of the radioactive materials in the reactor, some 2×10^{18} Bq, were released to the atmosphere.

Considerable environmental damage was done in the area surrounding the reactor, some 23 000 km^2 being designated as contaminated, while high levels of radioactivity were found in terrestrial animals, aquatic insects, fish and waterbirds. Radiocaesium in muscle from fish taken from freshwaters within 10 km of the reactor was in the range 6–192 Bq g^{-1} but was highly variable (Jagoe *et al.*, 1998). Nevertheless there are reports that wildlife is now thriving in the area, with for example 180 species of birds breeding, as abandoned farmland develops into wilderness (Connor, 2000). Birth defects and cancers have been recorded in children born in the Kiev region since the accident and the Ukrainian Health Ministry claimed in 1995 that some 120 000 deaths could be attributed to the accident (Hewitt, 2001), though estimates of mortality vary greatly. It appears that the greatest health risk is that of thyroid cancer. An overview of a rather confusing wealth of health science is provided by Moysich and Michalek (2000).

The radioactive cloud initially flowed over Scandinavia, but then moved southwards across central Europe to Britain (Eisler, 1995; Hewitt, 2001). The Mediterranean region was also contaminated and subsequently evidence of the accident could be found over much of the world. Heavy rain led to substantial depositions of radioactivity over parts of Scandinavia, central Europe and western and northern Britain. Of the cocktail of radionuclides deposited, only ^{137}Cs has a significantly long half-life, of 30 years. Caesium behaves like potassium and hence becomes widely dispersed in ecosystems. Livestock in some northern and upland areas became severely contaminated and some, to this day, remain unfit for human consumption, but much of the deposited radioactivity was leached or washed gradually into streams. The highest deposition of Chernobyl fallout occurred in areas with soft waters, where there would be high rates of transfer from water to fish.

In Sweden levels of radioactivity in freshwaters rose quickly. Fish species varied in the amount and rate of accumulation of radionuclides. For example in Lake Tröske, bream (*Abramis brama*), a bottom-feeding fish, accumulated 1000 Bq kg^{-1} ^{137}Cs as early as May, ten times greater than levels in pike (*Esox lucius*) and perch (*Perca fluviatilis*). By July, perch, which feed on zooplankton, had accumulated more than 8000 Bq kg^{-1}, whereas the fish-eating pike had accumulated only 4600 Bq kg^{-1}, a little higher than the bream at 3800 Bq kg^{-1} (Petersen *et al.*, 1986). Over the two year period 1986-87 the average concentration of ^{137}Cs in perch dropped from 9800 Bq kg^{-1} to 5040 Bq kg^{-1} but levels were still high in 1989 and there was great individual variation in concentrations in fish from the same lake. It was estimated that between 4000 and 7000 Swedish lakes had perch with a ^{137}Cs load more than 1500 Bq kg^{-1}, the present guideline for consumption, and angling virtually ceased (Håkanson *et al.*, 1989). Although there was a fairly rapid decline in ^{137}Cs in fish in the first four years after the accident, the subsequent decline has been very slow, owing largely to recycling within the lake ecosystem (Jonsson *et al.*, 1999) and complete recovery will take a very long time. Indeed, ^{137}Cs activity levels in pike could exceed the guidelines well into this century (Lundgren, 1993).

Because wild-caught freshwater fish feature little in the diet of most of the European population it was considered that the increase in radioactivity from this source due to the Chernobyl accident presented no cause for concern to human health. Indeed it has been calculated that even the critical group, that is those members of the population eating 100 kg annually of fish from the Baltic Sea, where contamination was greatest, would receive less than 0.08 mSv yr^{-1} (Camplin and Aarkrog, 1989). However, other species within the aquatic ecosystem are more exclusively fish-eating. Spraints (faeces) of otters (*Lutra lutra*) were collected from river banks in central Wales and from southwest Scotland. Central Wales was outside of the main deposition area for Chernobyl fallout. In the September following the accident, radioactivity in otter spraints from Wales was more than double that in a sample collected some 15 months before the accident and stored in a deep freeze. By the following January levels had returned to normal. By contrast, southwest Scotland received large quantities of Chernobyl fallout and average radioactivity was more than six times that in Wales (13 000 Bq kg^{-1}) and levels were still high in the following January, as has been found for other biological material from this area (Mason and Macdonald, 1988). Whether or not such high levels of radioactivity could adversely affect otters is unknown and it would be impossible to disentangle any mortality caused by radiation from other mortality factors, a major problem with many ecotoxicological studies.

The Chernobyl accident has stimulated a wealth of research in the last 15 years into radioactivity in the environment and on health, much of which is only now being published. The normal wastes released from nuclear power plants have not been shown to cause harm either to the natural environment or to humans. Compared with power stations burning fossil fuels and producing gases that cause acidification and global warming, nuclear technology would seem to be clean. This does, however, ignore the problem of the long-term disposal of high- and intermediate-level wastes and the decommissioning of redundant plants. It is planned that reactors will be left for a long time before decommissioning so that they become less radioactive. However, no full-size reactor has ever been fully decommissioned and the true cost of the process is unknown. A recent report on the decommissioning of the Dounreay nuclear plant in Scotland indicates that the task will cost £4 billion (US$5.8 billion) of taxpayers' money and will take at least 50 years to achieve. The half-life of plutonium, which fuelled Dounreay's reactors, is 24 000 years. And there remains the ever-present possibility of a major nuclear accident, the resulting release of radioactivity to the environment having 'the potential to be more insidious, more widespread, more prolonged, and less reversible than the effect of almost any other environmental disaster' (Depledge, 1986). The benefits of nuclear power need to be weighed extremely carefully against future and potential costs. Many governments are concluding that the risks are not worth taking.

OIL

Oil pollution incidents in freshwaters and their effects on living resources have received much less attention than those in marine ecosystems. The oiling of seabirds, for example, is commonplace and widely commented upon. Some accidents at sea have devastated marine life – the *Exxon Valdez* in Alaska in 1989, the *Apollo Sea* in South Africa in 1994, the *Sea Empress* in Wales in 1996, the *Erika* in

France in 1999 and the *Treasure* in South Africa in 2000 are five major incidents that resulted in massive mortality of seabirds. Accidents in freshwater ecosystems are much more localized and hence attract much less public outcry.

The Monongahela River flows into the Ohio River at the Golden Triangle in the city of Pittsburgh, Pennsylvania, thence down through Cincinnati to the River Mississippi. At Florette, some 37 km above Pittsburgh, a riverside storage tank was being slowly filled with no. 2 diesel fuel in early January 1988. The tank had recently been moved and reconstructed and the company had not carried out the normal testing before filling with oil. When almost full, the tank collapsed, releasing some 3.5 million litres of oil, most of which flowed into the Monongahela River and downstream to Pittsburgh. Within a day the water supply of 23 000 people was cut off and a further 750 000 people had their water rationed. More than a thousand families were evacuated, dozens of factories using water from the river stopped production, schools were closed and commercial traffic on the river came to a halt. Numerous fish, ducks and geese were killed and, as the oil continued downriver, drinking water intakes were closed. Access to the river was difficult because of steep banks, hindering clean-up operations, while the swift flow resulted in oil being forced under booms positioned to contain it. Only 380 000 l were recovered (Lemonick, 1988). Had the accident occurred in the summer, when fish and invertebrates were more active, the effects of the oil would have been much more severe.

In the marine ecosystem most oil pollution derives from tanker operations (cleaning of compartments etc.) and accidents, incidents at production installations being of less significance. Some 28 per cent of the world input of petroleum hydrocarbons to the sea is by way of rivers, however, suggesting that chronic pollution of freshwaters is widespread. In England and Wales in 1998, 30 per cent of all reported pollution incidents affecting rivers was caused by oils. Much oil in freshwaters is derived from petrol and oil washed from roads or other hard surfaces and from the illegal disposal of used engine oil. Other sources include irrigation pumps and boats, while accidents involving transporters, spillages from storage tanks and vandalism are also significant.

Detailed reviews of oil and its impact on freshwater ecosystems are provided by Vandermeulen and Hrudey (1987) and by Green and Trett (1989).

What is oil?

It is difficult to predict the potential toxic effects of oil because of its very complex chemical nature. Crude oil consists of thousands of different organic molecules, the majority of them hydrocarbons with between 4 and 26 atoms per molecule. There are also some sulphur and nitrogen compounds, and metals such as vanadium. Oils from different sources have very different compositions. The three major types of hydrocarbons are alkanes (e.g. ethane, propane, butane), cyclohexanes (naphthenes) and aromatics (e.g. benzene, toluene, naphthalene).

Crude oil is refined by a process of distillation, which separates off different fractions at increasing boiling points. At lower temperatures, the compounds used in the production of petroleum are separated off. At high temperatures, naphtha, which forms the basis of the petrochemical industry, separates and yet higher temperatures boil off diesel oil, bunker oil used to fuel ships and power generating stations, and tars. Further refinements are often necessary to produce commercial commodities. Oil products may also contain toxic compounds such as polynuclear aromatic hydrocarbons (PAHs), PCBs and metals, especially lead.

Effects on biota

Animals and plants may be affected by the physical properties of floating oil which prevents respiration, photosynthesis or feeding. Higher vertebrates, whose coats get covered in oil, lose buoyancy and insulation, while the ingestion of oil, frequently the result of attempts to clean the fur or plumage, may prove toxic. Many water-soluble components of crude oil and refined products are toxic to organisms, their eggs and young stages being especially vulnerable. There is also a range of sublethal effects.

The principal effect of oil discharges on the microbial community is one of stimulation, especially of heterotrophic organisms that utilize hydrocarbons. Shales *et al.* (1989) concluded that algae are less sensitive to the direct effects of oil than many other organisms, but they are especially sensitive to secondary effects. These may include an increase in primary production caused by the death, decomposition and nutrient release of sensitive species, an increase in primary production by nitrogen-fixing species or, in the absence of these organisms, a decrease in primary production.

The effect of aqueous extracts of various oils has been investigated on three species of floating duckweed *Lemna*. They proved generally tolerant of crude oils, but synthetic oils, such as raw coal distillate, were much more toxic, even a dilute 4 per cent aqueous solution resulting in a 50 per cent decrease in growth rate (Fig. 6.8). The toxicity was probably due to higher concentrations of aromatic hydrocarbons. Severe oiling of higher plants may reduce photosynthetic rates, either by interfering with the permeability of cell membranes or by absorbing the light required by the chloroplasts. Affected leaves may eventually turn yellow and die. The responses of plants to oil may be very species-specific, influenced especially by the thickness of the cuticle and hence the permeability of the leaves.

The toxicities of various oil derivatives to several algae, invertebrates and fish are shown in Fig. 6.9. The compounds tested were two phenols, two azaarenes from coal-derived oil and two polycyclic aromatic hydrocarbons. Snails were less sensitive than arthropods and fish were generally less sensitive than invertebrates. Within each pair of compounds the toxicity increased with increasing ring number of the molecule; for example naphthol (two rings) was 45 times more toxic to *Gammarus minus* than phenol (one ring).

A very wide range of differences has been revealed in the toxicity of oils to fish in short-term tests and these may depend on the type of oil, differences between

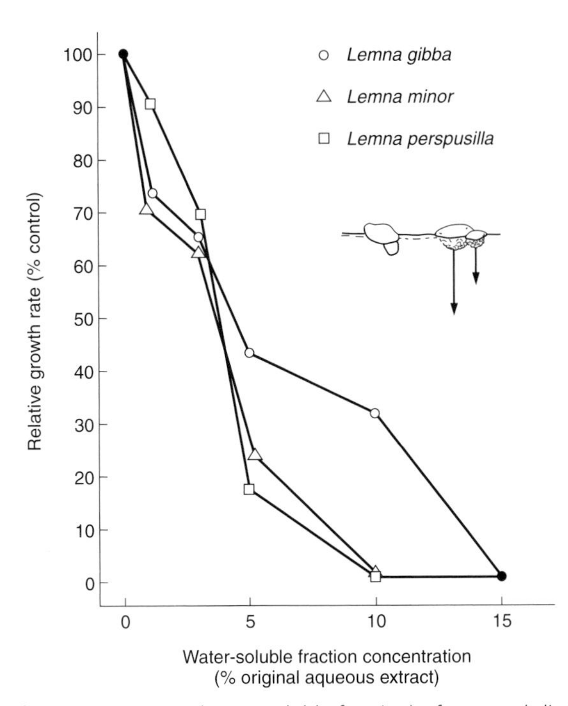

Fig. 6.8 Effects of aqueous extracts (water-soluble fraction) of raw coal distillate on frond multiplication rate of duckweeds (adapted from Green and Trett, 1989, after King and Coley, 1985).

the batches of the same oil, the changing nature of oil with time, the type of test system and the length of exposure (Hedtke and Puglisi, 1982). The aliphatic components of oils are relatively innocuous, while the monohydric aromatic compounds are generally toxic, increasing unsaturation of the molecules being associated with increased toxicity. Emulsifiers and dispersants, used to remove oil, contain surface-active agents that render membranes more permeable, thereby increasing the penetration of toxic compounds. In general terms the uptake of hydrocarbons by tissues and the histological damage due to exposure to hydrocarbons are both increased when an oil dispersant is mixed with the test oil. Most dispersants are quite toxic to fish and mixtures of dispersants with oil are much more toxic than oil alone. Frequently the oil : dispersant mixture is also more toxic than the dispersant alone, though this is not always the case.

Pearl dace (*Margariscus margarita*), collected from a site contaminated with diesel oil, showed many changes when compared with fish from a nearby reference site. There was an absence of young fish. The adult fish showed structural damage to the gills, liver and spleen and had low lymphocyte counts. They had an absence of those enzymes that break down pollutants, an absence of gut parasites, and they spawned later than fish from the reference site. It was considered that PAHs were

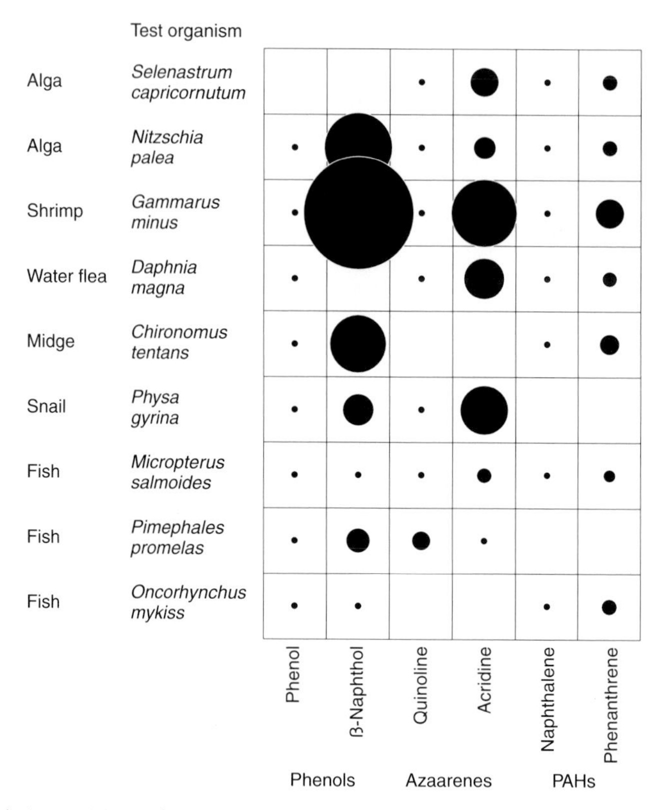

Fig. 6.9 Relative toxicities of single hydrocarbon compounds to freshwater organisms. The diameter of the circle is proportional to the toxicity. Comparisons can only be made between compounds within a class (i.e. between phenols, azaarenes or PAHs) and only within a species. Where boxes are empty, toxicities could not be compared (adapted from Milleman *et al.*, 1984).

responsible for the effects and would eventually lead to the extermination of this population (Khan, 1999).

There has been considerable research on the toxicity of PAHs. Early studies showed that anthracene was acutely toxic to bluegill sunfish (*Lepomis macrochirus*), especially when exposed to ultra-violet light. Oris and Giesy (1987) have investigated the phototoxicity of 12 PAHs to the larvae of the fathead minnow (*Pimephales promelas*). Six compounds exhibited acute toxicity, two slight toxicity and four showed no effect. Kazan *et al.* (1987) found that pyrene was not toxic even in ultraviolet light, but piperonyl butoxide, when present at non-toxic levels, synergized pyrene toxicity even in the dark. These experiments may be of considerable environmental significance because butoxide-like compounds are widespread.

Fish readily take up hydrocarbons but, when they are placed in clean water, the compounds quickly disappear, indicating that fish have a metabolic capability for removing hydrocarbons. Mixed-function oxidases, such as cytochrome P-450, in fish livers respond to foreign chemicals (xenobiotics). Fish react to PAH inducers by showing an increase in associated enzyme activities, resulting in the excretion of the foreign compounds (see review by Müller, 1987). Rapid metabolism has been reported for several PAHs.

Fish that live in waters receiving oily wastes develop an objectionable taste, known as *tainting*, which may result in serious economic losses to commercial fisheries. Wastes from outboard motor engines can also taint fish flesh. Taints are due mainly to unsaturated aliphatic hydrocarbons and some aromatic hydrocarbons. It is often difficult to pinpoint sources of taint in rivers receiving wastes from many sources.

Amphibians are likely to be vulnerable to the effects of oil because of their permeable skins and aquatic larval stages. In experiments with different concentrations of used engine oil, hatching success of the eggs of the tree frog *Hyla cinerea* was unaffected by concentration. However the growth of tadpoles was severely reduced at higher concentrations and at the highest concentration (100 mg°l^{-1}) no tadpoles metamorphosed (Mahaney, 1994).

Figure 6.10 illustrates the general effects of petroleum hydrocarbons on birds. Oil spills in areas where birds congregate in large numbers can be especially damaging to populations. The thermal conduction of oiled birds greatly increases, resulting in rapid heat loss. Ingested oil inhibits the movement of water and sodium across the gut wall, so that contaminated birds may die of dehydration. Growth rates, behaviour and egg production may also be adversely affected (Shales *et al.*, 1989). Birds may be contaminated though not incapacitated by oil, but they may

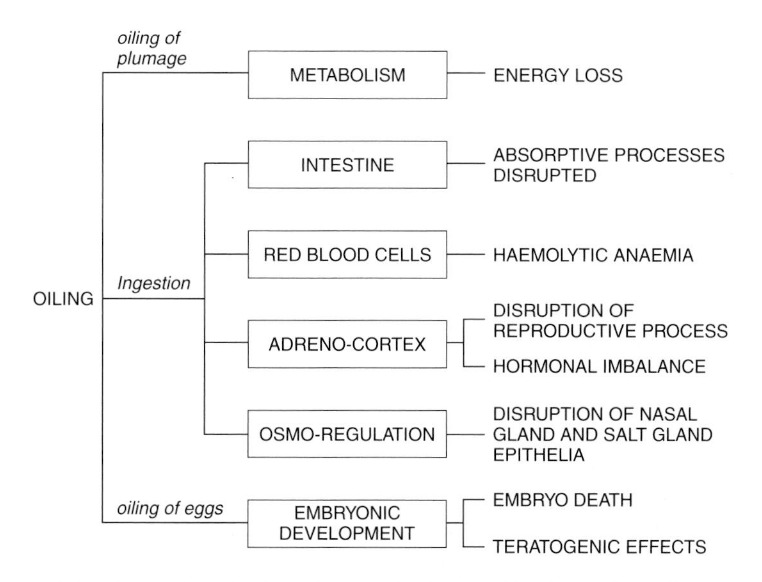

Fig. 6.10 Effects of petroleum hydrocarbons on birds (after Vandermeulen, 1987).

nevertheless fail to breed successfully because oil may be transferred to the eggs during incubation and prove toxic. Oil applied to the shells of mallard (*Anas platyrhynchos*) eggs drastically reduced hatching success, toxicity probably being due to the aromatic components of the oil. When 10 μl was spread over the egg shell, there was 100 per cent mortality. The dead duck embryos had a number of abnormalities, including deformed bills, reduction in liver size, incomplete ossification and incomplete feather formation (Hoffman, 1979). These laboratory experiments do have relevance to the field. Parnell *et al.* (1985) observed that, following a spill of no. 6 diesel oil in North Carolina, 24 of 98 nests of the brown pelican (*Pelecanus occidentalis*) were contaminated with oil transferred from the breast feathers of incubating adults. The hatching success of oiled eggs was significantly less than that of uncontaminated eggs.

Considerable effort is put into cleaning up and rehabilitating oiled birds but there is some question as to whether it is really successful. Rehabilitated American coots (*Fulica americana*) spent more time preening, bathing, eating and drinking, and less time sleeping, than non-exposed coots. Many failed to gain in weight and there was a high mortality (Anderson *et al.*, 2000).

The effects of the *Exxon Valdez* oil spill, two years after the accident, have been studied on a population of river otters (*Lutra canadensis*) which foraged in the intertidal zone (Duffy *et al.*, 1994). Otters abandoned ranges (measured in terms of usage of latrine sites) more than three times as often in oiled than in non-oiled areas, indicating a continuing population decline. Blood samples taken from live-trapped otters in the oiled area showed elevated levels of blood proteins and enzymes for up to two years after the spill, but these subsequently returned to normal. Porphyrins in the faeces of otters were also elevated (Blajeski *et al.*, 1996). There was some evidence of impaired immune systems, which could predispose otters to disease long after the oil had disappeared.

Because several components of oil and petroleum products, most notably PAHs, are known from laboratory studies to be mutagenic and carcinogenic, there has been much concern about their potential role in the environment, especially when they get into drinking water supplies. Proving a relationship between cancers and degree of pollution in the field is extremely difficult (p. 54) but given the high mutagenic and carcinogenic potential of a number of oil products, considerably more work is required to assess the true risk both to freshwater communities and to people.

Although the accidental pollution of freshwaters with oil occurs frequently and may cause severe mortality to aquatic life, the impacts on the aquatic community have been adequately described and reported on only a few occasions. Bury (1972) observed the effects of a spillage of some 9000 l of diesel oil, following a road tanker accident, into Hayfork Creek, California. Fuel entered the study area, 1.6–4 km below the site of the spillage, 36 h after the accident, the water turning brown and murky, with floating droplets of oil. Thousands of aquatic invertebrates were killed, as were 2500 fishes, a very large number of amphibian larvae and 36 aquatic snakes. Pond turtles (*Clemmys marmorata*) were found with eyes and necks swollen, damaged epidermis and uncoordinated movements. Most of the oil had flushed out of the stretch within three weeks, but the animal community by then had been severely damaged.

Guiney *et al.* (1987) studied the effects of a leak of 1310 barrels of aviation kerosene into a stream in Pennsylvania. Below the spill the benthic invertebrates and fish were eliminated. Recovery began within six months and was almost complete after a year. Thus although the short-term effects of this incident on the fauna were catastrophic, the community recovered relatively quickly. The factors that accelerated the re-colonization process included immediate and effective clean-up activities, plentiful unpolluted water upstream, invertebrate drift from upstream and immigration of fish from unpolluted sections of the watershed. Ecosystems that receive spills of crude oil, however, may take several years to recover, or indeed may never regain their former community of plants and animals.

Biodegradation and clean-up

Wherever possible, oil must be prevented from reaching the river in the first place. The River Rhine is a major shipping route and oil-contaminated bilge water is a potential source of pollution. Since 1965, Germany has established a bilge-draining facility. Currently eight bilge de-oiling boats pump out bilges while ships are moving, at no financial cost or wasted time to the transporter. It is estimated that this keeps 8–10 000 t of oil out of the river each year (Malle, 1994).

After an incident, oil that is deposited on the surface of water will disperse by physical and chemical processes, while some of the components will evaporate, or be lost by photochemical oxidation. As most of the remaining compounds are hydrocarbons they can be broken down by microorganisms. Brown (1989a) lists 21 genera of bacteria, 10 genera of fungi and 5 genera of yeasts that can degrade hydrocarbons. Complete degradation of a hydrocarbon may occur if a single strain of a microorganism uses that compound as its sole source of carbon and energy. Mixed populations of microorganisms may degrade up to 97 per cent of crude oil (Berwick, 1984). In general the substrate preferences of microorganisms for hydrocarbons are in the order aliphatics > heterocyclics > asphaltenes (Berwick, 1984; Brown, 1987). A number of factors influence the rate of biodegradation of oil, including oxygen availability, temperature, the type of oil spilled, the amount of emulsification and the availability of nutrients. Nitrogen and phosphorus are frequently limiting to microbial activity in freshwaters and the addition of these nutrients to lake-water samples increased the amount of mineral oil and hexadecane degraded over three weeks (Cooney *et al.*, 1985). A review is provided by Prince (1992).

The microbial breakdown of spilled oil is a natural process which slowly removes the often substantial residue left after physical and chemical clean-up technologies have been applied. The microbial process can be speeded up by seeding the oil spills (bioaugmentation) and inocula of microorganisms are commercially available to clean up spills. One product has been developed specifically to attack hydrocarbons in freshwaters and includes mutants that can degrade a variety of hydrocarbons. Once the substrate has been degraded the seeded microorganisms quickly lose their competitive ability and die out (Brown, 1989b). Nutrients are usually added to enhance the rate of breakdown. Scragg (1999) provides an overview of the bioremediation of oil-spills.

In areas where oil spills are likely to occur (oil loading and handling bays, garage forecourts etc.), interceptors can be built into surface drains to separate oil from water. Interceptor design and operation is described by Ellis (1989). When oil does pollute freshwaters, the clean-up consists of two stages – confinement followed by removal. Confinement involves the use of floating booms, or in very shallow waters the construction of temporary dams (Brown, 1989b; Ellis, 1989). The removal of the contained oil takes place in two stages: surface oil is skimmed off and the remaining traces removed using oil-absorbent devices. A variety of skimmers are in use and their operation is described by Ellis (1989). Sorbents include vegetable fibres and synthetic felted organic fibres, available in the form of cushions, blankets or loose chips. Some materials may hold up to 150 times their own mass of thick fuel oil. The chemical dispersal of oil is generally an undesirable option in the majority of spills in freshwater because of the highly toxic nature of dispersants (see above).

Oil-related industries have the potential to cause considerable environmental damage, at the exploration, drilling, transporting, refining and storage stages of operation. Countries in the developed world have therefore placed strict regulations on oil and related industries and the larger multinationals employ teams of scientists, including ecologists, to assess and reduce the impacts of their industry, often at great expense. Most of the problems in the developed world are caused by accidents at the facilities of both producers and users of oil products. Such accidents, unfortunately, occur with considerable frequency and very many could be avoided if adequate safeguards were implemented. It is surprising, in view of the many millions of dollars paid out for clean-up operations, compensation and fines in accidents such as the grounding of the *Exxon Valdez*, and also the very poor public image that such accidents project, that safety standards and accident procedures are often so lax within the industry.

Away from the developed West, environmental protection appears to be a very low priority of the oil industry in its exploration and exploitation activities, if the report of Read (1989) is typical. He described the exploration for oil by more than 20 foreign oil companies in 630 000 ha of forests in the upper Amazon basin in Ecuador. Initial exploration included building roads and trails, the use of dynamite, crop destruction, erosion, with attendant massive fish kills, and uncontrolled hunting by employees, with no concern for the indigenous population. At drilling sites, the surrounding land and waterways were polluted with drilling muds and oil wastes, while gas from the wells was burnt, rather than piped to cities where it could be used – Ecuador imports 50 per cent of gas for domestic consumption. Spills were described as commonplace and neither contained nor cleaned up. Almost twice as much oil was lost from the main pipeline as was spilled by the

Exxon Valdez, polluting hundreds of kilometres of the headwaters of the Amazon. Crops in flooded fields and cattle drinking from rivers died, while the indigenous peoples had to use polluted water for drinking and cooking. The network of roads allows access to speculators and colonists who destroy the forest and its wildlife and drive out the native peoples.

GLOBAL WARMING

The atmosphere has little effect on the passage of short-wave solar radiation (light energy). While some of this is reflected back by the clouds much passes through to be absorbed and converted to heat energy by the surface of the earth. The energy absorbed by the earth is radiated back at night as long-wave, infra-red radiation. A large proportion of this is absorbed by atmospheric gases to warm the atmosphere (the greenhouse effect). This process is essential to maintain the temperature of the earth, for it raises the average global temperature from −18 °C to 15 °C. However, human activities have been increasing the concentration of greenhouse gases in the atmosphere and hence raising the temperature of the earth (global warming). Greenhouse gases include water vapour, carbon dioxide, ozone, methane, nitrous oxide and the manufactured chlorofluorocarbons (CFCs) used in refrigeration systems. Of these, carbon dioxide contributes some 55 per cent to global warming, CFCs 24 per cent and methane 15 per cent. The burning of fossil fuels, releasing carbon that has remained dormant in the lithosphere for millions of years, together with deforestation, are responsible for most of the increase in atmospheric carbon dioxide. It has been estimated that temperatures may rise by between 0.2 and 0.5 °C per decade over the next century, and indeed the most recent predictions by the scientists of the Intergovernmental Panel on Climate Change (IPCC) suggest that the average temperature rise across the world may be as high as 6 °C by 2100. This will have major effects on both the quantity and quality of freshwaters, which will in turn affect natural ecosystems, agriculture and species. The rise in sea level that will result from the melting of permanent ice at the poles and in glaciers, and the thermal expansion of the oceans, could displace many millions of people as low-lying areas are inundated.

There are a number of changes in hydrological parameters likely to occur with global warming (Hayes, 1991; Graves and Reavey, 1996):

- decreased atmospheric humidity and increased evapotranspiration;
- decreased soil moisture;
- very variable rainfall, increasing incidence of floods and erosion;
- changes in seasonal snowfall and earlier snow-melt;
- changes in patterns of surface run-off, groundwater recharge and flow;
- reduced river flows and lake levels, and contraction of wetlands;
- increased incidence of droughts;

- reduced natural and artificial storage of aboveground and impounded freshwaters;
- changes in water quality, caused by drought, increased fire etc;
- increased saline intrusions into rivers from the sea;
- increased sea level with associated coastal flooding.

Most environmental scientists now accept the reality of global warming, though there are still detractors. As I write this in early November 2000, the UK is suffering a major storm and rainfall event which has brought transport to a halt, disrupted power supplies and flooded many towns and villages. It is the worst flood for 50 years, an extreme event but clearly not unique. However, it is the increased frequency of extreme events around the world that is providing the evidence for the onset of global warming and confirming the models of climate change.

Long-term data sets from a number of lakes have indicated a warming over recent decades (Gerdeaux, 1998). If surface water temperatures rise, the effects described above (p. 207) for waters receiving thermal discharges may become general over large areas. In the Experimental Lakes Area of Canada, water temperatures have risen by 2 °C over 20 years and the length of the ice-free season has increased by three weeks, earlier snow-melt leading to earlier and smaller spring stream flows (Schindler *et al.*, 1990; Schindler, 1997). There has been a decrease in water renewal time and an increase in the concentrations of most chemicals in the water. With a deepening of the thermocline there has been a decline in cold water species such as lake trout (*Salvelinus namaycush*) and the opossum shrimp (*Mysis relicta*).

Adrian and Deneke (1996) reported on a 20 year (1975–94) study of water temperature and plankton in a eutrophic lake, Heiligensee, in Berlin. The mean April water temperature increased by 2.6 °C over the period. There was a switch in the dominance of phytoplankton from diatoms and cryptophytes to cyanobacteria. At the same time the dominant zooplankton changed from the large-bodied *Daphnia galeata* to the small-bodied *D. cucullata*. In contrast the small-bodied *Cyclops kolensis* was replaced by the larger *C. vicinus*, which became abundant in winter, a season that in previous years it spent in diapause. The decline in *D. galeata* may have been caused both by a decline in food quality with the increase in cyanobacteria, and an increased predation pressure caused by *C. vicinus*.

It is thought that an increase in radiation intensity and in the depth of penetration of ultra-violet light (UV-B) in lakes may cause damage to planktonic animals. In oligotrophic lakes a substantial mortality of zooplankton was found, while the reproduction of the copepod *Diaptomus* was suppressed down to 6 m (Williamson *et al.*, 1994). It also damages the early life of some fish species (Battini *et al.*, 2000).

To investigate the probable effects of global warming on aquatic invertebrates a small stream was divided longitudinally at source, one channel being subject to a temperature increase of 2–3.5 °C, the other channel acting as a reference. In the experimental channel there was a decrease in the overall densities of invertebrates, especially chironomids, an earlier onset of insect emergence from the stream, increased growth rates, smaller size at maturity and precocious breeding in the amphipod *Hyalella azteca*, and altered sex ratios for the caddis *Lepidostoma vernale* (Hogg and Williams, 1996).

Fish are highly responsive to fluctuations in temperature (p. 208), effects being measured from the biological to the ecological level (Wood and McDonald, 1997). Fish communities are therefore likely to undergo marked changes with global warming. A number of endangered species of fish have very restricted ranges in desert pools and streams and are likely to become extinct with the drying out of their habitat (Carpenter *et al.*, 1992). Elsewhere, warm water species are likely to invade higher latitudes. Mandrak (1989) has predicted that 27 species, mostly Cyprinidae and Centrarchidae, could invade the Great Lakes of North America if conditions warm. The year-class strength of many species is dependent on temperature so that moderate warming may enhance productivity. At the same time there may be local extinctions among some stenotherms, or a northwards shift in distributions, possibly by some 500–600 km (Magnuson *et al.*, 1997). In Lake Constance, Germany, it has been predicted that whitefish (*Coregonus lavaretus*) will increase in numbers in the early stages of global warming, for the growth of the larvae is particularly responsive to temperature. However, whitefish are likely to decline in the later stages, because larvae develop near the lake bottom, where they require good oxygen conditions produced by lake overturn in late winter when the water temperature falls to 4 °C. If conditions become warmer, the lake temperature may not decline sufficiently to allow mixing. This will not only have consequences for the whitefish but will also reduce nutrient regeneration to the euphotic zone and hence affect primary production (Trippel *et al.*, 1991).

It is predicted that global warming may result in the extinction of those southern populations of Atlantic salmon (*Salmo salar*) in northern Spain and southwest France. At the northern end of the range, in northern Norway, feeding opportunities will increase so that salmon grow faster and migrate to the sea at a younger age (McCarthy and Houlihan, 1997).

Amphibians also respond to temperature. In southern England, over the period 1978–94, several species were shown to be spawning earlier or arriving earlier at ponds. The spawning time of the frog *Rana esculenta* and the newt (salamander) *Triturus vulgaris* became earlier at rates of 9–10 days per 1 °C increase in temperature (Beebee, 1995).

The sex of some animals, for example certain reptiles, is dependent on temperature during a critical period in the egg's incubation. In cool years more males hatch, in warm years more females. In a six year study on the Mississippi River, Illinois, it was found that hatchling male painted turtles (*Chrysemys picta*) became rarer in years when July temperatures were higher than average. It is suggested that the predicted temperature rises for the next century could be sufficient to eliminate male turtles altogether (Janzen, 1994).

Higher summer temperatures and lower rainfall will result in increased evapotranspiration and reduced soil water content. Many areas of intensive agriculture are already highly dependent on irrigation water and abstractions are likely to increase. Rises in domestic and industrial demand will lead to over-abstraction from both surface and groundwaters. In southeast England the headwaters of many rivers run dry in summer and, in drought years, there may be no surface flow even in winter (Fig. 6.11). Many wetlands may disappear. In more southerly countries the impacts are severe, with dire consequences for the aquatic biota. Indeed a lack of water could lead to major conflicts between countries in the future, the Middle East being one potential flashpoint.

Drought conditions in an upland stream were found to have stimulated the algal biomass, measured as chlorophyll, by 145 per cent, which may provide increased food for higher trophic levels and hence promote productivity within the stream (Freeman *et al.*, 1994). Macrophyte and invertebrate communities in streams appear to recover rapidly from the effects of drought once flows have returned to normal (Wood and Petts, 1994; Holmes, 1999a). In drought conditions, aquatic vegetation was found to spread across the river bed and take up and retain phosphorus (Boar *et al.*, 1995). Everard (1996) has argued that periodic droughts contribute to maintaining diversity because the varying conditions allow a greater range of species to survive, and this in turn will provide the ecosystem with the resilience necessary to survive the effects of long-term climate change.

In drought conditions poor water quality, a decreased oxygen concentration and the drying up of waterbodies can cause severe stress and mortality to fish populations (Everard, 1996). Many young Atlantic salmon were found to remain in shallow areas (riffles) during experimentally produced low flows. While this was

Fig. 6.11 Upper reaches of the River Stort, southeast England, dry because of over-abstraction (photograph by the author).

considered to be advantageous during normal low flows, prolonged drought events, including those predicted for global warming, could have severe implications for the salmon population (Armstrong *et al.*, 1998). In drought years brown trout were found to use deep pools in a stream as refuges, and they were absent from shallow pools where the temperature was above the incipient lethal level. Even in deep pools trout preferred the lower temperatures near the bottom, though some fish moved towards the surface at night when conditions cooled (Elliott, 2000).

As described earlier (Chapter 3) effluents already make up a substantial proportion of the low summer flows of many rivers. Effluents may become more concentrated, affecting all but the most resilient aquatic species. A decline in water quality in many rivers in England and Wales in the mid-1990s was considered to be due to low flows resulting from drought (Environmental Data Services, 1998). Higher concentrations in water of those metals and organic compounds that bioaccumulate could have ramifications throughout the food chain, far beyond the boundaries of the river.

Because streams obtain a substantial amount of their energy and nutrients from the surrounding terrestrial vegetation, any change in the riparian communities caused by global warming could affect the functioning of the stream ecosystem. The replacement of trees by grassland may not only reduce energy inputs but will also reduce shading, elevating water temperatures still further.

Storms are likely to be more frequent and more severe in many areas, leading to a greater risk of flooding. Fish may leave the river during floods, to be stranded and die as waters recede, while others may be washed out to sea in the torrent. Populations of small mammals and invertebrates in the floodplain can be destroyed and crops and livestock lost. The severe floods in West Bengal and Bangladesh in autumn 2000 left some 20 million people homeless.

This increase in extreme events results in the displacement of large numbers of people and in these conditions diseases such as cholera thrive (p. 84). Global warming may also allow the spread of diseases normally associated with the tropics to spread north. Of particular concern is malaria, which currently causes two million deaths annually.

The thermal expansion of water in the oceans, coupled with melting glaciers, is predicted to raise the average sea level by 12 cm by the year 2030 and as much as 80 cm by 2100. Many coasts are protected by sea-walls and the cost of upgrading all of these to cope with the increased tidal flow will be prohibitive. Large areas of semi-natural and agricultural land, claimed from the sea in the past, may therefore revert to salt marsh and tidal lagoons, while increased coastal protection will be targeted at areas of high population density. Low-lying islands, such as the Maldives, and delta areas such as Bangladesh are likely to be completely inundated. The rise in sea level will mean that saltwater will penetrate further upriver than nowadays, resulting in changes in the fauna and flora.

Natural, long-term fluctuations in climate make it difficult to show unambiguously that global warming is currently taking place, though the evidence is strong. Because of the extreme consequences predicted it is essential to follow the precautionary principle and to attempt to reduce emissions of greenhouse gases. The United States is the biggest emitter of greenhouse gases, with some 24 per cent of the total from just 4 per cent of the world population. In contrast the UK emits 2 per cent of the total. The United Nations Framework Convention on Climate Change (UNFCCC) was signed in Rio de Janeiro in 1992 and came into force in 1994, and contracting countries agreed to 'achieve stabilization of greenhouse gas concentrations in the atmosphere at a level that would prevent dangerous anthropogenic interference with the climate system ... within a time frame sufficient to allow ecosystems to adapt naturally to climate change, to ensure that food production is not threatened, and to enable economic development to proceed in a sustainable manner'. In December 1997 in Kyoto, Japan, the UNFCCC agreed legally binding targets for the reduction of greenhouse gases (the Kyoto Protocol). Under the protocol, the UK has agreed to reduce the emission of six greenhouse gases by 12.5 per cent below 1990 levels by 2008–12, and has a tougher domestic goal of reducing carbon dioxide levels to 20 per cent below 1990 levels by 2010. The EU as a whole agreed to reduce emissions by 8 per cent, the United States by 7 per cent and Japan by 6 per cent, while Australia was allowed an increase of 10 per cent. There were, however, several loopholes in the proposals. In particular surplus emissions may be traded with countries whose emissions are at present below the 1990 target. The former Soviet Union has carbon dioxide emissions currently more than 30 per cent below 1990 levels because of economic collapse following its break-up, allowing countries such as the United States to buy emissions credits and comply with their legal obligations without cutting energy consumption and greenhouse gas production at home. Furthermore, countries such as the United States and Canada wanted to count their vast forests as carbon sinks – existing forests contain large amounts of carbon and new plantations will absorb carbon as they grow – so that they could minimize cuts in energy use. A United Nations conference in The Hague in November 2000 attempted to close these loopholes and to develop sanctions for non-compliance in meeting targets but it broke up without agreement. In early 2001 the newly elected President Bush of the United States refused to implement the Kyoto Protocol, to the dismay and anger of many nations, who see the cutting of carbon emissions as critical to the long-term survival of the planet. A treaty was finally signed in Bonn in July 2001, with much watered-down targets and without the United States.

An increase in the generation of nuclear power represents one opportunity for reducing the emission of greenhouse gases but, as discussed above, the potential for devastating accidents makes it unlikely that this route would be publicly acceptable. The development of wind power is a most promising renewable energy source in which some countries, such as Denmark and the Netherlands, have already invested heavily. Hydroelectric, wave and tidal power also have potential for further development but all carry with them environmental concerns. Geothermal energy, utilizing the heat at the centre of the earth, and solar energy, e.g. capturing solar energy through panels to heat buildings, are further possibilities, the latter being cost effective in sunnier regions. The production of biogas from waste (p. 107) and the incineration of waste to produce electricity can also be developed further. Growing energy crops (biomass) to produce electricity or biofuels also has much potential. Scragg (1999) and the Royal Commission on Environmental Pollution (2000) provide more information on these alternative energy sources.

Along with a move towards renewable energy supplies there is considerable scope for reducing energy needs. Buildings can be made far more efficient in their use of energy, especially for heating and cooling, and some of the energy demand can be met within the building itself, for example by installing solar panels. Some 21 per cent of greenhouse gas emissions in the UK are derived from transport. Cars are seen as an essential form of transport to a majority of people in the developed world and, in a democracy, it may be impossible to curb car usage. The development of local integrated transport policies, a taxation system that penalizes vehicles with high carbon emissions, and the formulation of alternative fuel supplies will be necessary to reduce carbon emissions from transport.

This chapter is concerned with the impacts of global warming on freshwaters and one way of mitigating these is to develop measures for water conservation, a subject dealt with in the final chapter. The Royal Commission for Environmental Pollution (2000) considered that the UK must reduce its carbon dioxide levels by some 60 per cent of current levels by 2050 and suggested strategies for achieving this. Arnell (1996) discusses the impact of global warming on river flows and water resources. More general overviews of global warming are found in Graves and Reavey (1996), Houghton (1997) and Jarvis (2000), while the journal *Limnology and Oceanography* **41**(5) contains a series of papers on the likely impact on aquatic ecosystems.

CHAPTER 7

BIOLOGICAL POLLUTION

For thousands of years human beings have introduced species to regions where they do not naturally occur, often in error, often for food or fur or sport, sometimes in an attempt to eradicate other species, but also for cultural reasons, such as gardening. When an introduced species causes harm to an ecological system it can be considered as a biological pollutant under the definition of pollution given on p. 1. However, to describe invasive species as pollutants is perhaps not quite as simple as with the chemicals that we release into freshwaters. People have been colonizers for millennia. Are those species that hitched a lift with them on boats really any different from species that are transported over vast distances across oceans on rafts of vegetation? Our major crops and our livestock species evolved in very restricted areas of the world (Simmons, 1996) but have since been introduced widely over the continents. Their cultivation has led to widespread destruction of natural ecosystems, yet some of the habitats created are prized by conservationists, e.g. grazed chalk grasslands. There are major efforts to conserve weeds of arable crops, undoubtedly introduced with the early cereal seeds, now that herbicides have largely eliminated them. The brown hare (*Lepus europaeus*), introduced to Britain by the Romans, has a species action plan (p. 313) devoted to it because populations have declined. Some exotic species may have a positive conservation value (Lugo, 1997).

Natural habitats on small islands are generally much more vulnerable to invading species than those of continents (Elton, 1958), and lakes that have evolved in isolation over many thousands of years can be considered as islands. Successful invaders may be pioneer species with high reproductive rates, may be genetically variable, with high phenotypic variability, and habitat generalists with broad diets. They are often commensals of humans, succeeding best in habitats disturbed by people, or in early successional habitats with low species diversity, or with no predators, or, for predatory invaders, naïve prey with no experience of predation. Williamson (1996) has suggested that approximately 10 per cent of invaders become established and that 10 percent of established exotics become pests – the Tens Rule.

Table 7.1 Some rules for the initial establishment and long-term integration of invasive species into aquatic environments (from Moyle and Light, 1996).

Establishment	Integration
Most invasions fail	Most successful invasions are accommodated without major community effects
All aquatic systems are invasible	Major community effects are most often observed where species richness is low
Any species with the right physiological and morphological traits can invade, given the opportunity	Long-term success depends on a close physiological match between the invader and the system being invaded
Successful invasions are most likely when native assemblages are depleted or disrupted	Long-term success is most likely in aquatic systems highly altered by human activity
Invasibility of aquatic systems is related to interactions among environmental variability, predictability and severity	Invaders are much more likely to extirpate native species in aquatic systems with either extremely high or extremely low variability or severity

Table 7.1 lists rules that appear to govern invasions into aquatic ecosystems during the initial establishment of species and their long-term integration into the community.

MACROPHYTES

Major problems have been caused by the introduction of aquatic plants. Three South American floating species, water hyacinth (*Eichhornia crassipes*), water fern (*Salvinia molesta*) and water lettuce (*Pistia stratiotes*) have spread through much of the tropics, where they block rivers, interfere with fishing and hydroelectric power generation, shade out native plants, and deoxygenate the water beneath their dense mats. *Eichhornia* first appeared in Australia in 1894 and by 1976 there were infestations in every region except Tasmania. The plant reproduces vegetatively by offsets, doubling its population every six days (Arthington and Mitchell, 1986). Following the closure of the dam on the Zambezi River in Africa in 1959, and the filling of Lake Kariba behind it, *Salvinia molesta* built up large populations, occupying 21.5 per cent of the lake's surface (over 1000 km^2) at its peak in 1962. The introduction of a grasshopper, combined with a decline in available plant growth nutrients as the lake matured, led to a decrease in *Salvinia*; it eventually occupied less than a tenth of the area that it covered at its peak (Marshall and Junor, 1981). More recently a weevil introduced from Brazil has been used throughout the tropics: it can reduce *Salvinia* by 99 per cent within a year (Room, 1990).

Fig. 7.1 Giant hogweed (*Heracleum mantegazzanium*), an invasive plant of watersides and damp meadows (photograph by Chris Gibson).

In the United States there are more than 20 invasive aquatic plant introductions, alligator weed (*Alternanthera philoxeroides*) and purple loosestrife (*Lythrum salicaria*) being especially invasive (Benson, 2000). In Europe considerable concern has been expressed over five invasive alien plants. The swamp stonecrop (*Crassula helmsii*) is an amphibious aquatic introduced from Australia as an oxygenation plant for garden ponds. The species was first recorded as naturalized in England in 1956 and is now widespread. Growing readily from fragments, it is easily dispersed by wildfowl, anglers and indeed anyone dabbling in ponds. Living in both water and the damp margins it forms monocultures, totally suppressing native species within a few years. Most methods of control have proved ineffective, the herbicide diquat offering the best chance of success, though it does, of course, also kill native species (Dawson, 1994; Leach and Dawson, 1999). More recently, another escapee from ornamental ponds, the floating pennywort (*Hydrocotyle ranunculoides*) from North America is invading waterways in southern England. It can grow at 30 cm per day and forms floating mats which can cause localized flooding by choking drainage ditches, and kill fish and invertebrates by de-oxygenating water.

The other three major invasive pests are all species of river banks: Himalayan balsam (*Impatiens glandulifera*), Japanese knotweed (*Fallopia japonica*) and giant hogweed (*Heracleum mantegazzianum*). All have escaped from gardens to form dense stands, making access to the water difficult along extensive lengths of river in some places and they are currently increasing in range in Britain and in continental Europe (Dawson and Holland, 1999; Pysek and Prach, 1995). They suppress the native flora, including grasses, and when they die back in winter, a bare riverbank remains which is susceptible to erosion. The giant hogweed (Fig. 7.1) may reach 5m in height, with leaves of 1 m in length and its sap causes dermatitis if it gets on skin exposed to sunlight. Aspects of the ecology and control of these species are discussed in de Waal *et al.* (1994, 1995).

In Florida a tree from Australia, *Melaleuca quinquenervia*, was deliberately introduced in 1906 to help drain the Everglades. With a rapid growth it is effective at lowering local water tables but has proved to be highly invasive and can outcompete any of the native plants of the Everglades. It produces monocultures of no wildlife value in an area prized for its biodiversity. Some 19 000 ha consist of pure stands of *Melaleuca*, with up to 600 000 ha lightly to moderately infested and, with no natural enemies, it could eventually invade much of southern Florida (Bush, 1997).

Invasions of disease organisms may also cause widespread damage. *Phytophthora* disease of alder (*Alnus glutinosa*), a dominant riverside tree in northern Europe, was first recorded in southeast England in 1993. It has since killed some 66 000 trees (Environment Agency, 2000). Were there to be widespread mortality, the impact of the loss of alders on stream ecology could be severe, for they provide habitats in submerged roots, shading and leaf-fall to fuel food webs, while insects fall from the canopy to feed fish.

INVERTEBRATES

The zebra mussel (*Dreissena polymorpha*) is a small bivalve native to the Black Sea. It has invaded the American Great Lakes from Europe, probably having been brought in with the ballast of a ship, and is one of at least 139 alien species to colonize the Great Lakes since the opening of the St. Lawrence Seaway and the Erie Canal, linking the lakes to the Atlantic Ocean (Mills *et al.*, 1994). First recorded in Lake St. Clair in 1985, it became a major pest in the Great Lakes in only three years, reaching peak densities of over 300 000 m^{-2} and in places making up to 70 per cent of the macroinvertebrate biomass. It is set to spread over much of North America. In the Great Lakes it has clogged water intake and distribution networks, and changed the ecology of the lakes by filtering large populations of plankton and thus transferring nutrients from the water column to the benthos. Water clarity is much greater in the spring, algal production being transferred from the plankton to the benthos, though in some as yet unknown way the mussels cause summer blooms of the cyanobacteria *Microcystis*. *Dreissena* have altered habitats for fish and fouled fishing gear and navigation buoys (Claudi and Mackie, 1994; Nalepa *et al.*, 2000). The cost of mitigating damage in the Great Lakes is so far estimated at US$5 billion. The invasion of zebra mussels into the Hudson River appears to have reduced the mean summer dissolved oxygen in the water by some 14 per cent (Caraco *et al.*, 2000). *Dreissena* has accelerated regional extinction rates of freshwater bivalves by some 10 times. Since 1990, the species has been spreading in the Mississippi River basin, which supports 60 endemic species of mussels, and mass extinctions are predicted (Ricciardi *et al.*, 1998). Zebra mussels have, however, also become an important prey item of several species of diving ducks. Waterfowl populations could therefore benefit from this introduction (Hamilton and Ankney, 1994).

Spencer *et al.* (1991) have described the effects of introducing opossum shrimps (*Mysis relicta*) on fish and their predators. The shrimps were put into Flathead Lake, Montana, to provide additional food for kokanee salmon (*Oncorhynchus nerka*). The shrimps ate the main food of the salmon, zooplankton, without themselves becoming a major part of the fish diet because they retreated to the lake bottom during the day and escaped predation. As shrimp populations rapidly increased, the salmon population sharply declined. This forced bald eagles (*Haliaeetus leucocephalus*) and grizzly bears (*Ursus horribilis*), which exploited the fish as they spawned in the tributaries of the lake, to seek food elsewhere.

Crayfish are considered a delicacy and some species have been widely farmed or introduced to supplement wild fisheries. The noble crayfish (*Astacus astacus*) was a prized species in mainland Europe. Alien crayfish were imported from North America to Italy in the 1860s and with them came crayfish plague, a disease caused by the fungus *Aphanomyces astaci*, to which European species have no immunity. The infected animals show a darkening of the exoskeleton, followed by degeneration of the muscles. In the later stages, the animals often become diurnal, seem to walk on tip-toe, and eventually fall on their backs. The fungus then emerges through the exoskeleton and releases large numbers of spores into the water. The disease spread rapidly through Europe and later the Middle East, largely helped by people moving infected animals and equipment around the continent.

Crayfish were then widely introduced from North America into Europe to compensate for the loss of native species but, carrying plague, they made the situation worse. The striped crayfish (*Oronectes limosus*) was introduced into Germany in 1890 and has since spread through much of mainland Europe, but its flesh is considered inferior to native species. The signal crayfish (*Pacifastacus leniusculus*) was introduced into Europe in the 1960s and the red swamp crayfish (*Procambarus clarkii*) into southern Spain in the 1970s, where it now causes considerable damage to rice fields and river banks.

Crayfish plague was first recorded in England in the early 1980s, and it has resulted in widespread declines in the native or white-clawed crayfish (*Austropotamobius pallipes*), which is now protected by law. Over half of the localities in England and Wales from which the species was known in the 1970s are in catchments from which plague has been reported (Holdich, 1991) and surveys during the 1990s have revealed a marked decline in distribution. The disease almost certainly came in with signal crayfish introduced for farming but which escaped into the river systems and are now widely distributed in southern England. Signal crayfish also eat native crayfish and exclude them from shelter, adding to the problem. Feral populations of Turkish crayfish (*Astacus leptodactylus*), noble crayfish, red swamp crayfish and possibly striped crayfish also exist at a small number of sites in England.

The introduction of crayfish into Europe has not been entirely bad. In southern Spain the red swamp crayfish has resulted both in a thriving export industry in an area suffering acute economic depression (Holdich, 1991), and in a thriving population of otters (*Lutra lutra*) (Adrián and Delibes, 1987).

FISH

Fish have been widely introduced for a variety of reasons – for food, for sport and to control disease vectors such as mosquitoes. They may be accidentally introduced with the ballast water of ships, or they may escape from ornamental ponds, or be released from aquaria. Anglers have been especially remiss by moving fish from water to water, country to country, both for sport and as live bait (p. 311), with complete disregard for the ecological consequences (Maitland, 1995; Litvak and Mandrak, 2000). About 56 per cent of the fish fauna of New Zealand and 32 per cent of the United Kingdom are exotics. Some 114 species of fish have been introduced into Europe and around 61 non-indigenous species have become established in the United States (Benson, 2000).

Successful introductions may have a number of impacts. They may cause local extinctions of native species, they may stunt the growth of native fish through competition, and they may alter the trophic ecology of the habitat. They may also

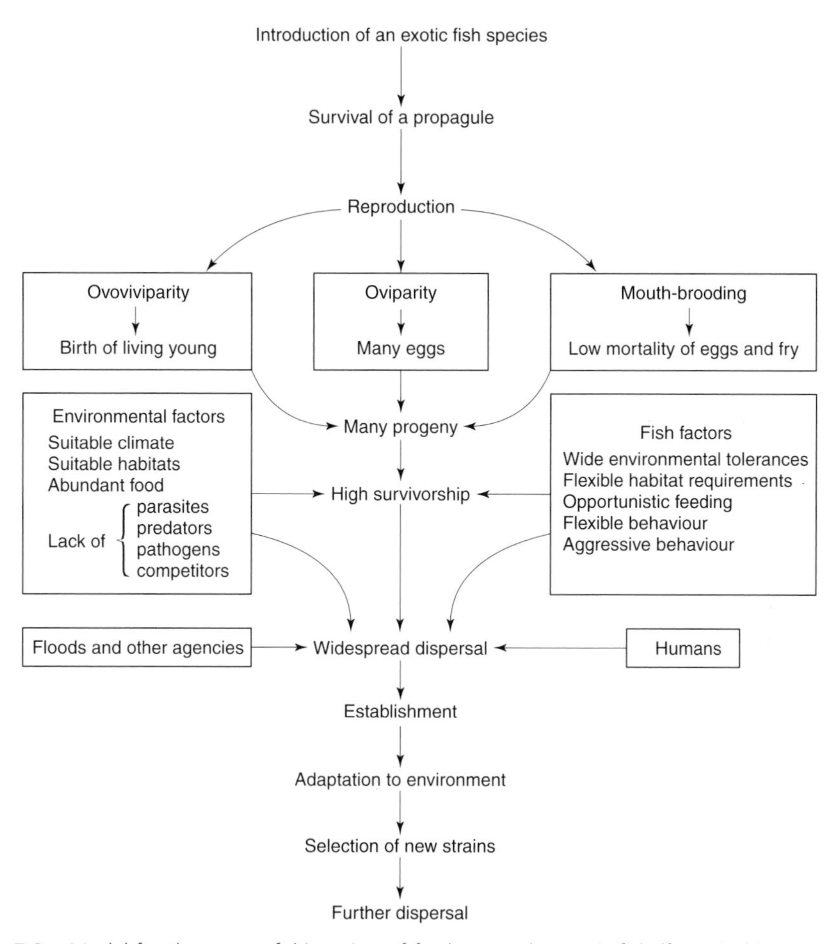

Fig. 7.2 Model for the successful invasion of freshwaters by exotic fish (from Arthington and Mitchell, 1986).

introduce disease. For example, the monogenean parasite *Gyrodactylus salaris* was introduced into Norway with farmed salmon (*Salmo salar*), where it escaped to cause massive mortality in wild populations and in some cases totally eradicated them. The nematode *Anguillicola crassus*, introduced into Europe in Japanese eels (*Anguilla anguilla*) has spread rapidly and had a major impact on eel stocks.

Figure 7.2 presents a general model of how attributes of fish interact with the environment to result in a successful invasion, based on experiences in Australia. Successful species generally produce large numbers of eggs, or produce either live young, or mouth-brooded eggs and young, which increases their survival rate. They are generally tolerant and flexible in their requirements, thriving in an environment free of predators, competitors, parasites and disease. Their dispersal is aided both by floods and human assistance.

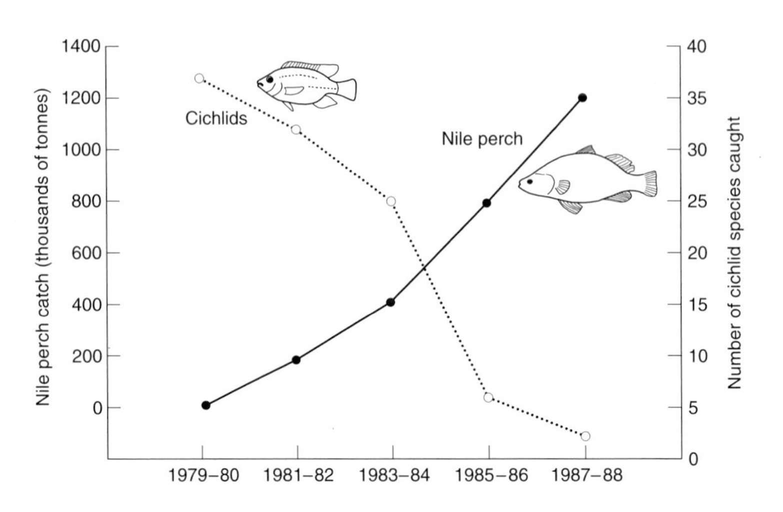

Fig. 7.3 Catch of Nile perch (*Lates*) and number of cichlid species caught in trawls each season in Lake Victoria in the late 1980s (from data in Witte *et al.*, 1992).

A classic example of the ecological consequences of an unplanned invasion is that of the sea lamprey (*Petromyzon marinus*), an ectoparasite of fish that kills its host within a week. It spawns in freshwater. In 1829 a canal was opened linking Lake Erie to Lake Ontario and hence all of the Great Lakes to the Atlantic Ocean, the Niagara Falls previously forming a barrier to navigation. Sea lampreys invaded slowly at first but there was an explosive population increase in the 1930s, causing the commercial catch of lake trout (*Salvelinus namaycush*) to decline more than 30-fold and bankrupting fishing communities in Lakes Huron and Michigan (Mills *et al.*, 1994). This apparently simple effect of predator on prey is probably not the whole story however, for pollutants in the Great Lakes may have played some role in the decline of the trout.

One of the most dramatic changes in biodiversity following the introduction of a fish is seen in the East African Lake Victoria. Up until the 1980s the lake supported some 300 species of endemic cichlid fish. Overfishing of large *Tilapia* species led to a sharp decline in fish yields during the late 1950s. To offset this the Nile perch (*Lates niloticus*), a large voracious predator, was introduced in the early 1960s. The perch increased only slowly at first but there was a dramatic population increase in the early 1980s, concurrent with a marked decline in the number of species of cichlids (Fig. 7.3). Some 200 species have disappeared or become seriously endangered. There were widespread changes in the food web of the lake. The large populations of detritivorous/planktivorous cichlids were severely affected, and there were major losses of those species specializing on fish, molluscs and insects.

However, it is not just simply a case of an exotic predator restructuring the food web. The sudden explosion in the population of *Lates* occurred at a time of increasing algal blooms and depleted oxygen levels at depths below 25 m. It was thought that initially predatory cichlids kept juvenile *Lates* under control. Overfishing of native species both released Nile perch from predation, allowing it to become the dominant predator, and depleted populations of plantivorous cichlids, which accelerated the process of eutrophication. Low oxygen forced the remaining native species into shallower waters, where they were more vulnerable to predation by both perch and fishermen, leading to a further acceleration in the effects of eutrophication – a positive feedback was created, leading to a collapse in biodiversity. A further twist to the story is that a cyprinid, *Rastrineobola argentea*, which feeds extensively upon zooplankton, increased in parallel with the increase in *Lates* and the decline of the zooplankton feeding cichlids. Densities of the larger zooplankton declined, leading to larger blooms of algae and more production descending to the bottom of the lake to add to de-oxygenation problems.

What may be seen as an ecological catastrophe has had benefits to the protein-starved local communities, for the fishery is three or four times larger than it was, and provides much greater employment. It is not known whether the Nile perch population is sustainable in the long term with the changing ecological conditions in the lake. Detailed discussions of the impact of *Lates* in Lake Victoria can be found in Goldschmidt *et al.* (1993), Lowe-McConnell (1994), Gophen *et al.* (1995) and Seehausen *et al.* (1997).

Most introductions have had less spectacular but nevertheless serious impacts on native fish. The predatory zander (*Stizostedion lucioperca*) was first introduced into British waters in 1963 and has since spread. In general, despite fears to the contrary, they appear to have had little effect on fish populations. In canals their impact is linked to the level of boat activity. Where boat traffic is light the water is clear, macrophytes have high density and diversity, and zander do not establish populations large enough to affect fish populations. In heavily used canals the water is murky, there are few macrophyte refuges for small fish, and zander, which are adapted to locating prey in the dark, build up large populations and reduce the density of other species (Smith and Briggs, 1999) – another example of an invader doing well in ecologically degraded conditions.

HIGHER VERTEBRATES

The cane toad (*Bufo marinus*) was introduced into Australia in 1935 to control beetle pests of sugar cane, and it has both spread rapidly and reached higher population densities than in its native South America (Lampo and De Leo, 1998). Its eggs and larvae are highly toxic to predators. Crossland (2000) used replicated artificial ponds to examine the impact of cane toads on the predatory native tadpoles *Lymnodynastes ornatus* and *Litoria rubella*. *Lymnodynastes ornatus* populations survived poorly in the presence of *B. marina*, but this helped the survival of the later breeding *L. rubella*, which are normally eaten in large numbers by *L. ornatus*. There were therefore both negative and positive effects of cane toads on native amphibian populations.

There have been some 72 introductions of waterfowl into the UK, though the

Fig. 7.4 Canada goose (photograph by Sheila Macdonald).

majority have not successfully established. The mandarin duck (*Aix galericulata*), deliberately introduced into England, has filled a vacant niche (a tree-hole nesting species of woodland pools) and this population is probably now of international significance for conservation. The Canada goose (*Branta canadensis*) was introduced in the seventeenth century into waterfowl collections but it was not until the middle of the twentieth century that the population began to grow rapidly (Fig. 7.4). They are said to be aggressive to other waterfowl species but there is little evidence that any population has declined because of them. They cause local contamination of lakes and erosion at places where they gather, primarily where they are fed, but of course people get pleasure in feeding them. They cause local and minor agricultural damage. In many parts of England, however, this is the only large, spectacular flocking species to delight birdwatchers, a benefit much greater than the alleged damage they do.

The ruddy duck (*Oxyura jamaicensis*), a native of North America, was first recorded breeding in the wild in the UK in 1960, having undoubtedly escaped from the collections of the Wildfowl and Wetlands Trust. After a slow start the population has grown to a maximum of 3600 birds (Fig. 7.5). This attractive species has become a popular member of the British avifauna. Ruddy ducks have, however, begun to disperse southwards, being first recorded in Spain in 1983. Here breeds a closely related and threatened species, the white-headed duck (*Oxyura leucocephala*), whose population has been increasing from a low of 22 birds in 1977 to some 2400

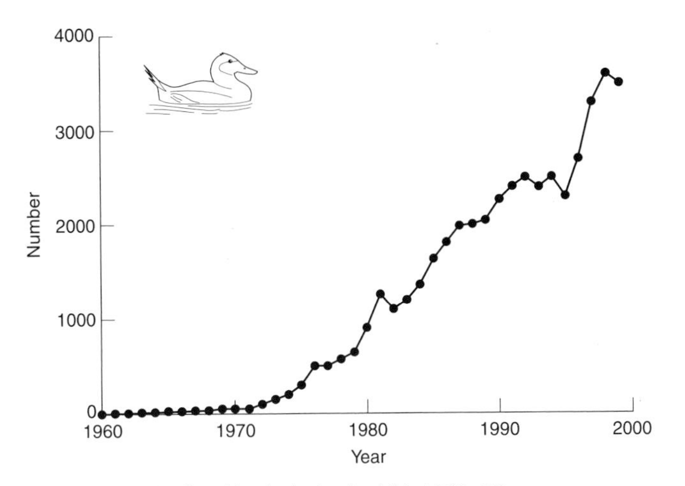

Fig. 7.5 Population growth of ruddy ducks in the UK, 1960–99.

in 2000, due to considerable conservation effort in Spain. The problem is that the species readily hybridize to produce fertile offspring, and female white-headed ducks apparently prefer to mate with male ruddies. The concern is that the white-headed duck will disappear as a species in Spain and that ruddy ducks or their hybrids will disperse eastwards to affect white-headed ducks in their strongholds in Kazakhstan. There is now a European strategy to eradicate the ruddy duck from the continent.

The eradication programme raises some interesting points. DNA evidence suggests that the two species have been separated for at least two million years and hence are separate species, but the ducks clearly do not recognize this. There is no clear-cut definition of a species, except perhaps that of biological species, which do not produce fertile offspring if they reproduce together. In any event, the white-headed duck genes will survive within the hybrid. The initial clamour for a cull was led by powerful bird conservation organizations but once the British government, reluctantly, had accepted the need for one, the conservation organizations have been strangely silent. There has been an air of secrecy about the whole matter and discussion appears to have been effectively suppressed within their magazines and newsletters (Lawson, 1996, 1998), presumably because of the distaste felt by very many of their members for this kill. To many, this is an ethical, not just a conservation issue. The majority of British ruddy ducks live among flocks of other waterfowl in protected areas, where no shooting is allowed. Conservationists have always been strong proponents of the polluter pays principle (p. 303). There is no doubt from where the British ruddy duck population originated, a conservation organization, and the general dangers of non-native invaders had been cogently discussed (Elton, 1958) at the time of their escape, but no action was taken when it could have been both successful and non-controversial. The control programme has an estimated cost of £10 million, to be borne by the taxpayer, not the polluter. It would seem much more cost-effective to target action at those ruddy ducks, if it is deemed necessary, that cross into the Iberian peninsula.

Of 23 species of mammals in Europe that are more or less associated with water, seven are introduced, all but one (Chinese water deer, *Hydroptes inermis*) due to the activities of the fur industry (Mason, 1995). Three are considered to have become pests – coypu (*Myocastor coypus*), muskrat (*Ondatra zibethicus*) and American mink (*Mustela vison*).

In Britain, the coypu had established a feral population by 1944 and quickly became widespread in eastern England, where it caused damage to river banks and crops, as well as to the indigenous wetland flora. Systematic control began in 1962 and an eradication programme was started in 1981, which exterminated the coypu population by 1989 (Gosling and Baker, 1991). Muskrats established feral populations in England but were quickly exterminated. Both species persist on the continent of Europe.

The American mink was first recorded breeding in the wild in Britain in the 1950s and the species is now widely distributed along watercourses, as it is in a number of areas on the European mainland. Mink have been accused of depredations on livestock, on native wildlife, including competition with otter (*Lutra lutra*) and European mink (*Mustela lutreola*), and on fisheries. The occurrence of remains of domestic animals and reared game in the diet of mink is rare. Their impact on native prey, including fish, is likely to be slight (Dunstone, 1993). The one exception is the water vole (*Arvicola terrestris*), whose population appears to decline in the face of expanding mink numbers. The water vole used to be a common sight along waterways but in many areas it is now very rare. There is little doubt that mink have precipitated the decline. For example, a population (already small) of 20 voles on a 1.6 km stretch of a tributary of the River Thames disappeared within a year following the colonization by mink (Barreto and Macdonald, 2000). However the ultimate cause of the local extinction of water vole populations appears to be agricultural intensification, with the conversion of old grasslands into improved grasslands or arable land, leading to a loss of structural diversity. Added to this is the channelization of rivers. Water voles are forced to live in a broken, narrow belt of degraded riverside habitat, where they are highly vulnerable to predation or other adverse influences (Barreto *et al.*, 1998). Habitat restoration is essential to reverse the decline in vole numbers: in a diverse habitat, water voles and mink should be able to coexist.

Claudi and Leach (2000) provide a collection of recent papers on introduced freshwater organisms in North America.

GENETIC POLLUTION

The ruddy /white-headed duck hybridization described above is an example of genetic pollution but in this section the impact of genetically modified organisms on populations is considered. The current concern is with the effects of farmed salmon on their wild cousins. Wild salmon are known to have excellent homing abilities, returning to their natal stretch of water to breed after having grown and matured at sea. These reproductively isolated populations have become genetically fine-tuned to their particular environment. Salmon are farmed in cages in sea lochs and other sheltered marine environments. Cross-breeding of salmon stocks takes place to produce fish suitable for the industrial farming which takes place in cages. A rapid growth rate, large harvestable size, resistance to disease and a placid nature are all desirable characteristics. Unfortunately escapes from these farms are all too frequent, while some deliberate stocking takes place, so that the salmon in some rivers are predominantly derived from farmed stock. Interbreeding between wild salmon and farmed stock is likely to upset the genetic integrity of native populations, and may be one reason for their current decline. Following an escape of farmed salmon into a river in Northern Ireland in 1990, the wild population was found to be genetically very different from the pre-escape population. The later presence of an allele not previously detected indicated that further escapes of farmed salmon had taken place (Crozier, 2000).

Of even greater concern is likely to be the development of genetically engineered or transgenic fish, that is fish that have been modified by genes from another species. While legislation may insist on their secure maintenance, experience of escapes from fish farms makes their eventual appearance in the environment almost inevitable.

CONCLUSION

At the outset of this chapter, it was suggested that, if humans are considered as part of the ecosystem, then introductions may be little different from natural colonization events. Most species fail to establish in new environments and many that do have little ecological impact. However, some develop into major pests having enormous effects on species, ecosystems and economics. The effects of biological pollutants differ from most other types of pollution in one important aspect: once established it will be almost impossible to eradicate them. In some instances biological control may work, as with *Salvinia* described above, but such control is usually effective only in simple ecosystems. Most aquatic systems, even if damaged, are still complex.

The only effective way to limit biological invasions is to legislate to prevent them. Many countries regulate both the import of fish and the movement of native fish between localities, but the rules are easy to circumvent. There seems to be little regulation of the aquarium or pond trade, the latter being the source of many aquatic plant introductions. For those species that surreptitiously hitch a lift, such as zebra mussel, no regulations are likely to be of much use. Further biological pollution by invading species is therefore certain. As I was completing this chapter, the *Guardian* newspaper was reporting the imminent invasion of British rivers by a 'killer shrimp', *Dikerogammarus villosus*. Spreading across Europe from the Black Sea, via the Rivers Danube and Rhine, it apparently has all the characteristics of a successful invader – tolerance of environmental conditions, rapid reproduction, catholic diet – and attacks most invertebrates with impunity, leading to the fear of local extinctions.

CHAPTER 8

BIOLOGICAL ASSESSMENT OF WATER QUALITY

To manage an aquatic environment receiving polluting substances we need information concerning:

- the pollutants entering it and their sources, quantities and distribution;
- the effects of these pollutants within it;
- trends in concentrations and effects, and the causes of these trends;
- how far these inputs, concentrations, effects and trends can be ameliorated and the methods and costs.

The first stage in this management is to carry out a survey, which is *a programme of measurements that defines a pattern of variation of a parameter in the environment*. As an example, we may be concerned about the release of zinc into a river in an effluent from a factory. Our initial survey may involve measuring the concentrations of zinc in the river sediment at a number of stations downstream from the factory, together with sampling the fauna and flora at these stations. The survey will inform us only of the situation at one point in time.

The next stage will be surveillance and research, which will enable us to learn more about a problem before any policy decisions are made. Surveillance is defined as *the repeated measurement of a variable in order that a trend may be detected*. In our example we may measure the concentration of zinc in the sediments at three monthly intervals to determine how sediment loadings vary. Similarly, the animals and plants will be sampled to see if the original observations are repeated. The research function will be to examine the pollution process in more detail, using experimental and analytical techniques. For instance, the survival of fish in concentrations of zinc flowing from the factory could be studied in the laboratory, or experimental streams (mesocosms) might be used to study whole communities. Furthermore, the tolerance of organisms to concentrations of zinc lower than that in the effluent might be studied as a guide to fixing a standard for the zinc level.

From the surveillance and research programme, and taking into account economic considerations, a policy for managing the pollution

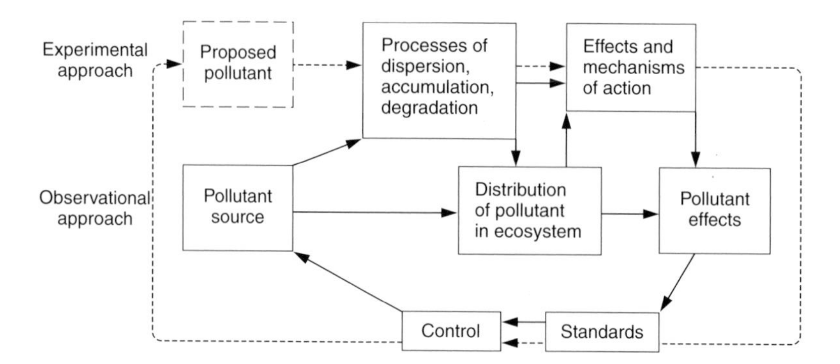

Fig. 8.1 Flow diagram of pollution and its control (from Edwards, 1975).

might be decided. For example, the level of zinc in the effluent might be reduced by 75 per cent by installing a treatment plant in the factory. It will be necessary to see that this reduction has the desired effect – an improvement in the state of the receiving river – and also to ensure that the effluent's quality is maintained. These observations on *performance in relation to standards* are known as monitoring.

It should be noted that these terms are used rather loosely and the boundaries between them are indistinct. Furthermore, there is not necessarily an ordered sequence in the procedure. The monitoring programme might produce useful research data which can be used to redefine the policy, which may then require modifications in the monitoring strategy.

The strategy outlined above for examining pollution includes both experimental and observational approaches (Fig. 8.1). The experimental approach involves the simulation of the pollutant's behaviour and its action on resources in experimental systems, whereas the observational approach examines the distribution of pollutants and their pattern of effects on natural resources. The integration of these two approaches is essential if a management policy for a particular pollutant or pollution source is to be effective. Both approaches have their disadvantages. The experimental simulation does not take into account the complexity of the normal pollution situation in which a variety of factors influence the way a pollutant affects its target. This very complexity, however, makes the interpretation of observational data exceedingly difficult.

WHY BIOLOGICAL SURVEILLANCE?

When carrying out a pollution surveillance programme, why not merely measure the levels of pollution at key points in the transfer pathway, especially if for economic reasons, we are only interested in people and their food supplies? Pollutant levels can be measured at the sites of discharge and the sites of abstraction from the watercourse for potable supply, irrigation etc., and we need not concern ourselves with problems in the river. Moreover, the concentrations of chemicals can be measured more or less accurately and repeatedly, using standard methods. Biological materials are notoriously difficult to sample and are inherently variable. There are, however, considerable advantages in biological surveillance.

Animal and plant communities respond to intermittent pollution which may be missed in a chemical sampling programme. For example, a river may be sampled at station A on the first Tuesday in every month and analysed back in the laboratory for 10 chemical determinands, one of which is zinc. Such a routine may be necessary for logistical reasons where a large number of sites are under surveillance. On the second Wednesday in every month a factory immediately upstream of station A may discharge an effluent containing zinc. By the time the next water sample is taken, the zinc will have disappeared downstream and the chemist will not detect it. A biologist, however, sampling at monthly intervals alongside the chemist, will record an unexpected depression in the diversity of the biological community, as some species may be eliminated and many individuals killed by the zinc. Some of the species missing may be known to be especially sensitive to zinc; they are acting as *sentinels* or *indicators*. As replacement of organisms at station A will take time, being by immigration and reproduction, the polluting incident will be apparent for several weeks or even months after the event.

The chemist may get round the problem of periodic sampling by installing an automatic analyser at the station in permanent contact, via telemetry, with a central control where alarms can sound if significant changes in water quality occur. These are expensive, however, they deal with only a few determinands and they are liable to fail under the often rigorous and unpredictable conditions in the field.

This brings us to the next advantage of biological surveillance. Biological communities may respond to new or unsuspected pollutants in the environment. It would obviously be uneconomic and impracticable regularly to determine concentrations of the 1500 or so known pollutants, and the water industry routinely tests for about 30. If, however, a change in a biological community is detected and gives cause for concern, then a detailed screening for pollutants, and indeed of chemicals hitherto not considered as contaminants, can be made. The decline in western grebes at Clear Lake, California (p. 70), gave the first evidence that organochlorines were biomagnified in food chains to have detrimental environmental effects.

The example of the grebes demonstrates another advantage of biological monitoring, namely that some chemicals are accumulated in the bodies of some organisms and these levels can reflect the environmental pollution levels. Chemicals may accumulate in the body over long periods of time, whereas at any particular point in time the pollutant may be present at too low a level in the water to be detected without the concentration of large volumes of liquid. In this way many organisms may be used in surveillance programmes.

BIOACCUMULATORS AS SENTINEL ORGANISMS

The concentration of pollutants in the tissues of organisms was discussed in Chapter 2 (p. 39). Where pollutant levels in water are very low, the analysis of tissues can bring detection within the scope of most instruments and operators. This is especially advantageous for collaborative surveillance programmes involving laboratories with widely differing facilities and expertise. Whereas the analysis of a water sample will record the level of a pollutant at a particular time, the bioaccumulator will reflect the level of pollution over much longer periods. Furthermore, dried tissues can be stored before analysis, whereas water samples need almost immediate analysis, particularly as transformations of material due to microbial activity may occur. Organisms are often monitored to assess the risk to the public of exposure to material with which they are likely to come into contact, *critical material* such as food. Fish are of particular importance in freshwaters.

Several criteria need to be fulfilled before organisms can be considered as satisfactory sentinels to alert us to the presence of contaminants. They should be relatively sedentary so that they reflect only local pollutant levels. They need to be readily identifiable and in sufficiently large numbers to ensure genetic stability. Ideally, they should also be large enough that low concentrations of pollutants can be detected within individuals, though analysis of combined samples is often satisfactory. The life cycles should be long enough to ensure that there is a good balance of age groups throughout the population during the monitoring period. The degree to which organisms concentrate pollutants will vary with both the pollutant and the species so, for large-scale surveys, a single, widespread species is needed. Although the use of bioaccumulators has considerable appeal, the results are often difficult to interpret. The total pollutant content and concentration may vary with the age of an organism, its size, weight and sex as well as the time of year, the sampling position and the relative levels of other pollutants in the tissues (Phillips, 1993; Phillips and Rainbow, 1993).

In addition to measuring pollutant levels in organisms collected from a site, sentinels can also be placed into the environment in cages and accumulation measured over defined periods of time. This approach can be very valuable in comparing pollution levels between sites and at different times of year.

Bioaccumulators have so far mostly been used to monitor the coastal marine environment because of the economically important fisheries and shellfisheries in these regions. Intertidal seaweeds, such as *Fucus*, and bivalves, especially mussels (*Mytilus* spp.) have received particular study because they satisfy most of the criteria given above.

Macrophytes

For freshwaters, bryophytes (mosses and liverworts) have received special attention. They have a number of advantages for monitoring heavy metal pollution, being easily sampled and identified, abundant and widespread. They are generally tolerant of high levels of pollutants and concentrations of metals in moss tissues correlate strongly with concentrations in water and sediment (Goncalves *et al.*, 1992). Their populations are stable over many years and are homogeneous; once dried, bryophytes can be stored indefinitely. The different moss species are dis-

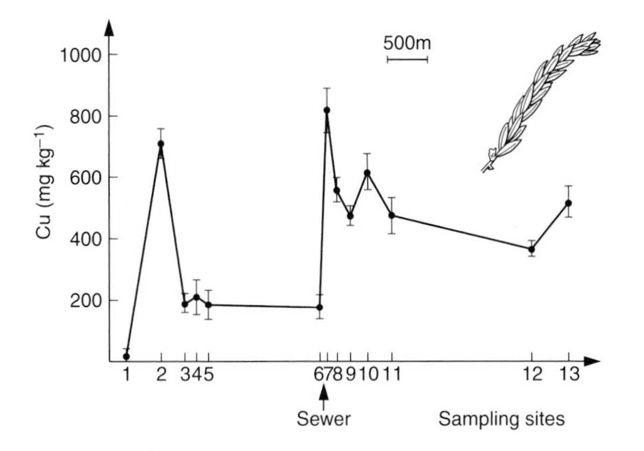

Fig. 8.2 Concentrations of copper (mg kg^{-1} dry weight ± standard error of the mean) in the moss *Cinclidotus* down the River Bienne, France (after Mouvet, 1985).

tributed along a well-defined gradient of immersion. Methods for processing mosses are considered by Wehr *et al.* (1983) and the use of moss bags, in which mosses from unpolluted sites are transferred to sites of interest, is described by Kelly *et al.* (1987), Lopez *et al.* (1994) and Bruns *et al.* (1997).

Figure 8.2 shows the concentration of copper in the moss *Cinclidotus danubicus* down the River Bienne in northern France. Two sharp peaks revealed points of discharge of copper which were hitherto unknown. Further investigations led to the discovery of copper in effluents between sites 1 and 2, while a sewerage discharge between sites 6 and 7 included wastes contaminated with metals from factories upstream.

Metals not detected in water can be measured in bryophytes, suggesting that these plants can be used to monitor episodic pollution. Metals and organochlorines have been shown to accumulate very rapidly following an accident so there is effectively no time lag, while high levels are present in bryophytes up to 13 days after an accident, when the pollutants have effectively disappeared from the water (Mouvet *et al.*, 1993). The analysis of metals (and possibly also accumulating organics) in bryophytes should become routine where discharges of these contaminants are likely.

Mosses are more or less passive samplers. There has been recent interest in using non-biological accumulators for organic compounds. One such is the 'passive, *in situ* concentration-extraction sampler' (PISCES) which has been used to locate point sources of PCBs in the Black River, New York (Litten *et al.*, 1993). The sampler consists of standard plumbing units, sealed with Teflon and filled with hexane. Samplers were suspended in the river at various points and hydrophobic compounds, such as PCBs, were taken up by diffusion, simulating uptake across the gills of a fish. The point source of PCB pollution was the discharge of a paper mill.

The large filamentous alga *Cladophora glomerata* has been suggested as a useful monitor of heavy metals (Whitton *et al.*, 1989) while higher plants have also been used to monitor both heavy metals (Nasu *et al.*, 1984; Reimer and Duthie, 1993) and organochlorines (Lovett Doust *et al.*, 1993, 1994). Higher plants, however, are less ubiquitous than bryophytes, being scarce, for example, in oligotrophic waters or in shade. Having a more complex structure they also exhibit different levels of accumulation between organs as well as marked interspecific differences (Reimer and Duthie, 1993).

Bivalves

Bivalve molluscs have been extensively used for biomonitoring in coastal environments, and to a more limited extent in freshwaters. The small bivalve *Sphaerium corneum* has been utilized to detect the organochlorine pesticide dieldrin, which was used as a moth-proofing agent for textiles and which entered streams in treated sewage effluent. Most of the dieldrin was accumulated by direct partitioning of residues from water into tissue fats, rather than via particulate food, and a steady state was achieved in laboratory studies after a 20 hour exposure at 10 °C, a bioaccumulation factor of 1000 being recorded. The time to attain a steady state decreased as water temperature increased (Boryslawskyj *et al.*, 1987).

The freshwater mussel *Quadrula quadrula* has been used to monitor copper wastes from an electroplating works. Animals were placed in cages at several distances below the point of discharge (Fig. 8.3). At 0.1 km below the discharge the

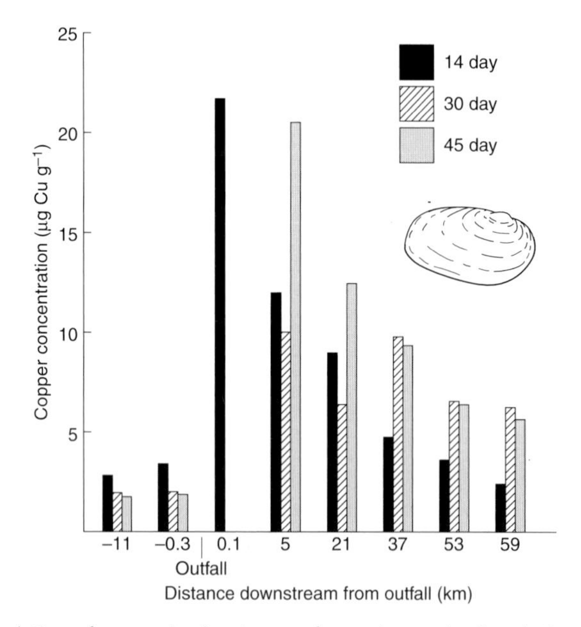

Fig. 8.3 Accumulation of copper in the tissues of caged mussels, *Quadrula quadrula*, exposed in the River Muskingum, United States (from Foster and Bates, 1978).

mussels accumulated 20.6 µg Cu g^{-1} in 14 days and died. The amount of copper accumulated with time decreased at greater distances from the outfall. At no time did the copper in the effluent exceed the legal limits, so the sentinels were indicating how very low levels of the metal were having adverse effects on the stream fauna. The native mussel fauna was absent for 21 km below the outfall because of levels of copper. As mussels are very slow to recolonize, the stability of the native fauna had been seriously altered and recovery would be slow.

The freshwater mussel *Dreissena polymorpha* has been used for many years to monitor heavy metals in the Rivers Rhine and Meuse in the Netherlands. Mussels were both collected on site and transplanted from clean sites to the rivers. Since 1976 cadmium concentrations have markedly decreased in mussels exposed in the Rhine but increased in those in the Meuse. Copper concentrations have not changed (Kraak *et al.*, 1991). The clam *Corbicula fluminea* was transplanted into the River Lot in France at sites below a disused zinc mine. The concentrations in tissues of cadmium, but not zinc, reflected the concentration gradient in the river (Andres *et al.*, 1999).

The crustaceans *Asellus aquaticus* and *Gammarus pulex* have also been used in cages in streams to determine accumulation and heavy metal burdens. Zinc was the most rapidly bioaccumulated metal, concentrations stabilizing at 4–5 times the background level after 5–6 weeks (Shutes *et al.*, 1992).

Fish

Because fish are the main critical material in freshwaters, i.e. they are the major component of the food chain leading to humans, there is clearly much merit in using them as sentinels. The main problem with fish is that they are highly mobile, so that levels of contamination cannot easily be related to specific sources of pollution. Notable exceptions are eels (*Anguilla* spp.), which are generally highly sedentary during their extended period of development in freshwaters, sometimes remaining for up to 20 years before beginning their long journey to oceanic spawning grounds. Eels are carnivores and hence high in the food chain, while they spend much of their time in contact with sediments, from which they may absorb contaminants. These features make eels potentially useful indicator organisms for both metals and organic contaminants (Beumer and Bacher, 1982; Mason, 1987; Brusle, 1991; Pieters and Hagel, 1992). Mercury in eels is generally highly correlated with length, so that standardization of the size of fish is desirable, though there is less correlation between eel length and cadmium and lead concentrations (Barak and Mason, 1990a). Mercury in the flesh represents long-term accumulation and if liver concentrations are higher than those in the flesh, it is likely that the water is severely contaminated (Barak and Mason, 1990b). Cadmium and lead levels in eel livers, but not flesh, also reflect environmental levels of these metals (Barak and Mason, 1990b).

Eels are the only freshwater fish species exploited commercially in significant numbers in eastern England. Most are exported to continental Europe, but in the

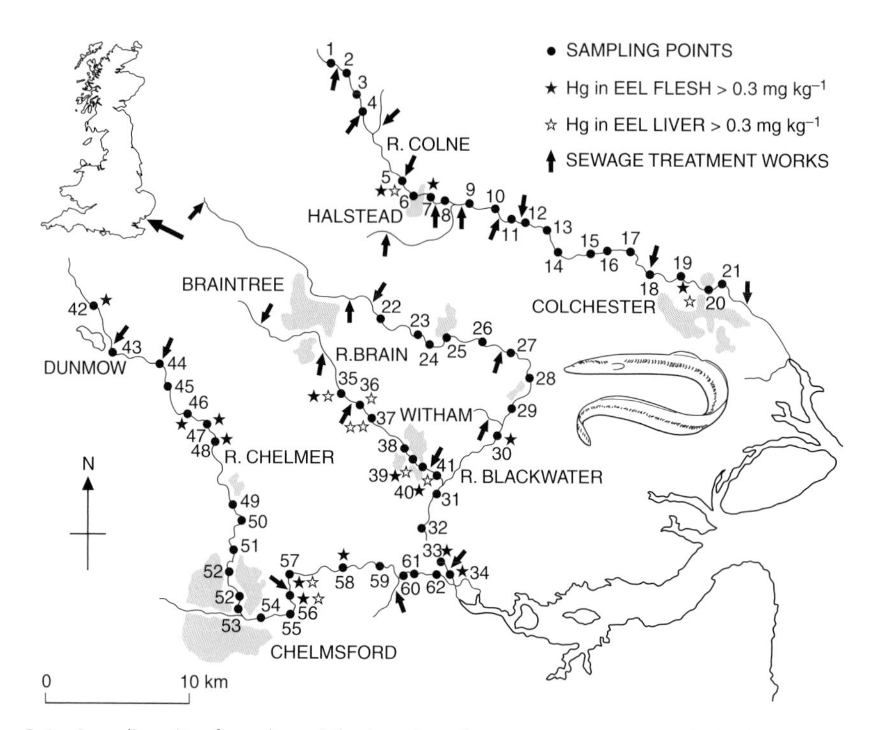

Fig. 8.4 Sampling sites for eels and the location of sewage treatment works in rivers in Essex, eastern England. Sites are indicated where the mean mercury concentration in eel flesh (filled stars) and livers (open stars) was greater than 0.3 mg kg^{-1}.

East End of London and neighbouring seaside towns 'jellied' eels are a local delicacy, eaten in considerable quantities by those who have acquired the taste. Eels were collected from more than 60 sites in rivers draining into the Colne/Blackwater estuary, eastern England, and analysed for heavy metals (Mason and Barak, 1990). Routine analysis of water had shown no problems with metals. Figure 8.4 shows those sites where mean mercury concentrations in eel flesh or liver exceeded 0.3 mg kg^{-1}, the level above which fish are considered unsafe to eat, according to a Directive of the European Union. Seventeen sites (33 per cent) had mean levels of mercury in eel flesh greater than 0.3 mg kg^{-1}. The River Brain was the most contaminated, with concentrations as high as 0.64 mg kg^{-1} in flesh, and high concentrations also in liver, indicating continued chronic pollution. Increased mercury in eels tended to occur below point discharges of effluents from sewage treatment works, some of them serving quite small villages. Eels with high mercury concentrations also tended to have raised concentrations of lead and cadmium. The analysis of eels detected a potential heavy metal problem which had been overlooked by routine chemical surveillance.

Pike (*Esox lucius*) have also been used as biomonitors; as top predators they have the potential to accumulate large amounts of contaminants. Mercury was analysed in pike from 220 Swedish lakes and it was found that concentrations were related to local sources, atmospheric deposition and acidification (Björklund *et al.*, 1984). An analysis, over time, of pike from Swedish waters has shown declines in both DDT and PCBs as the use of these compounds has been restricted (Olsson and Reutergardh, 1986). Bream (*Abramis brama*) has also been suggested as a good indicator of contamination status of freshwaters (Scharenberg *et al.*, 1994).

BIOLOGICAL SURVEY

The objectives of a particular study must be clearly defined before the work programme is begun. A biologist working within the water industry will mainly be involved in large-scale survey, surveillance or monitoring studies, necessitating visits to many, often widely scattered localities during the course of a year, frequently for different purposes. Severe constraints will be placed on the amount of sampling that can be undertaken at any particular site and a compromise will have to be reached between the sampling effort and the amount of data required to produce meaningful results. It will not be feasible to sample the entire biota and a decision must be made as to which group of organisms will provide the most information for solving a particular problem.

Ideally a sample should be compared with detailed information from past samples for that site but frequently the biologist is involved only after a pollution incident has occurred. It may be possible to find an acceptable reference site, for example upstream of the incident, or a local tributary, but the assumption then has to be made that the biological community was identical in both the unpolluted control and the polluted site before the incident took place. As an example, 28 shallow lakes in the Broadland area of eastern England were surveyed and many of them were found to have a depauperate flora and fauna and were suffering severe eutrophication (p. 134) (Mason and Bryant, 1975). Despite the fact that the area had been renowned for decades for its biodiversity and importance for nature conservation, no previous survey had been carried out so that the extent of the ecological degradation had to be pieced together from published natural history notes, unpublished reports and the personal recollections of local naturalists. In no cases were the survey methods of previous observations recorded. Of these 28 broads, 3 were still biologically rich and could be considered as possible reference sites but it was known that in 2 of these lakes marked changes had occurred (severe pH fluctuations in one and changes in the dominance of macrophytes in the other). It was not known whether these changes were due to natural factors or human influences. The biologist investigating pollution events is often faced with these dilemmas.

The ecology and sensitivity to pollution of many of the organisms used in water quality assessments are still very inadequately known, even in those countries with a long tradition of research in freshwater biology. The key organisms in the ecology of one particular catchment may be insignificant in a neighbouring catchment of differing geology or flow regime. Pollutants affect organisms differentially at different life stages or different times of year, while natural changes in population size may underlie any of the effects of pollution. Disentangling the effects of pollution from natural events is often very difficult.

Biologists in the water industry have also to explain their findings to non-biologists. They therefore frequently resort to simple indices of water quality, single numbers which may help in communication but which result in a considerable loss of biological information. The danger is that they themselves may come to rely too heavily on interpreting change in terms of indices, such that more subtle analyses of the data, giving increased biological understanding, will not be attempted. Indeed they may not be allowed time by their managers to undertake more penetrating approaches to the understanding of pollution problems.

Despite these many difficulties, biological assessments of water quality have proved very successful and even mild, intermittent pollution, frequently missed by routine chemical sampling, has been detected. The discussion below will deal chiefly with survey and monitoring techniques, rather than with the undertaking of specific research projects.

SAMPLING STRATEGY

The definition of objectives is the essential first stage in the design of a sampling programme. There are three prime objectives in surveillance and monitoring studies for water quality:

- environmental surveillance, where the objective is to detect and measure adverse environmental changes, such as the effects of unknown or intermittent pollutants, or to follow the improvement of conditions once a pollutant has been removed;
- establishing water quality criteria in which causal relationships between ecological changes and physicochemical parameters are determined;
- appraisal of resources which may involve a large-scale survey to assess general water quality.

Alternatively, particular water resource problems, involving nuisance species, or the impact of new developments, may be investigated. Such problems will also include the management of fisheries or the conservation of threatened species.

In designing a sampling programme it is essential to include randomly allocated replicates and controls. A preliminary sampling should be undertaken to evaluate the sampling design and examine options for subsequent statistical analysis; this often saves considerable time later in the programme by emphasizing design faults at the beginning of the study. The efficiency of the sampling device should be determined at the beginning, as should the size and number of samples in relation to the size, density and spatial distribution of the organisms being sampled.

Surveys may be either extensive or intensive. Extensive surveys aim to discover what species are present in an area, usually with a measure of relative abundance, and are especially used where the water quality over many sites is being monitored or compared. Such surveys have been criticized, or even considered valueless, because they are too superficial to detect or interpret subtle environmental changes, such as alterations in species dominance due to biological interactions, so that it is impossible to disentangle natural changes from those caused by pollution. Such a view is undoubtedly too pessimistic and there is ample evidence of the ability of faunal surveys to detect pollution without a detailed foreknowledge of the ecology of a site. It is true, however, that sampling design is frequently given inadequate consideration in extensive surveys.

Intensive surveys usually aim to determine population densities. The main considerations in designing a quantitative survey are the dimensions of the sampling unit, the number of sampling units in each sample and the location of sampling units within the sampling area. Populations of organisms are usually highly aggregated so that a large number of samples are required to obtain a population estimate that is statistically meaningful (Peckarsky, 1984). For example, 98 samples were required to obtain an estimate of mean density (±40 per cent), with 95 per cent confidence limits, for the limpet *Ancylus fluviatilis* (Edwards *et al.*, 1975). Such sampling intensity would clearly be impossible in an extensive survey and even in an intensive survey the rarer species will be inadequately sampled. Edwards *et al.* (1975) have shown that, in sampling a riffle, only 44 per cent of the species taken were common to both of any two random samples. Nevertheless, for studies of pollution it is the difference between means over time, or the difference between sites, which is important (Resh and McElravy, 1993), so accuracy in determining populations of individual species is not required. Six samples may be sufficient to provide estimates of the total number of individuals in a community with confidence limits ± 40 per cent of the mean (Canton and Chadwick, 1988).

The number of samples required for a specified degree of precision can be readily calculated if an estimate of the mean abundance is made from a pilot survey:

$$D = \frac{1}{x} \sqrt{\frac{s^2}{n}} \tag{8.1}$$

where D is the index of precision, x is the mean, s^2 the variance and n the number of samples. A standard error of ±20 per cent of the mean is usually acceptable for ecological studies so that the number of samples required to obtain this level of precision ($D = 0.2$) is:

$$n = \frac{s^2}{0.2^2 x^2} \tag{8.2}$$

If a series of samples is taken over time it must be remembered that the number of samples required to maintain this level of precision will change as the population size and degree of aggregation changes. For example, on collecting 30 random benthic samples of the tubificid worm *Potamothrix hammoniensis* in a shallow lake in July, the standard error of the mean with a population density of 10 130 m^{-2} was 16 per cent but after the death of adult worms during the late summer, the standard error in October increased to 45 per cent of the population mean of 660 m^{-2} (Mason, 1977a).

A small sampling unit is generally preferable to a large unit because more samples can be handled for the same amount of effort and the statistical error in estimating the mean is reduced. Many small units cover a greater range of habitats than a few large units so that the population estimate will be more representative of the sample area. Sampling units must be selected at random from within the sampling area for the sample to be representative of the whole population. To prevent all of the sampling units falling randomly in one part of the sampling area, or if the location to be sampled shows environmental pattern, stratified random sampling is often used. The sampling area is divided into smaller areas (strata), usually of equal size, and sampling units, divided equally among the strata, are selected at random from within each stratum on each sampling occasion. Sampling design and strategies are discussed by Resh and McElravy (1993) and Waite (2000).

THE CHOICE OF ORGANISMS FOR SURVEILLANCE

It is usually impossible to study the entire biota present in a sampling area because of the constraints of time and of the wide variety of sampling methods required for different groups of organisms. A survey or monitoring programme must therefore be based on those organisms that are most likely to provide the right information to answer the questions being posed.

The use of a single species as a water quality indicator is generally avoided because individual species show a high degree of temporal and spatial variation due to habitat and biotic factors and these confuse any attempt to relate presence, absence or population level with water quality. Similar constraints may limit the value of ratios of species or groups, such as the *Gammarus*/*Asellus* ratio (Hawkes and Davies, 1971). Nevertheless the *Gammarus*/*Asellus* ratio has been shown to perform well when compared with biotic indices (p. 274) in assessing organic pollution and it is extremely simple to use (Whitehurst, 1991). When using a single species as an indicator, considerable care is needed in identification, because similar species may show very different reactions to pollution. For indicator species to be worthwhile, they must be able to register subtle, rather than gross and obvious effects of pollution.

The use of assemblages of organisms allows this more subtle approach. To be suitable for a broad survey or monitoring programme a biological system requires the following features:

- the presence or absence of an organism must be related to water quality rather than to other ecological factors;

- water quality must be reliably assessed and expressed in a simplified form, but the system must be sufficiently quantifiable so that comparisons can be made;
- the assessment should indicate water quality conditions over an extended period of time, not just at the time of sampling;
- the assessment should relate to the point of sampling, not the watercourse as a whole, to locate sources of pollution rather than describe the general water quality of the catchment;
- a minimum of time and manpower should be required for sampling, sorting, identification and data processing.

Numerical abundance at some sites, widespread distribution and a well-documented ecology are also important factors to take into account in selecting a group of organisms for water quality assessment.

Bacteria

Despite the ubiquity of bacteria in aquatic ecosystems and the large populations developed, little attention has been given to the use of indigenous bacteria in the assessment of pollution, with the exception of the biochemical oxygen demand (p. 81). This may be due to the techniques required to enumerate and identify them, though these processes have been made easier by the use of flow cytometry and immunological methods such as ELISA (enzyme-linked immunosorbent assay). Because of their morphological, physiological and genetic characteristics, microbial communities could act as excellent early-warning systems for pollution. In contrast, by the time a change has been detected in assemblages of larger organisms, which react much more slowly to pollution events, it may be too late to reverse the damage. As examples of their role as indicators, bacteria that metabolize petroleum hydrocarbons are ubiquitous in aquatic ecosystems but are present in much higher numbers in environments exposed to oil, while the number of bacteria resistant to heavy metals increases even in the presence of very low concentrations. Genes for resistance to heavy metals are often carried on independent pieces of bacterial DNA, called plasmids, which can be transmitted from one bacterium to another. Because the acquisition of specific plasmids may render bacteria tolerant to several metals, bacteria could prove useful indicators of metal pollution. The use of recombinant DNA technology will considerably aid the use of bacteria in pollution assessment but they will probably be used mostly as biomarkers and biosensors for specific pollutants (p. 57). Scragg (1999) provides an overview of the role of microorganisms in pollution assessment.

Algae

Algae have been popular organisms for the assessment of water quality, especially eutrophication. For lakes, a compound quotient has been developed, where the number of species of Chlorococcales, cyanobacteria, centric diatoms and Euglenophyta are divided by the number of Desmidaceae in a water sample. Values of the quotient less than 1.0 (rich in desmid species) indicate oligotrophy, while values of 5.0 or more are indicative of hypertrophic conditions. In Loch Leven, Scotland, compound quotients of the order of 1.6–1.9 were recorded in the early part of this century but 50 years later, following cultural eutrophication, the quotient had increased to 7.2 (Brook, 1994). The compound quotient may be useful for synoptic surveys but the development of a more subtle index, suitable for the routine management and monitoring of lakes and reservoirs, is proving elusive. Methods for the use of attached, epilithic algae in eutrophication assessments of lakes are being developed (Danilov and Ekelund, 2000).

For assessing the eutrophication status of rivers in Scotland an Algal Abundance Index (AAI) has been developed (Marsden *et al.*, 1997) which may be more widely applicable:

$$\mathrm{AAI} = \frac{2(\text{number of abundant records}) + \text{number of common records}}{\text{number of site visits}} \times 10 \quad (8.3)$$

Abundance or cover is assessed on a semi-quantitative five-point scale – abundant, common, present, rare, absent. The AAI can be related to general levels of pollution, an AAI of less than 20 indicating oligotrophic conditions, while an AAI of more than 70 is associated with hypereutrophication. The method is simple enough to be used alongside routine invertebrate sampling.

More sophisticated approaches have included the development of the Trophic Diatom Index (TDI) (Kelly and Whitton, 1995; Kelly, 1998). In the field the surfaces of five representative boulders are vigorously scrubbed with a toothbrush to remove diatoms. In the laboratory, counts of at least 200 diatoms are made of 86 taxa (genera or species) which are readily identifiable and for which environmental data are available. Each algal taxon is assigned a sensitivity indicator value from 1–5 depending on the concentration of phosphorus in the water at which it was most abundant, and an indicator value from 1–3 which gives greater weight to those taxa known to be good indicators of phosphorus. The TDI is then calculated as:

$$\mathrm{TDI} = (\mathrm{WMS} \times 25) - 25 \quad (8.4)$$

$$\mathrm{WMS} = \frac{\Sigma a_j s_j v_j}{\Sigma a_j v_j} \quad (8.5)$$

where WMS is the weighted mean sensitivity, a_j is the proportion of species j in the sample, s_j is the pollution sensitivity (1–5) of species j, and v_j is the indicator value of species j (Kelly, 1998). The value of the TDI ranges from 0 (very low nutrient concentrations) to 100 (very high nutrient concentrations). There is a good correlation between the TDI and phosphorus concentration but results are influenced by the presence of organic pollution from sewage treatment works. This can be partly allowed for by calculating the percentage of pollution-tolerant diatoms

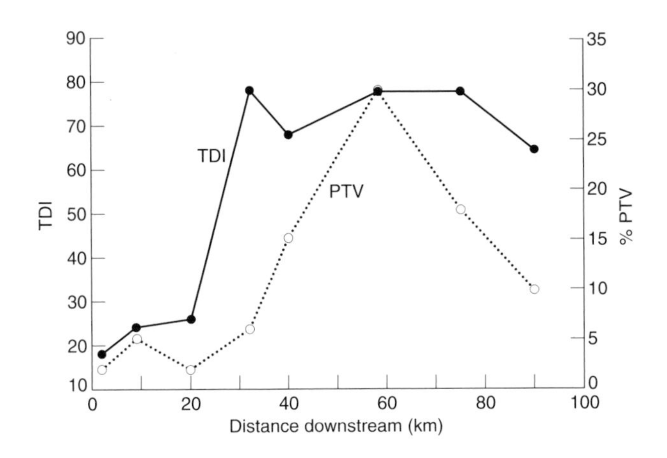

Fig. 8.5 The Trophic Diatom Index (TDI) and percentage of Pollution Tolerant Valves (% PTV) at sites down the River Derwent, Derbyshire, England, in July 1999 (from unpublished data of Mark Barnett).

(valves) in the sample and when this is high, caution is required in interpreting the TDI in terms of nutrients alone. A table of taxon sensitivities and indicator values for calculating the TDI is provided in Kelly (1998). An example of the use of TDI is given in Fig. 8.5. The River Derwent rises in moorland and the TDI was low in the upper reaches. It increased rapidly after the first 20 km due to both the releases of treated sewage effluent from small towns and run-off from livestock production and the river began to eutrophicate. The percentage of pollution-tolerant valves increased to 30 per cent at site 6, indicating a source of organic pollution. Both TDI and PTV were correlated with nitrate levels in the River Derwent, but only PTV was correlated with phosphate concentrations.

Reviews of the use of diatoms in water quality assessment are provided by Round (1991) and Kelly and Whitton (1998).

Macrophytes

Macrophytes are conspicuous and relatively easy to identify in the field and the relationship between water quality and plant distribution has been studied in some depth (Haslam, 1987). In Great Britain a large-scale survey has led to macrophyte vegetation being classified into four broad groups, representing an environmental gradient from lowland eutrophic to upland oligotrophic rivers. Within these there are ten river community types, further divided into 38 subtypes (Holmes *et al.*, 1998). Altitude, gradient and catchment and soil type are the major determinants of the groups and these will, of course, influence the use of vegetation in pollution assessment.

In Britain a method of using macrophytes to indicate eutrophic status of rivers has been developed (Holmes, 1999b). Macrophyte surveys of 100 m lengths of river are made, the cover of each taxon present being scored on a 1–9 scale. Each species on the list is given a species trophic rank (STR) according to its perceived tolerance of nutrient enrichment. Highly tolerant species are given a score of 1, while those highly intolerant of enrichment score 10. A mean trophic rank (MTR) for the site is calculated by multiplying the cover value of each taxon by its STR. The sum of these is then divided by the sum of the cover scores for the site and this score is multiplied by 10 to produce an MTR on a scale of 10–100.

One of the potential problems with using macrophytes as indicators is that they obtain their nutrient supply from both water and sediment. Those species that obtain most of their nutrients from the sediment may respond differently to reductions in nutrient loading, while rivers with similar nutrient loadings but different amounts of sediment may also show different responses to changes in nutrient status as measured by the MTR (Kelly and Whitton, 1998). It is suggested that the inclusion of information on functional groups of macrophytes (root : shoot biomass, total length of shoots, roots etc.) may improve the predictive value of macrophyte indices (Ali *et al.*, 1999). Macrophytes are also seasonal, as well as being strongly influenced by management practices (e.g. for flood prevention) and river use.

Other approaches to using river macrophytes as indicators of water quality have been developed in France (e.g. Robach *et al.*, 1996). Bryophyte communities have also been used (Vanderpoorten, 1999), though an index has not been developed. There are also systems for using macrophytes to assess the water quality of lakes (e.g. Palmer *et al.*, 1992; Lehmann and Lachavanne, 1999).

Fish

Fish, being highly mobile, can avoid intermittent pollution incidents, while their capture requires considerable manpower. Fish populations are widely manipulated by managers. They are, however, easy to identify, their ecology and physiology are relatively well known and, as some are at the top of the food chain, they may reflect changes in the community as a whole. Karr (1991) has developed an Index of Biotic Integrity (IBI) (see below, p. 276) for fish communities in the United States. This includes measures of abundance, total species richness, the numbers of various fish groups (darters, sunfish, suckers), the numbers of sensitive and tolerant species, trophic composition and a measure of fish condition. A total of 12 metrics are given values and added together to produce the IBI, which ranges from 12 to 60 and is considered a measure of the health of the entire fish community.

MACROINVERTEBRATES AND WATER QUALITY ASSESSMENT

In running waters macroinvertebrates have been used most widely for water quality assessments and the remainder of this chapter will be concerned largely with them. There are a number of reasons for preferring benthic macroinvertebrates to other groups (Metcalfe, 1989; Metcalfe-Smith, 1996). The sampling procedures are relatively well developed and need low manpower resources, and there are identification keys for most groups. Macroinvertebrates are reasonably sedentary, with comparatively long lives, so that they can be used to assess water quality at a single site over a long period of time. The group is heterogeneous and a single sampling technique may catch a considerable number of species from a range of phyla, so that at least some species or groups are likely to respond to a particular environmental change. Macroinvertebrates are also generally abundant.

They have some disadvantages. Their aggregated distribution means that, to obtain a quantitative sample of a site, many samples must be taken (p. 261). The muddy, depositing substrata of the lowland reaches of rivers, or of lakes, are often dominated by chironomids and tubificid worms, which are difficult to identify, while the water in these situations is frequently deep, making sampling difficult. The insect species in the community may have emerged as adults and hence be absent for part of the year, so that care needs to be taken in interpreting the results of monitoring.

Sampling for macroinvertebrates

The simplest method of sampling, suitable for shallow waters over eroding substrata, is the kick sample. The operator faces downstream and holds a standard pond net vertically in front, with the bottom against the substratum (Fig. 8.6). The substratum upstream of the net is then vigorously disturbed with the feet and the dislodged invertebrates flow into the net. In shallow waters stones can also be turned over by hand in front of the net. By attempting to disturb a known area, or by kicking for a fixed period of time, this method can be made semi-quantitative for relative abundance estimates. The technique is rapid and inexpensive and is particularly suited to faunal surveys or extensive surveillance programmes. There are obvious potential problems in comparing results between sites of different flow regimes, substratum types and so on, and between individual operators, which will be inevitable in extensive surveys. Furse *et al.* (1981) examined this variation. Using a standard pond net (a frame of 230 mm × 250 mm, with a 900 μm mesh net of depth 275 mm, on a 1.5 m handle) they found that a 3 min kick sample of all available habitats at a site collected 62 per cent of families and 50 per cent of species that could be attained by sampling for 18 min. They considered that their results justified the use of a 3 min kick sample with a standard pond net for extensive surveys. This method is now used routinely in the UK – the operator takes a kick sample for 3 min following a zigzag path across the stream to ensure that all habitats, including those on the margins, are sampled.

A variety of samplers has been designed for the quantitative collection of invertebrates (Hellawell, 1986). The Surber sampler, combining a quadrat with a net (Fig. 8.7a), and cylinder samplers (Fig. 8.7b) are widely used. The Surber sampler consists of a net with a hinged frame attached to its lower margin, which can be

Fig. 8.6 A kick sample being taken with a standard pond-net (photograph by the author).

pushed into the substratum and locked into place. The frame quadrat encloses 0.09 m^2 and stones and gravel within this area are lifted and stirred so that invertebrates are dislodged into the net. Side-wings on the net help to reduce the loss of animals around it.

In shallow waters, open-ended cylinders can be pushed into the substratum to enclose a known area, a typical example being the Neil sampler, which has two openings near the base. The upstream opening is covered with a coarse metal mesh which allows a through-flow of water but excludes drifting invertebrates. The downstream opening has an attached net for collecting animals. The openings of the Neil sampler have sliding doors which are opened after vigorously stirring the enclosed substratum, and the sampler is traditionally made of stainless steel or aluminium. A considerably cheaper version can be made by using a 50 cm length of 25 cm diameter plastic sewer piping. The movable doors can be dispensed with and a

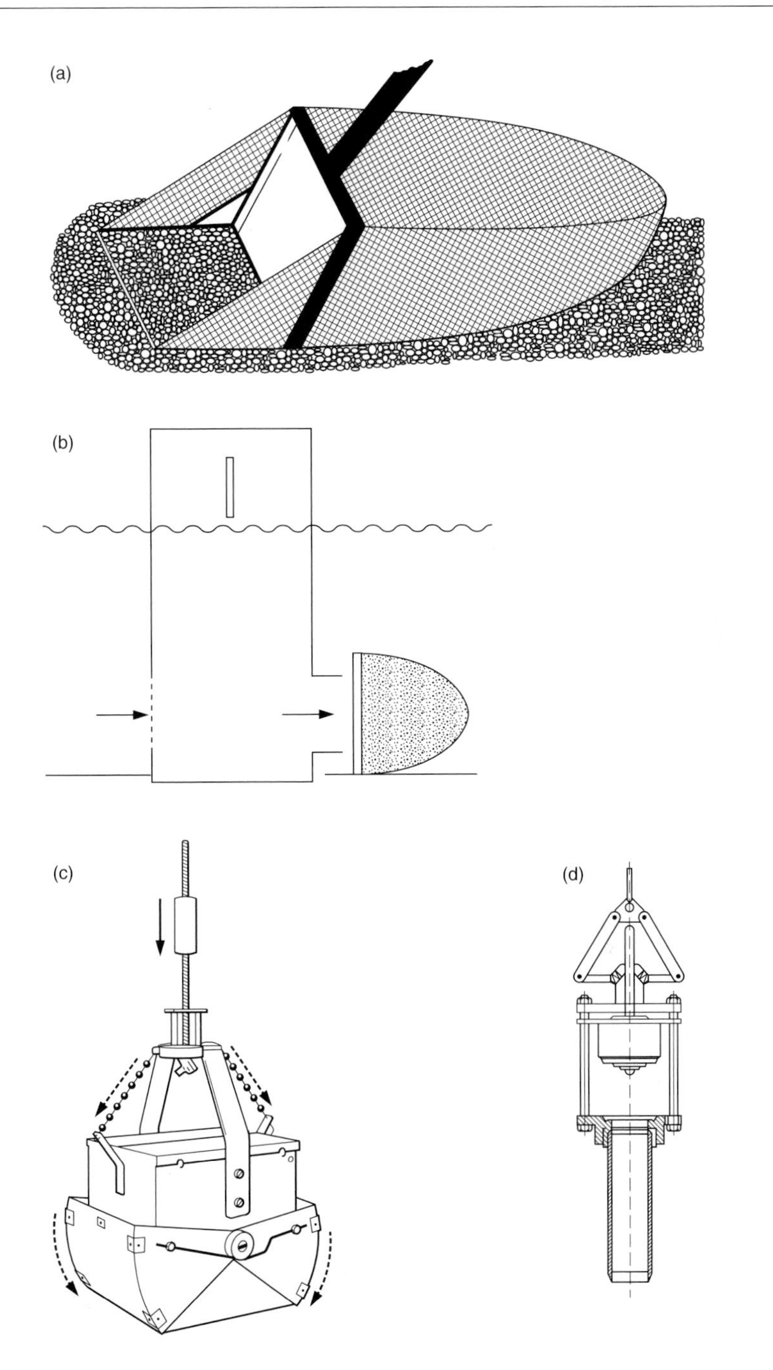

Fig. 8.7 Types of samplers for benthic invertebrates: (a) Surber sampler; (b) cylinder sampler; (c) Ekman grab; (d) corer.

standard pond net can be held over the downstream aperture as the substratum is stirred (Fig. 8.7b).

For shallow, still waters, such as ponds or the muddy edges of reservoirs, a cylinder sampler without inflow and outflow apertures can be used. The sediment may be removed with a plastic beaker or a small, fine-mesh hand net and this method proved highly satisfactory for sampling the benthos of shallow lagoons (Mason, 1986). In water less than 2 m deep, sediment can be collected from a boat using a pond net and this can be calibrated against a standard sampler (Mason, 1977a), which results in a considerable saving of time.

For sampling deeper waters of lakes and rivers a variety of grab and core samplers has been devised (Downing, 1984). The jaws of grabs (Fig. 8.7c) close beneath a known area of substrate, the mechanism being operated using a messenger released from the surface. For very fine sediments, corers (Fig. 8.7d) are usually favoured. These have a smaller diameter than grabs (less than 15 cm) and consist of an open-ended Perspex tube. The top of the tube has a valve which allows the free passage of water during the descent of the corer but which is closed, either automatically or with a messenger, when the sampler is lifted, to prevent the loss of sediment. Multiple units, which take several samples at once, have been developed.

Artificial substrata

Considerable interest has been shown in the use of artificial substrata for the collection of macroinvertebrates. A variety of substrata has been tried (Flannagan and Rosenberg, 1982; Rosenberg and Resh, 1982), including wire mesh trays filled with stones placed on the river bed, or wire mesh baskets filled with crushed limestone (Fig. 8.8a) or bark chippings from coniferous trees, placed on the river bed or suspended at various depths in the water. Artificial substrata may be mimics of plants, consisting of nylon ropes attached to a plastic or wire base. Units of multiple hardboard plates (Fig. 8.8b), suspended in the water, have also been used.

Artificial substrata are, of course, initially devoid of invertebrates, so that it is important for a representative invertebrate community to have developed before they are retrieved. Using modified Hester–Dendy multiplate samplers exposed for 60 days in a stream, it was found that the number of individuals present peaked at 39 days and then declined, whereas new taxa were still colonizing at the end of the exposure period (Meier *et al.*, 1979). Using a similar collector in a canal, the number of species present was found to have increased very little after three weeks (Cover and Harrel, 1978), while, with artificial macrophytes in a shallow lake, an equilibrium community developed after 35 days (Mason, 1978). Species disappeared and new ones were added in substrata exposed for long periods while overall populations increased, affecting diversity. It is generally agreed that an exposure period of six weeks will enable a reasonably representative community to develop.

There are several advantages in using artificial substrata for the assessment of water quality. The samples can be readily processed because there is usually little extraneous material, such as silt. The effect of natural substrata is reduced and the

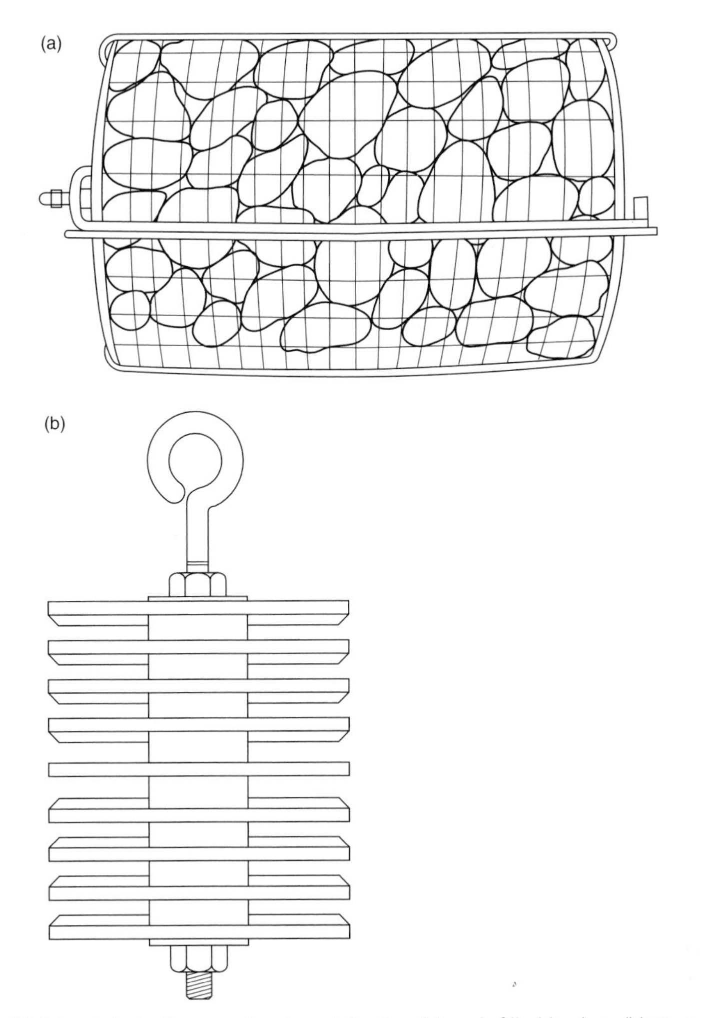

Fig. 8.8 Artificial substrata for sampling invertebrates: (a) rock-filled basket; (b) Hester–Dendy multiplate sampler.

samples can be used in sites where more conventional techniques are difficult to apply, for instance in deep, swiftly flowing rivers. A higher level of precision is obtained and comparison between sites is made easier. Disadvantages are that the sample obtained may not be representative of the community present in the natural substratum of that site, for example the fauna of stony riffles may colonize artificial substrata positioned in mud and substrata may be selective. They only collect invertebrates colonizing during the period of exposure and this makes comparisons with other techniques difficult. They are subject to vandalism and to loss during spate conditions. However, if the aim of a study is to assess water quality, rather than to examine the local macrofauna, then artificial substrata have much to recommend them because, even if the sample collection is richer than in the surrounding river bed, as it frequently will be if artificial substrata are placed on muddy river beds, it shows that a diverse fauna can live in those water quality conditions. As Green (1979) has succinctly remarked, 'the health of canaries can be a good indicator of the safety of coal mines, even though canaries are not natural inhabitants of coal mines'. For extensive comparative surveys, however, it is necessary to standardize the artificial substrata and their placement within a waterbody, but this is easier than with traditional sampling methods.

Drift and emergence samplers

Drift and emergence samplers are reviewed by Peckarsky (1984) and Davies (1984). Emergence traps collect only a proportion of the community (i.e. insects) at certain times of the year, while the emergence periods for different species vary, so the technique is of little use in the assessment of water quality. The collection of drifting invertebrates is subject to similar problems in that different species have different predilections to drift, and daily, or longer-term, variations and other factors make it difficult to relate drift to the macroinvertebrate community in a river bed. One component of drift, however, the exuviae (pupal skins) left behind by hatching chironomid pupae, has been used in the assessment of water quality (p. 120). Exuviae are collected with a pond net from streams or lakes, especially where concentrations of scum or flotsam occur. As surface samples are taken, the type of waterbody does not affect collection. The exuviae are easily separated from detritus by sieving. Identification requires considerable expertise. The technique can provide a rapid biological assessment of an entire catchment from a single set of field collections taken on a single day at a good time of year (late May to September). The method also reveals points of change in the fauna due to changes in water quality. It is most effective during the summer and has little value in the detection of sporadic pollution incidents at other times of year (Wilson, 1994).

Comparison of sampling methods

Some knowledge of the efficiency of the various sampling techniques is essential, especially where comparisons are made of the fauna between sites using different methods. Hughes (1975) compared a Surber sampler, a modified Neil sampler, which was also used in conjunction with an electric shock pulser, and an artificial substratum. The Surber and Neil samplers gave similar results, the electric shock method was highly selective and was the least consistent. The artificial substratum collected the most species and numbers and was the most consistent, but Hughes considered that it failed to represent the fauna in the surrounding river bed and so preferred the Surber and Neil samplers.

When sampling sediments beneath deeper waters, the accuracy of the sampling device will depend on the substrate to be sampled. The efficiency of seven grabs in sampling small plastic pellets (model invertebrates) from different substrata was tested in a large tank. Three grabs had low efficiencies in all substrata. The remainder were adequate in a fine gravel substratum, if the minimum acceptable efficiency was 50 per cent, but only two grabs sampled efficiently when small stones were present and none of the grabs was adequate for sampling when stones greater than 16 mm diameter were present (Elliott and Drake, 1981).

Flannagan (1970) compared a range of core samplers and grabs for sampling sediments beneath deep waters, core samples also being taken by a diver. The Ekman grab and multiple corer gave good quantitative estimates of total biomass when compared with the diver. The Ekman grab collected fewer oligochaete worms and the corer fewer chironomids. Overall the multiple corer seemed to perform better and was the preferred sampler in Hellawell's (1986) review.

Sorting samples

Experienced personnel may sort samples in the field, especially if the taxonomic precision required is low, for example to the family level only, and if estimates of population size are not needed. Data may be entered into a portable PC and indices of water quality could be calculated on site.

Sorting samples in the laboratory, especially muddy sediments, is tedious and time-consuming. Coarse detritus can be removed using a 4 mm sieve, but a smaller sieve to retain invertebrates has to be a compromise between retention efficiency and the time available for sorting, bearing in mind that hand-sorting will be inefficient if large amounts of detritus remain with the sample. Some 87 per cent of chironomids and 82 per cent of tubificid worms were retained by a 500 μm aperture sieve and, using a 250 μm sieve, the retention efficiency was 98 per cent and 99 per cent respectively (Mason, 1977a). The smaller mesh sieves readily become clogged with clay particles so that wet sieving becomes very time-consuming. For routine surveillance programmes a 500 μm aperture sieve should be adequate.

A variety of flotation techniques has been tested to separate invertebrates from the substratum using solutions such as calcium chloride, carbon tetrachloride or sucrose. These are likely to add another source of loss to the sorting programme and can, in themselves, be time-consuming and messy. The most effective method is probably to transfer the contents of a 500 μm sieve, using a gentle jet of water, to a white tray, which has been divided into squares, and then sort each square systematically under a good light. A stain, such as rose bengal, can be added to help sorting, but where possible the sorting of a live sample is best because the movement of animals aids detection.

DATA PROCESSING

Biologists in the water industry, as already stated, must communicate their findings to managers and the public, who are usually unfamiliar with ecological techniques.

A single figure describing the biological impact of water quality at a site is therefore a tempting way to present data and is generally preferred by water managers. However, the use of a biotic or diversity index to describe the biological impact of water quality reduces the amount of information extracted from the data. The ready calculation of indices also tends to replace more sophisticated analyses which would enable the relationships between organisms and the measured physico-chemical parameters of water quality to be better understood. This would eventually place the biological management of freshwaters on a sounder footing. Biotic and diversity indices can be used for overall assessments of water quality and their ability to detect changes in quality will be illustrated later in this chapter, but it must be stressed that information, and indeed the expertise of biologists, is tragically wasted if indices become the be-all and end-all of biological surveillance programmes, as they so often are in the water industry.

Biotic indices

A large number of biotic indices have been developed to assess water quality (Metcalfe-Smith, 1996). A biotic index takes account of the sensitivity or tolerance of individual species or groups to pollution and assigns them a value, the sum of which gives an index of pollution for a site. The data may be qualitative (presence–absence) or quantitative (relative abundance or absolute density). They have been designed mainly to assess organic pollution.

Saprobic Index

The first biotic index, devised in the early years of the twentieth century, was the Saprobien system which recognized four stages in the oxidation of organic matter – oligosaprobic, α-mesosaprobic, β-mesosaprobic and polysaprobic. The presence or absence and relative abundance of indicator species in the zones is recorded.

Table 8.1 The Saprobic Index of Pantle and Buck (1955)

	s value		*h* value
Oligosaprobic	1	Occurring incidentally	1
β-mesosaprobic	2	Occurring frequently	3
α-mesosaprobic	3	Occurring abundantly	5
Polysaprobic	4		

The Saprobic Index ranges are:

1.0–1.5	Oligosaprobic	no pollution
1.5–2.5	β-mesosaprobic	weak organic pollution
2.5–3.5	α-mesosaprobic	strong organic pollution
3.5–4.0	Polysaprobic	very strong organic pollution

The indicator species are mainly microorganisms (bacteria, protozoans, algae and rotifers) but include some macroinvertebrates and fish. A value (h) is given to express the abundance of each organism in the different Saprobien groups as well as a value (s) for the Saprobic grouping (Table 8.1). Sládeček (1973) provides a detailed list of saprobic values. The Saprobic Index is calculated as:

$$\mathrm{SI} = \frac{\Sigma(sh)}{\Sigma h} \tag{8.6}$$

The Saprobic Index and derivatives are widely used in continental Europe (e.g.

Table 8.2 Macroinvertebrate metrics for calculating an Invertebrate Community Index

Category	Metric	Definition	Response
Species richness	Total number of taxa	Measures macroinvertebrate biodiversity	Decrease
	Number of EPT	Total taxa of Ephemeroptera (E) (mayflies), Plecoptera (P) (stoneflies) and Trichoptera (T) (caddis)	Decrease
	Number of Ephemeroptera taxa	Number of mayfly taxa to genus/species	Decrease
	Number of Plecoptera taxa	Number of stonefly to genus/species	Decrease
	Number of Trichoptera taxa	Number of caddis to genus/species	
Composition measures	% EPT	% of this group in collection	Decrease
	% Ephemeroptera	% of mayfly larvae in collection	Decrease
Tolerance/intolerance measures	Number of intolerant taxa	Richness of sensitive taxa	Decrease
	% tolerant taxa	% tolerant invertebrates	Increase
	% dominant taxon	Dominance of most abundant taxon	Increase
Feeding measures	% filterers	% of FPOM filterers	Variable
	% grazers + scrapers	% of this feeding guild	Decrease
Habit measures	Number of clinger taxa	Number of insects which attach to surfaces	Decrease
	% clingers	% clinging insects	Decrease

FPOM, fine particulate organic matter.

Sládeček 1979; Friedrich, 1990; Johnson and Goedkoop, 2000) but the approach has found little support in Britain or North America.

Index of Biotic Integrity

This is the approach being adopted by the Environmental Protection Agency (EPA) of the United States. Biological integrity is defined as *the ability of an aquatic ecosystem to support and maintain a balanced, integrated, adaptive community of organisms having a species composition, diversity and functional organization comparable to the natural habitats of a region* (Karr and Dudley, 1981; Barbour and Yoder, 2000). An Index of Biotic Integrity (IBI) is constructed by combining several biological indicators (metrics) into a summary index of a single number. Generally an IBI is created by combining a minimum of seven metrics from one biological assemblage. Each metric is given a score in relation to an environmental stress, e.g. we might expect species richness to decline from 5 (no pollution) to 1 (severe pollution). The IBI should approximate to a straight line, declining as environmental stress increases.

The use of fish in an IBI has already been described (p. 266). A typical suite of metrics for determining an Index of Biotic Integrity for invertebrates – the Invertebrate Community Index (ICI) – is given in Table 8.2. They include measures of richness, species composition, pollution tolerance or sensitivity, feeding guild and habit. The table includes only the most frequently used metrics for streams. There are others, and different suites of metrics for still waters and wetlands. Accessing the EPA web-site and searching under IBI will provide the reader

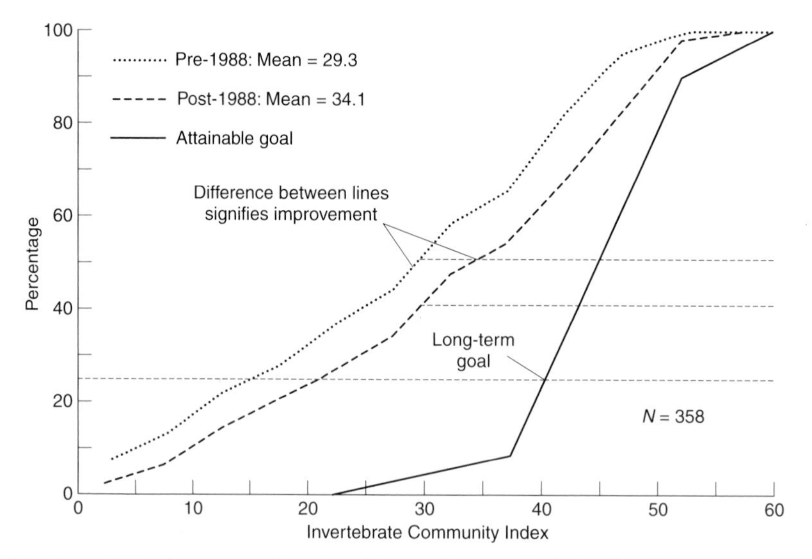

Fig. 8.9 Cumulative frequency of an Invertebrate Community Index for Ohio streams (adapted from Barbour and Yoder, 2000).

with a wealth of detail. Figure 8.9 illustrates cumulative frequencies for an ICI for Ohio streams and shows an improvement in condition (biotic integrity) with post-1988 surveys compared to pre-1988 surveys. Nevertheless the conditions are still short of the long-term goal. Karr (1999) and Karr and Chu (1999) provide reviews of the IBI approach.

BWMP Score

In Britain, the once widely used biotic indices, such as the Trent Biotic Index and Chandler Biotic Score, have been largely superseded by the Biological Monitoring Working Party (BMWP) Score. Its development is described by Hawkes (1997). The approach is much simpler than those discussed above, combining measures of taxon richness and sensitivity to pollution. The BMWP Score was designed to give a broad indication of the biological condition of rivers throughout the United Kingdom. Sites are sampled, where possible, by a 3 min kick/sweep with a standard pond net, all major habitats (bottom substrata, vegetation, margins etc.) within the site being included. Where it is not possible to use nets, artificial substrata can be substituted. Identification of taxa is to the family level only and no account is taken of abundance. Each family is given a score, between 1 and 10, depending on their perceived susceptibility to pollution, with those taxa least tolerant, such as families of mayflies and stoneflies, given the highest scores (Table 8.3). The BMWP Score is the sum of the scores of each family present in a sample. Values greater than 100 are associated with clean rivers, while heavily polluted rivers score less than 10. The total BMWP Score can be divided by the number of taxa to produce the Average Score Per Taxon (ASPT), which is independent of sample size (a larger sample is likely to include more families, thus inflating the BMWP Score if not standardized).

The performance of the BMWP Score using invertebrate samples from 268 sites in 41 river systems in England and Wales, collected using the standard method, has been examined (Armitage *et al.*, 1983). The ASPT was less influenced by season than the BMWP Score, so that samples taken in any season will provide consistent estimates of ASPT. ASPT was also less influenced by sample size than the BMWP Score, so that more information can be obtained for less effort. Physical and chemical data, and physical data alone, were used to predict BMWP Scores and ASPT using multiple regression techniques. On average a higher proportion (65 per cent) of the variance was explained in equations used to predict ASPT than in those used to predict BMWP Scores (22 per cent). The authors suggested that the ratio of observed to predicted ASPT could be used to indicate the possible influence of pollution on the macroinvertebrates. Somewhat different conclusions were drawn from a Spanish study, where ASPT, but not BMWP, was found to be dependent on temperature and hence season. Water quality explained 54 per cent of the variation in BMWP, better than for the English study, possibly because a wider range of water qualities was investigated. ASPT and BMWP were highly correlated (Zamora-Muñoz *et al.*, 1995).

Table 8.3 The Biological Monitoring Working Party (BMWP) Score

	Families	Score
Mayflies	Siphlonuridae, Heptageniidae, Leptophlebiidae, Ephemerellidae, Potamanthidae, Ephemeridae	
Stoneflies	Taeniopterygidae, Leuctridae, Capniidae, Perlodidae, Perlidae, Chloroperlidae	
River bug	Aphelocheiridae	10
Caddis	Phryganeidae, Molannidae, Beraeidae, Odontoceridae, Leptoceridae, Goeridae, Lepidostomatidae, Brachycentridae, Sericostomatidae	
Crayfish	Astacidae	
Dragonflies	Lestidae, Agriidae, Gomphidae, Cordulegasteridae, Aeshnidae, Corduliidae, Libellulidae	8
Caddis	Psychomyidae, Philopotamiidae	
Mayflies	Caenidae	
Stoneflies	Nemouridae	7
Caddis	Rhyacophilidae, Polycentropidae, Limnephilidae	
Snails	Neritidae, Viviparidae, Ancylidae	
Caddis	Hydroptilidae	
Mussels	Unionidae	6
Shrimps	Coriphiidae, Gammaridae	
Dragonflies	Platycnemidae, Coenagriidae	
Water bugs	Mesoveliidae, Hydrometridae, Gerridae, Nepidae, Naucoridae, Notonectidae, Pleidae, Corixidae	
Water beetles	Haliplidae, Hygrobiidae, Dytiscidae, Gyrinidae, Hydrophilidae, Clambidae, Helodidae, Dryopidae, Elminthidae, Chrysomelidae, Curculionidae	
Caddis	Hydropsychidae	5
Craneflies	Tipulidae	
Blackflies	Simuliidae	
Flatworms	Planariidae, Dendrocoelidae	
Mayflies	Baetidae	
Alderflies	Sialidae	4
Leeches	Piscicolidae	
Snails	Valvatidae, Hydrobiidae, Lymnaeidae, Physidae, Planorbidae	
Cockles	Sphaeriidae	3
Leeches	Glossiphoniidae, Hirudidae, Erpobdellidae	
Hoglouse	Asellidae	
Midges	Chironomidae	2
Worms	Oligochaeta (whole class)	1

Various biotic indices have been compared during a biological surveillance programme of a chalk stream in southern England (Pinder *et al.*, 1987; Pinder and Farr, 1987a,b; Pinder, 1989). It was concluded that the ASPT was the best indicator of water quality over the range of conditions encountered. At one site, below a small discharge of partially treated sewage, the ASPT was depressed, while the BMWP increased compared with a site further upstream. It was concluded that the discharge was resulting in mild enrichment of the stream, rather than causing pollution, and that there may be some advantage in evaluating both ASPT and BMWP scores. A Spanish version of BMWP (but not ASPT) was found to be a good indicator of the impact of drainage from coal mines on stream fauna (Garcia-Criado *et al.*, 1999).

The BMWP score has been used to assess the effect of a spillage of chlorpyrifos, an organophosphorus insecticide, on the macroinvertebrate fauna of the River Roding, to the northeast of London. A road accident resulted in the release of about 500 litres of the insecticide into a tributary and, within two days, it had reached the tidal reaches of the river, 26 km downstream. Some 90 per cent of the

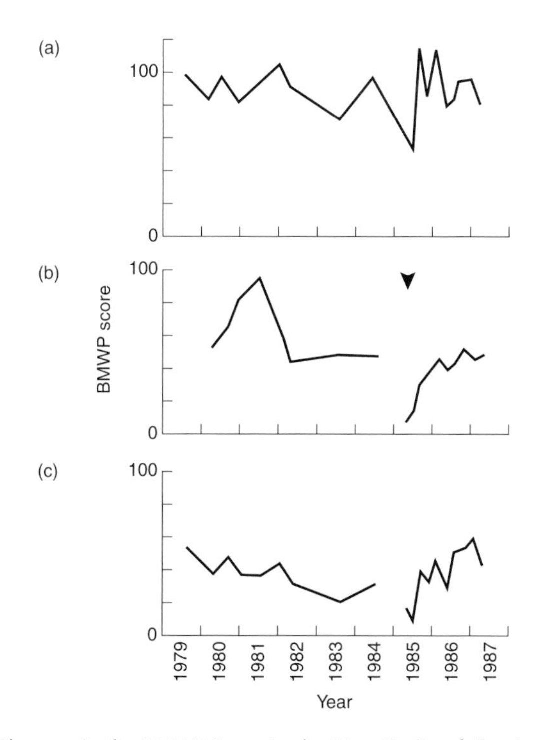

Fig. 8.10 Changes in the BMWP Score in the River Roding following a spill of the insecticide chloropyrifos: (a) control site above the spillage; (b) 4 km downstream; (c) 23 km downstream of the spillage. The arrow marks the spillage (adapted from Raven and George, 1989).

fish biomass and all of the aquatic arthropods were destroyed over a 23 km stretch of river. There was little effect on molluscs and annelids. Figure 8.10 illustrates the BMWP score for a control site and affected sites 4 km and 23 km downstream of the spillage. Soon after the spill, in April 1985, the BMWP score fell sharply at both sites as sensitive arthropod taxa, including those families that contributed most to the score, were eliminated. The effect was most severe at the upstream site (b in Fig. 8.10), which had better water quality before the accident and received the pulse of pesticide in a more concentrated form. The BMWP then gradually recovered as recolonization occurred, largely by drift from unaffected sites above the spillage. Chironomids began to recolonize 13 weeks after the spillage and *Asellus aquaticus* also recovered quickly, probably because of the greatly reduced predation by fish. Most other arthropods had recolonized within 79 weeks, but the beetle *Oulimnius tuberculatus* and the mayfly *Caenis moesta* were still absent 108 weeks after the spillage.

Modifications to the BMWP score and similar methods, including the incorporation of abundance ratings, have been developed (Walley and Hawkes, 1996, 1997; Cao *et al.*, 1997). However, for routine use on a large scale, biotic indices must be simple. Studies have shown that the BMWP score and its derivative, the ASPT, are generally reliable indicators of water quality. How these can be used to develop a water quality index is described on p. 283.

Diversity indices

Biotic indices have been developed largely to measure responses to organic pollution and may be unsuitable for detecting other forms of pollution. Stoneflies, for example, have a prominent role in most biotic indices owing to their great sensitivity to a decrease in oxygen in the water, but they are more tolerant of metals and may be abundant in rivers receiving substantial quantities of metal wastes from mine tips. Diversity indices are used to measure stress in the environment. It is considered that unpolluted environments are characterized by a large number of species, with no single species making up the majority of the community. Maximum diversity is obtained when a large number of species occur in relatively low numbers in a community. When an environment becomes stressed, species sensitive to that particular stress will be eliminated, thus reducing the richness of the community. In addition, certain species may be favoured (e.g. as a result of the reduction of competition or predation) so that they become abundant compared with other members of the community. Most species diversity indices take account of both the number of species in a sample and their relative abundances but the sensitivity of individual species to particular pollutants is not allowed for. Diversity indices are frequently calculated along with biotic indices.

The most widely used indices of diversity are those based on information theory, the most frequent measure being the Shannon index, which assumes that individuals are randomly sampled from an indefinitely large population:

$$H' = -\Sigma \frac{N_i}{N} \ln \frac{N_i}{N} \tag{8.7}$$

where N is the total number of individuals of all species collected, and N_i is the number of individuals belonging to the *i*th species, i.e. the number of individuals in a given taxon is divided by the total number of organisms in the collection, the resulting ratio being multiplied by the logarithm of that ratio; the results of this calculation for each taxon are added together to provide the diversity index. The Shannon diversity index has values of H' which usually fall between 1.5 and 3.5, rarely rising above 4.5 (Magurran, 1988). Lower values of H' are generally characteristic of polluted conditions, where a few pollution tolerant species dominate the community. Higher values are recorded from unpolluted waters. Different base logs may be used in diversity indices so care is needed in comparisons.

The abundance of organisms is obviously important in assessing the effects of pollution, but it can make the interpretation of diversity indices very difficult, especially when water quality over time at a particular site is being examined or when diversity indices from a wider geographical area are being compared from samples taken at different times of year. There may be marked changes in the seasonal abundance of animals. Over two years, Mason (1977b) examined the diversity of monthly samples of macroinvertebrates collected from a hypertrophic lake, devoid of submerged plants, and a clear-water, eutrophic lake with rich macrophyte growth. Diversity was generally lower at the hypertrophic site, but in June of both years the diversity index was lower at the unpolluted site, owing to the presence of a very high population of the chironomid larva *Tanytarsus holochlorus*, which developed rapidly and then emerged from the lake.

When sampling is infrequent, as with most surveillance programmes, the appearance of seasonally abundant species could result in the misinterpretation of water quality conditions using diversity indices. Sampling method, the area sampled, the time of year and the level of identification all influenced the diversity index (Hughes, 1978), while the seasonal variations in the index at a site were found to be greater than differences between sites along a river (Murphy, 1978; Pinder and Farr, 1987a). None of the four diversity indices compared by Pinder *et al.* (1987) was found to produce values that were independent of season, size of sample and level of identification. Such independence is a necessary requirement to facilitate comparisons taken at different times, by different operators at different sites. Extreme care is obviously needed in the interpretation of diversity indices. Mason (1977b) concluded that the number of species (*S*) alone gave a more consistent indication of the difference in eutrophic status of two lakes and Winner *et al.* (1975) drew the same conclusions in a study of streams polluted with copper. A number of studies have suggested that *S* is a better and more realistic indicator of diversity than information statistics (Green, 1979).

Pinder and Farr (1987b) compared the performance of two diversity indices (the Shannon and Simpson) with biotic scores. Contrary to theory, as water quality deteriorated in their study stream, diversity tended to increase. The ASPT, in contrast, decreased in relation to water quality and was recommended as the best method of analysis over both diversity indices and other biotic indices, for it detected relatively small changes in water quality.

Biotic indices, including ASPT, have been designed chiefly to detect organic pollution and there may be some merit in continuing to use diversity indices where other pollutants, such as metal discharges from mines or acidification, are significant. The most frequently used method, the Shannon–Wiener Index (equation 8.7) has been much criticized and Magurran (1988) recommends that it should be replaced by the log series index.

Multivariate analysis

While biotic and diversity indices reduce information from samples into a single value, multivariate techniques retain information on the taxa within a sample, which may be of great importance in determining the real effects of changes in water quality on the community. Multivariate analyses can be carried out on both presence–absence and quantitative data but it has been argued that, as abundance is easily influenced by extraneous factors, presence–absence data give a less ambiguous measure of association. Multivariate techniques can identify discontinuities present within communities which can be related to environmental change. They can be used to generate hypotheses about the causality of distribution but the relationship of distribution to environmental features must then be studied using experimental techniques. Manly (1994) and Waite (2000) provide good general introductions to multivariate techniques in ecology.

Two powerful techniques for analysing community data are detrended correspondence analysis (DECORANA) and two-way indicator species analysis (TWINSPAN). DECORANA is an ordination technique that arranges sites into an objective order, those sites with similar taxonomic composition being placed closest together. An axis score is produced which can be used to relate the ordination to environmental factors. TWINSPAN classification arranges site groups into a hierarchy on the basis of their taxonomic composition, while species are classified simultaneously on the basis of their occurrence in site groups. The technique also identifies indicator species that show the greatest difference between site-groups in their frequency of occurrence. Waite (2000) describes the techniques.

TWINSPAN has been used to classify the running waters of Great Britain on the basis of their macroinvertebrate fauna. Over 600 sites have been sampled using standard kick-sampling methods (p. 277). The DECORANA ordination showed that the highest correlations (Axis 1) were with substratum type and alkalinity, probably representing variations between different river types. Axis 2 correlated with variables related to distance downstream and discharge category, a measure of the long-term average flow of the river. Many rivers showed a progressive change in TWINSPAN group downstream so that TWINSPAN groupings could be assigned to particular stretches, provided they had no atypical physical or chemical features (Wright *et al.*, 1989).

It is possible to predict the probability with which a given species or family will be captured at a particular site using environmental data. A software package, the River Invertebrate Prediction and Classification System (RIVPACS) has been developed (Wright *et al.*, 1993, 1994), currently in its third version of RIVPACS III (Moss *et al.*, 1999; Wright, 2000). The main use is the provision of a target macroinvertebrate assemblage for a site against which the effects of changes in water quality on the assemblage can be assessed. Up to 12 environmental variables are used in predictions of the invertebrate assemblage: distance from source (km), mean substrate type (phi units), altitude (m), discharge category (9 groups in $m^3 s^{-1}$), mean water width (m), depth (cm) and slope ($m\,km^{-1}$), latitude (°N) and longitude (°W), alkalinity (mg $CaCO_3\,l^{-1}$), mean air temperature (°C) and air temperature range (°C). Predictions can be made for three seasons (spring, summer, autumn) separately or combined (winter is excluded because of the difficulties of sampling). The predicted target assemblage of macroinvertebrates can be used to generate expected BMWP or ASPT scores against which to assess the results of field surveys.

RIVPACS forms the basis of a biological general quality assessment (GQA) used by the Environment Agency in England and Wales. A target fauna for a site is set using RIVPACS and the site is then monitored against this target using the GQA. The GQA is determined by combining two parameters, an ecological quality index based on the ASPT measure (EQI ASPT) and on the number of taxa present (EQI N-Taxa). The general quality assessment has six categories (Table 8.4). The GQA is determined from the lowest of the two grades. Thus, if the EQI ASPT was c and the EQI N-Taxa was d, then the biological GQA would be assigned d (fair). It is also possible to include a statement of the confidence in the quality grade assigned to a stretch (Hemsley-Flint, 2000). A description of the biology of these classes is given in Table 8.5. A similar approach is being developed in Australia, the AUSRIVAS system (Davies, 2000). Wright *et al.* (2000) contains a number of recent papers on RIVPACS and similar approaches to pollution assessment.

The use of RIVPACS requires considerable expertise and generally the sorting and identification of macroinvertebrates must be done in the laboratory. In some

Table 8.4 The lower limits of the grades for EQI ASPT and EQI N-Taxa used in deriving the Biological General Quality Assessment for England and Wales.

Grade	GQA limits for EQI ASPT	GQA limits for EQI N-Taxa
a very good	1.00	0.85
b good	0.90	0.70
c fairly good	0.77	0.55
d fair	0.65	0.45
e poor	0.50	0.30
f bad	<0.50	<0.30

Table 8.5 Definitions of the biological GQA grades used in England and Wales

	Grade	Definition
a	Very good	Biology similar to or better than that expected for an average, unpolluted river of similar size, type and location. High diversity of families, with several species in each, and generally no dominance of a single family.
b	Good	Biology shows minor differences from Grade a and falls a little short of that expected of an unpolluted river of this size, type and location. Small reduction in pollution-sensitive families, moderate increase in individuals in pollution-tolerant families such as oligochaetes and chironomids. Indicates first signs of organic pollution.
c	Fairly good	Biology worse than expected for an unpolluted river of this size, type and location, with an absence of many pollution-sensitive families, a reduction in numbers of individuals, and an increase in individuals within pollution-tolerant families.
d	Fair	Big differences from expected for an unpolluted river of this size, type and location, with sensitive families scarce and with few individuals. A range of pollution-tolerant families, some with high numbers of individuals.
e	Poor	Biology restricted to pollution-tolerant families, some families with large numbers of individuals. Sensitive families rare or absent.
f	Bad	Small number of highly pollution-tolerant families only (oligochaetes, chironomids, leeches and *Asellus*) which may be present in high numbers. In severe cases even these may be absent.

circumstances a more rapid appraisal may be preferred. In western Britain there is a widespread problem with organic pollution from livestock farms (p. 83), there being many point sources, often episodic in nature, affecting fisheries and the quality of water used for drinking and recreation. In west Wales TWINSPAN has been used to generate indicator species of organic pollution which have been adopted into a simple flow chart (Fig. 8.11). Invertebrates are collected in a 1 min kick sample and the assessment is made on site. The procedure is simple enough to be used by non-biologists after a short period of training and it allows the rapid pinpointing of sources of pollution with the minimum of resources (Rutt *et al.*, 1993).

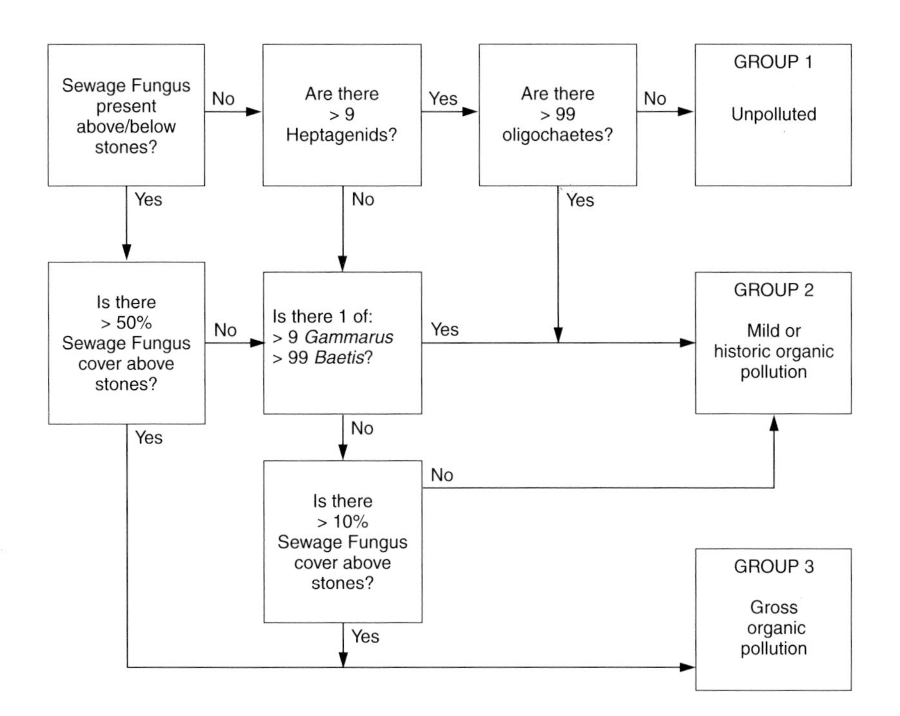

Fig. 8.11 Flow chart, derived from TWINSPAN indicator species, for assessing organic pollution from farms in west Wales (adapted from Rutt *et al.*, 1993).

CONCLUSIONS

Only a few of the many approaches to processing data in routine surveillance programmes have been presented here and new developments and modifications of old methods are continually being tried. There have been attempts to derive universal indices for monitoring water quality and attempts will be made in the future. Such an index would suit the politician, who could compare river health on a national, or indeed continental, basis. But rivers are highly variable and no one index is likely to be universally applicable. It is no use applying a system on a particular river which patently does not work just because it is the nationally, or internationally, agreed method. Indices of water quality will have to be tuned to local conditions. The RIVPACS approach, by comparing a sampled biological community with one predicted from environmental data for a river of that type, size and locality, overcomes many of these problems. Other approaches will be needed for assessing particular pollution problems, for example the use of the Trophic Diatom Index for eutrophication. The meiofauna may be especially useful in assessing metal pollution (Burton *et al.*, 2001). Locally derived indices may be of considerable value for solving local problems, as described above for west Wales.

The methods we have examined relate the changes in the biological community to pollution but there are many natural influences which must not be forgotten. Rundle *et al.* (1995) describe how the liming of acidified streams increased pH and reduced aluminium levels to those typical of circumneutral streams but the macroinvertebrate community did not change in the predicted manner. Colonization by acid-sensitive species was minimal, and recovery did not occur over a period of five years. The reasons were unknown but it seems likely that some other factor on which the invertebrates depend did not respond to the liming and was hence limiting recovery. Natural stochastic events may also have impacts on the river biota which may be much greater than pollution events. A severe flood, for example, may completely alter the bed of the river, move boulders, dead wood and accumulations of sediment and rip out beds of macrophytes.

Multivariate techniques are potentially powerful tools in pollution research and monitoring for they can be used to explore the relationships between biological communities and water quality characteristics, and to generate hypotheses that can be tested experimentally. Such techniques, however, require taxonomic penetration to species level for a wide spectrum of organisms, which is the reversal of current trends in the water industry towards simple biotic indices, and which will require an increased deployment of resources for biological surveillance (Edwards, 1989). Biotic indices are likely, therefore, to remain the major tool in everyday management of water quality.

In conclusion, a biotic index for day-to-day use should have the following characteristics (Extence *et al.*, 1987):

- The system should be based on established methods and it should be possible to calculate results retrospectively for historical data.
- The method should be as simple as possible to use, both in the field and in the laboratory.
- Non-specialists should be able to easily appreciate the meaning of any grading or index rating.
- The index should use as much information as practically possible from the sample, as it is the whole community and not just key groups that respond to variations in water quality.
- The index should ideally be applicable to all river types, whether they be fast or slow flowing, habitat-rich or habitat-poor.
- It should be possible to associate index values with water quality classes, and existing or potential river stretch uses, and thus check for compliance with targets.
- The index should be cost effective.

CHAPTER 9

MANAGING WATER RESOURCES

The evolution of rivers is intimately connected to the evolution of human society. The majority of the capital cities of the world are close to water, and we need water for drinking, for washing, for fish. Waterways were once the main transport routes, because roads were so appalling. Water also provided a main source of power to grind corn and produce flour. But rivers also periodically cause devastating floods so there is a long history of trying to tame them.

The prime task of the water industry, considered here in the broad sense, is to manage the hydrological cycle for the benefit of users. The main sources of water are from reservoirs, and from natural lakes, river flows and groundwater. Other sources, such as desalination of seawater or rainmaking by cloud-seeding, may be important in some parts of the world. The water industry develops these resources and manages them to provide water, wherever possible, in the quantities required by domestic, industrial and agricultural users. It must be of an acceptable quality for these purposes, requiring varying degrees of treatment. Water is also used extensively for recreation and amenity and the water industry is responsible for the development and regulation of these demands. Included here are fisheries. In underdeveloped countries fish may be the prime source of protein to the local community. The conservation of biodiversity should also be a requirement of the water industry.

Society produces a vast array of waste products and water provides an effective means of disposing of many of these. The water industry must ensure that the disposal of wastes causes the minimum of damage to resources. Water may be the source of electric power, using dams and turbines. The water industry is also responsible for reducing damage caused by flooding, which can be highly destructive of life, industry and agriculture. Navigation must be maintained on many watercourses.

Water resources must be managed in a sustainable way, sustainability being the capacity for indefinite continuance (Everard, 2000). Society depends upon nature's *goods* (clean water, food etc.) and *services* (water regulation and supply, waste treatment etc.). The global value of freshwater services is estimated annually in terms of trillions of (10^{12}) US dollars.

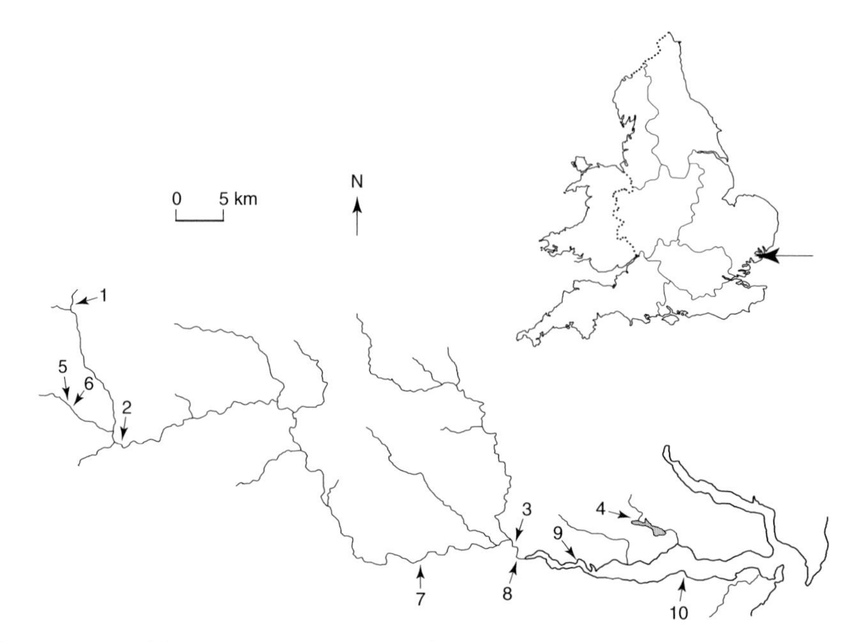

Fig. 9.1 Map of the River Stour – numbers refer to sites discussed in text.

As an example of a managed watercourse we can consider my local one, the River Stour (Fig. 9.1), with a long history of use. The total length of the river and its tributaries is some 340 km, 20 km of which are tidal. By English standards it is a large river but it is tiny on a world scale. The water classification of the river is good (46 per cent), fair (51 per cent) or poor (3 per cent). Most of the river supports good cyprinid fisheries. The catchment is primarily farmland and by the early medieval period it was one of the richest agricultural areas in the world, as can be seen by the large scale of its village churches (Fig. 9.2a). The river has therefore been managed for a very long time almost from source, many of the headwaters being channelized. At Kirtling Green (1 on Fig. 9.1) the river receives water from the Ely–Ouse Transfer Scheme (p. 306), so downstream it is both swifter and deeper than it would previously have been, especially in winter when it is effectively in permanent spate. Some of this water is taken out at Wixoe (2 on Fig. 9.1) and pumped to the top of the River Blackwater in an adjacent catchment. The remainder of the transferred water is abstracted at the lower end of the Stour (3) and pumped, along with water abstracted from the bottom of the Blackwater, to two large water storage reservoirs, which serve the populous south of Essex. There is also a water supply reservoir, Alton Water (4) on a tributary of the Stour, which receives water pumped from an adjacent catchment. Alton Water has been zoned for other uses, including nature conservation, angling and a variety of water sports. There is a cycle path around the perimeter, a café and a number of picnic sites.

The town of Haverhill was developed rapidly in the 1950s and 1960s from a small village to take an overspill population from London. Built in the floodplain, it was prone to flooding, the problem being solved by the development of a flood park (5; Fig. 9.2b) upstream. The tributary itself has been channelized through the town as a concrete drain (6; Fig. 9.2c) – the population effectively turned its back on this troublesome river, an occurrence all too common with urban developments in the middle of the twentieth century. Fortunately this view is changing and many towns are seeing urban regeneration with the river as the central feature, shops, restaurants, housing and parks now facing the water. Rivers bring the countryside into the town and also act as corridors for wildlife.

Industry in Haverhill is also responsible for a sickly, sweet smell from an unknown compound that is apparently formed from the combining of industrial effluents in the sewage treatment works. On some days this smell pervades the length of the River Stour.

An alternative to a floodpark is a flood relief channel (7; Fig 9.2d), which cuts off a meander in the river, flows only when river levels are high, and in this case prevents flooding of the village of Nayland. In other parts of the catchment floods are managed naturally, with extensive areas of water-meadows taking excess water (Fig. 9.2e); they also often provide a public open space

The river has a number of large water-mills associated with it (8; Fig. 9.2f). Usually the wheel was situated in a mill-race, a channel carefully constructed to provide an even flow of water year-round. During low flow, much of the river water passes through the mill-race, while in periods of spate surplus water spills over a weir to follow the main river channel. The mechanisms controlling the water and mill wheels were highly engineered, skills that proved invaluable in the Industrial Revolution – it is no accident that the great textile factories of the Victorian age were known as mills. Because both grain and flour were transported by horse and cart, the pond below the mill was an essential watering-hole. The mills in the Stour valley are nowadays used as educational field centres (as Fig. 9.2f), hotels, or highly sought after private apartments or houses.

The River Stour was once commercially navigable as far as Sudbury but now there are only canoes and rowing boats. Angling and riverside walks are the main recreational pursuits, while there are places close to villages where ducks and swans congregate to be fed (Fig. 9.2g), providing many children with their first experience of wildlife.

Central to our life, rivers also play a major role in our culture. The Stour valley produced three artists of world renown – John Constable, Thomas Gainsborough and Alfred Munnings. The area is marketed as Constable Country, the artist having painted many landscapes of the river valley, and supports a lucrative tourist industry, including many amateur painters. A recent anthology, *The River's Voice* (King and Clifford, 2000), contains almost 200 poems in English from the fifteenth century to the present day, concerned with rivers. There are many similar examples in music. Rivers and the arts enrich our lives.

The tidal reach of the River Stour is bound by the grazing marshes of Cattawade (9; Fig 9.2h), a Site of Special Scientific Interest (SSSI), 1 of 39 in the catchment.

a)
b)
c)
d)
e)
f)

Fig. 9.2 Views of the River Stour catchment. Letters refer to sites discussed in text (photographs by the author and Sheila Macdonald).

However, it currently suffers from a lack of water, caused by abstractions upstream, which cause the lower river to run backwards sometimes during the summer.

The river widens into the 40 km long Stour Estuary (10). Most of the estuary is largely unspoiled (Fig. 9.2i) and holds internationally important populations of waterbirds, totalling some 48 000 in winter. Because of this it has a number of conservation designations – it is an SSSI, it is protected under the Ramsar convention on wetlands of international importance, and under European Union law it is both a Special Protection Area (SPA) and a Special Area for Conservation (SAC). It is also an Area of Outstanding Natural Beauty for its landscape. Despite this its future is not entirely secure. On opposite banks at the mouth of the estuary are the large container ports of Harwich and Felixstowe (Fig. 9.2j), both expanding and taking valuable inter-tidal land. The dredging at the entrance to the port, to allow the largest of container ships to berth, causes slippage of the mudflats upstream and is largely responsible for a 40 per cent loss of salt marsh over the last 25 years. Sea level rising, because of global warming and the natural sinking (isostasis) of the east coast, could threaten the entire estuary in the future. The management activities on the Stour are typical of the majority of rivers in developed areas.

The water industry has little control over some key points in the water cycle, notably precipitation and severe floods. This is not to say that human influence is not profound on these parts of the hydrological cycle. The destruction of forests, particularly those of the tropics, is a case in point. Forests generate much of the rainfall, up to 50 per cent, for example, in the Amazon basin, and much of this precipitation is lost with forest clearance, a process occurring at a rate of some 64 000 to 204 000 km^2 per annum in the tropics. Forests also act as sponges, absorbing water and regulating its release to rivers. Felling results in rapid run-off, followed by drought, so that agricultural areas lower in the catchment alternately suffer devastating floods, followed by periods of water shortage, resulting in tremendous loss of life and agricultural production. Floods erode valuable soil, depositing it in rivers and estuaries to damage fisheries and interrupt navigation. In Nepal, for example, some 37–75 tonnes of soil per hectare are stripped by monsoon rains from deforested mountainsides each year. Around 3 billion tonnes of soil are carried annually by the rivers Ganges and Brahmaputra and deposited in the Bay of Bengal (Lean *et al.*, 1990).

The large-scale destruction of rainforests in the developing world, often by multinational organizations from the developed world seeking short-term profits, may also have major repercussions for world climate. More solar heat may be reflected from land cleared of forests, altering global patterns of air circulation and wind currents and possibly decreasing rainfall in equatorial and temperate lands. Forests also act as sinks for carbon dioxide but, with large-scale burning, they become sources. An increase in atmospheric carbon dioxide will, through the greenhouse effect, raise the temperature of the earth (p. 229). Global warming could be catastrophic for the carefully managed water resources of the developed world, emphasizing that resource conservation must, in the long term, be viewed on a global scale. Some 50 per cent of the world's population is directly affected by the way watersheds are managed.

WATER MANAGEMENT

In England and Wales, the provision of water and the treatment of sewage is carried out by ten water companies, while a number of smaller organizations provide only water. Some of them are now part of much larger, multinational organizations, and they carry out many business activities other than those of water supply and treatment. The water industry is regulated by the Office of Water (OFWAT).

The task of monitoring river quality and setting standards is undertaken by the Environment Agency (EA), established in 1996 by combining several regulatory

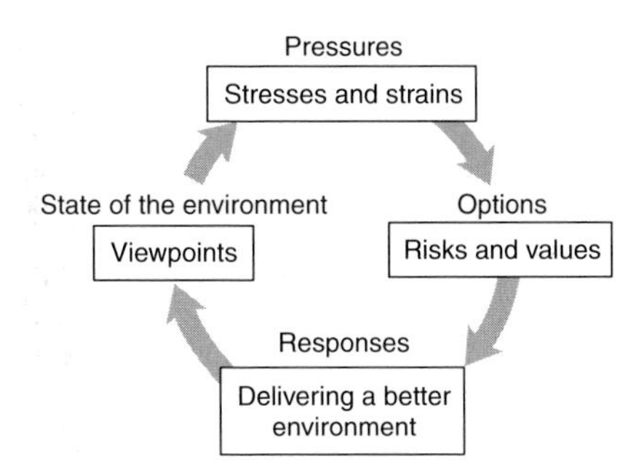

Fig. 9.3 The Environment Agency's four frameworks.

authorities, and under the direction of central government through the Department of Environment, Food and Rural Affairs (DEFRA). The principle of sustainability is enshrined within the Agency's vision: *a better environment in England and Wales for present and future generations.* Following on from the 1992 Earth Summit in Rio de Janeiro, the philosophy is to think globally but act locally. The Agency's strategy identifies areas for management by reference to four frameworks (Fig. 9.3). The first framework assesses the state of the environment at any one time. The second considers the pressures that are affecting the environment, while the third examines the various options for taking action. The fourth enables the Agency to manage the environment in such a way as to deliver the right responses to improve it. The environmental strategy has 11 themes:

- addressing climate change;
- managing water resources;
- delivering integrated river-basin management;
- flood defence;
- managing freshwater fisheries;
- enhancing biodiversity;
- regulating industry;
- improving air quality;
- managing waste;
- conserving the land;
- business development.

Each of these themes has a number of action points. Those for integrated river-basin management, for example, include improving water quality, improving the quality of river and wetland habitats and landscapes, improving economic and recreational navigation etc. Business development includes involving local communities in the development of local action plans. Some of these themes will be expanded below.

In Scotland the Scottish Environmental Protection Agency (SEPA) fulfils a similar role to the EA. In Northern Ireland the Environment and Heritage Service is responsible for water matters. The European Environment Agency, based in Copenhagen, provides information and analysis in the European environment. One of its tasks is to produce a triennial state of the environment report. Its Internet site gives access to environmental agencies throughout the continent, so that the approaches of different countries to environmental management can be compared.

Water resource management in the United States is complicated. Catchment management is rarely possible because the majority of large rivers cross state boundaries. The use of surface waters in the United States is allocated according to Riparian and Appropriation doctrines. The Riparian doctrine, which applies to states east of the Mississippi River, gives the right to the reasonable use of water, undiminished in quantity and quality, to the owners of land that borders the stream. The doctrine protects owners against unreasonable withdrawals or use of water that may reduce their rights to quality or quantity. The Appropriation doctrine applies in those states west of the Mississippi, where water is generally scarcer, and is based on a first-come, first-served principle, though the water taken must be put to beneficial use. In theory, therefore, upstream users could take so much water that rivers run dry downstream, an all too familiar occurrence in many drier parts of the world where dams are built for hydroelectric power or water supply. There are a very large number of organizations in the United States involved in various aspects of water resources at the national, regional and local levels, leading to many conflicts and inefficiencies in overall resource management. Nevertheless, over 90 per cent of water supply units are under direct municipal management and there are hardly any private sewage treatment and disposal companies. The Environmental Protection Agency (EPA) was established in 1970. It has the responsibility to do research, set standards, monitor emissions and enforce the law on emissions for water, air and solid waste. There are increasingly restrictive standards and stricter enforcement, leading to a greater control over water quality. This is seen by Black (1987) as the only way, in an entrepreneurial society which dislikes legislation and long-term planning, of maintaining a healthy aquatic environment. Gore (1997) reviews water quality legislation and management in the United States.

EA, SEPA and EPA all have comprehensive web-sites where details of their strategies, activities and data can be found.

In developing countries, the most serious threat from water pollution is that to health (p. 84). Uncontrolled industrial discharges are often the first evidence of a deterioration in water quality. Developing countries are keen to attract foreign investment, while manufacturers from the developed world may be lured by the low environmental standards, in addition to cheap labour. A vicious circle therefore develops. Even where quality standards are set, they may not be enforced because the government pollution officers are frequently poorly trained and poorly paid and hence have a very low morale. Similar problems may beset the efficient running of sewage treatment facilities, where these have been built into the municipal sewerage system. Unfortunately often they are not and many large cities pump untreated sewage into rivers. The provision of wholesome water for drinking and washing is an essential first step in the improvement of the quality of life for many communities in the developing world.

BIOLOGISTS IN THE WATER INDUSTRY

The construction of reservoirs, the drainage of land, the building of treatment works and so on, are largely the province of engineers, and engineers once dominated the water industry, initially to the virtual exclusion of other disciplines. Many large-scale works have been executed in the past without advice on the ecological consequences, which have frequently been severe. Large reservoirs today, for example, are usually designed with multiple uses in mind, that is, as well as the generation of electricity or the storage of water, reservoirs can also be used for irrigation, controlling floods, fisheries, wildlife conservation and recreation. Multiple use can help mitigate some of the adverse factors consequent on the building of a reservoir.

A realization of the importance of biologists to the management of water resources has resulted in a marked increase in their employment over the past three decades. There are also, of course, many biologists outside the mainstream water industry whose brief, wholly or in part, is concerned with aquatic resources, such as wildlife or landscape conservation, or fisheries.

There are a number of objectives in the management of water resources in which the biologist has an important role.

- The classification of water resources. Water resources are increasingly classified according to the uses to which they are put and the quality necessary for a particular use must be defined. For example, certain industrial processes, raw water for potable supply and game fisheries require water of the highest quality, whereas low-quality water may be acceptable in a waterway used primarily for shipping. An assessment of the biological resources is necessary in an initial classification.
- The collection of baseline data. These data will allow any changes in water quality caused by the development of a resource to be detected, together with changes that may interfere with the present and planned usage of water.
- Water quality surveillance. This is routine work carried out to determine the effectiveness of wastewater management programmes or carried out in relation to specific uses, such as water abstraction, fisheries or recreation.
- Specific investigations. These may involve determining the effects of a specific pollution incident and the subsequent recovery of the freshwater community, the effect of a new impoundment on water resources downstream, the development of communities within impoundments, and so on.

- Forecasting. The forecasting of changes consequent on variation in the intensity of use of a resource or on altered pollution inputs is essential for the rational exploitation of water resources. The biologist can provide essential input for the development of predictive models for forecasting both in the short and long term.

For the tasks outlined above the biologist will be working as a member of a multi-functional team, but there are two further areas in which the biologist has a dominant role:

- Fisheries. The development and maintenance of commercial and recreational fisheries, the support of angling as a pastime, and the development of other water-based recreation.
- Wildlife conservation. Many aspects of water use, including pollution and various forms of development, are inimical to wildlife. The biologist's role is to assess wildlife resources, predict the impact of development and ensure that vulnerable species are conserved within wetland habitats.

Described below are some activities and problems that require biological skills.

CATCHMENT MANAGEMENT

The most efficient way to manage aquatic resources is at the level of the catchment or river basin, that is, management of the river all the way from its various sources until it discharges into the sea – *integrated catchment management*. In this way resources are better used, for they are considered together in relation to regional needs. Water supplies are rationalized over a larger area, allowing greater integration, while effluent disposal can be considered in relation to water supply. This reduces the likelihood of conflicts between those providing water and those disposing of effluents and it allows the setting of environmental standards that reflect the characteristics of the particular basin. Land-use within a catchment will also markedly influence water and habitat quality within the river and hence its biota (Allan *et al.*, 1997). For example, erosion is much greater from cultivated lands than from natural cover and the increased sediment loads settle as beds of silt within the river, often some distance downstream, changing the ecological communities that occur.

One of the major difficulties with integrated catchment management is that very many rivers cross national boundaries, where poor water management in one country can cause problems in countries downstream (Higler and van Liere, 1997). This is not a problem in England and Wales, where the Environment Agency's approach to catchment management planning is through the Local Environment Agency Plan (LEAP), of which there are some 50. A LEAP:

- provides for the implementation of functional strategies;
- identifies present and future uses;
- sets objectives and standards for each use;

- identifies interaction and potential conflicts between uses;
- sets out an action plan to achieve the agreed uses;
- allocates responsibility for achieving actions together with an investment framework.

The planning process begins with the formation of a multifunctional team under a project manager who identifies all the current and future uses in the catchment. This may result in the publication of a separate Environmental Overview, ideally in conjunction with the relevant local government body. Whether separate or as part of the draft LEAP, the overview will include a detailed statement of the actual conditions within the catchment and how they fall short of expectations. A series of activity plans are then developed, containing options to address environmental shortfalls, with start dates and likely costs. The process is overseen by a local Area Environment Group (AEG), whose members represent a range of interests and who are independent of the EA. A draft LEAP is sent out for wide consultation among catchment user groups and the general public. After taking note of the opinions of consultees, a final LEAP includes the strategy for the catchment and a series of prioritized activities to achieve the strategy. As an example, the South Essex LEAP, covering a catchment of 1841 km^2 on the north shore of the Thames estuary east of London, much of it built-up and industrialized, identified 12 major issues and 31 activities. Wherever possible partner organizations are identified to implement individual strategies. Plans are reviewed annually for progress, with a full review planned every five years.

For LEAPS to achieve success, precise environmental requirements for each activity will need to be determined. This should entail a detailed scientific investigation at the local level. For example, an issue might be the development of a sustainable population of migratory Atlantic salmon (*Salmo salar*). The water quality in the river will need to be good throughout, with high oxygen content and low levels of pollutants. It will require a flow regime to stimulate migration. There will need to be free access from the sea to the spawning grounds, which will require suitable gravel beds. For this one activity it is clear that a detailed scientific study, catchment-wide, is needed, especially if the re-establishment of a salmon fishery, lost through environmental degradation in earlier years, is the aim.

This catchment management approach has an important political and social dimension, accepting that river improvements will not take place if those living within the catchment – the stakeholders – are not involved. However, the ecological basis for any action may be weak. The ecology of a river is intimately bound up with the landscape and land-use processes within its catchment, all of which will differ from catchment to catchment, and to some extent from year to year. Management adds further variability to ecological processes within the river. Every river is therefore likely to be different and will require an ecological study before actions are taken. Very often, however, the only data available are on flow regimes, routine water samples for a limited range of determinands, twice-yearly invertebrate samples from a limited number of sites, and possibly a five-yearly fisheries survey, a cheap and cheerful approach to monitoring. Activities based on such limited information may prove very hit-and-miss. The new Water Framework Directive (see below, p. 302) may considerably strengthen the LEAP process.

Werritty (1997) reviews procedures for integrated catchment management, while Ashby *et al.* (1998) provide an example of stakeholder involvement in watershed management in Colombia. Gardiner (1996) and Schrage and Enderlein (1996) discuss the role of Environmental Impact Assessment in catchment management.

WATER QUALITY STANDARDS

This book has been concerned with the release of pollutants to freshwaters that are potentially hazardous to the environment and ultimately to ourselves. It is unlikely that the discharge of potentially dangerous substances will ever be completely eliminated so that we need to have an assessment of the risk caused by exposure before we can set standards to prevent, or at least minimize, any damage. To make a risk assessment, information is needed on the toxicity of the pollutant under examination and the likely exposure of an organism to it. The procedure may also involve the perceived risk to society and the sensitivity of the ecosystem receiving the pollution. Higher standards may be required for more sensitive ecosystems, or for pollutants thought by society to be especially harmful (Jarvis, 2000).

Management decisions are based on the comparison of water quality data with criteria and standards. The following definitions are generally accepted:

- **Criteria**: scientific requirements on which a decision or judgement may be based concerning the suitability of water quality to support a designated use.
- **Objectives**: a set of levels of water quality parameters to be attained in water quality management programmes, which also involve considerations of costs and benefits.
- **Standards**: legally prescribed limits of pollution that are established under statutory authority.

Figure 9.4 illustrates an approach for deriving water quality standards to protect aquatic life from toxic pollutants. The initial stage is to review the published information on the toxic effects of a pollutant. Laboratory and field studies are kept separate and both are evaluated for scientific competence. The screened information on laboratory effects is used to assess the minimum adverse effects concentration. This is then reduced by an arbitrary safety factor, which takes account of the severity of the effect, the length of the exposure period, and the influence of environmental factors, such as pH. This tentative standard is then compared against field observations and, if it conflicts with them, the data are re-appraised to determine the reason for the conflict. If the conflict cannot be reconciled, the field observations take precedence and a more stringent standard is adopted. The tentative

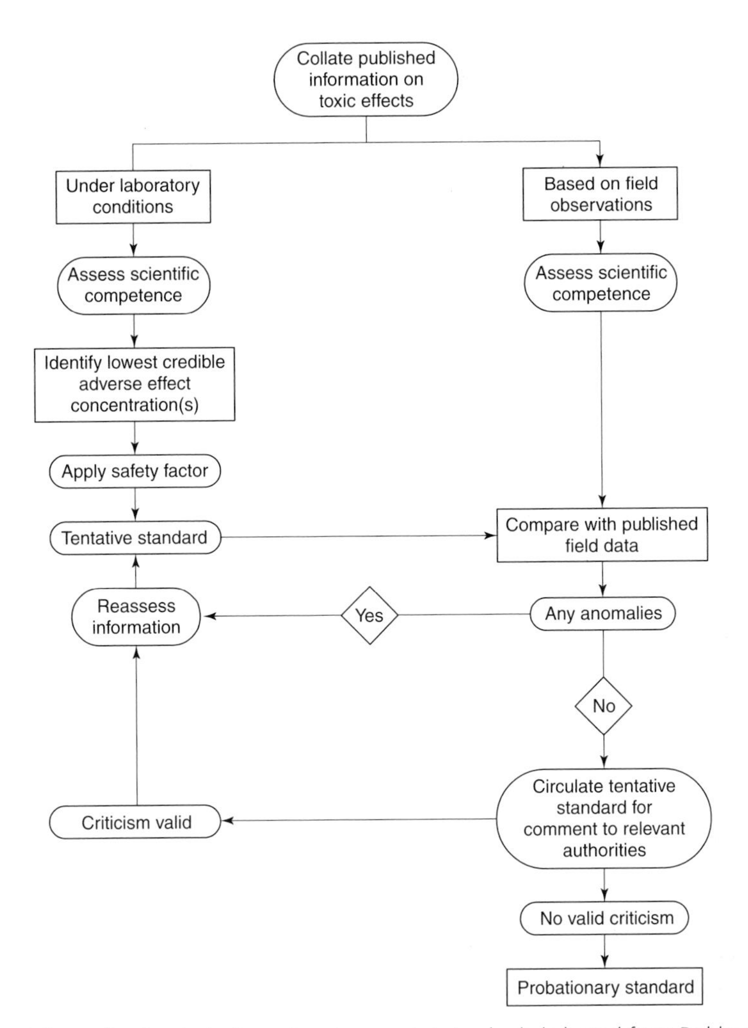

Fig. 9.4 Scheme for the derivation of environmental standards (adapted from Dobbs and Zabel, 1996).

standard is then circulated to the relevant authorities for comment and any valid criticisms will lead to a reassessment before it is finally adopted as a probationary standard.

The application of the safety factor will depend on the quality of the available toxicity data. If only acute toxicity data are available, then an arbitrary safety factor of 100 may be applied. If data on chronic toxicity are available, this may be reduced to 10. To protect wildlife from the effects of biomagnifying pollutants in the Great Lakes region, the lowest adverse effect levels (LOAELs) were determined, either from field observations or from controlled laboratory studies in which domestic species were fed known quantities of Great Lakes fish. The standard, the no observable adverse effect level (NOAEL), was defined as 10 per cent of the LOAEL, i.e. allowing a safety factor of 10. It was considered that this would protect the most sensitive species (e.g. bald eagle *Haliaeetus leucocephalus*) on which there were no data. The water quality criterion for total PCBs to protect the most sensitive species was calculated as 0.1 pg l^{-1} (i.e. 1×10^{-16}) (Ludwig *et al.*, 1993). Such effects-based standards are being increasingly developed, using direct toxicity assessments (DTAs) to control effluent discharges by toxicity rather than chemical concentration (Wharfe and Heber, 1998).

In England and Wales those industries with the greatest potential to pollute are regulated under a new Pollution Prevention and Control Act (1999) by the Environment Agency. The mechanism of action is by way of integrated pollution prevention control (IPPC). Control of waste disposal is often considered in terms of the transfer of waste to the disposal route posing least environmental hazard. Before the release of a potentially harmful substance is authorized, it must be shown that attempts to minimize releases have been made under the BAT principle (best available technology). The emphasis is on recycling and recovery to prevent release to the environment. If it is not possible to prevent releases then they must be minimized and rendered environmentally harmless, having regard to the best practicable environmental option (BPEO). This approach aims to limit the potential waste disposal routes so that environmental protection is ensured in the long term.

Within the European Union there are Environmental Quality Objectives (EQOs) for polluting substances, levels of water quality that should be achieved to be suitable for the agreed use of a particular stretch of water. To meet these objectives Environmental Quality Standards (EQSs) are set. An EQS is that concentration of a substance that must not be exceeded if a specified use of the environment is to be maintained. EQSs should:

- allow successful completion of all life stages of aquatic organisms;
- not produce conditions that prevent aquatic organisms from living in parts of the habitat where they should be present;
- not allow bioaccumulation, in wildlife and humans, of harmful substances;
- not produce conditions which alter the functioning of ecosystems.

EQSs are placed on individual discharges that will leave a river with water of the necessary quality. All discharges will not, therefore, be of the same standard and where the assimilative capacity of a river is high, relatively little expenditure may be required to treat some discharges. Many consider that individual quality standards are difficult to enforce and that they always tend towards the minimum values required to meet the objectives, imposing the minimum cost on the polluter. They are most suitable for biodegradable effluents, such as treated sewage, whose effects can be minimized by dispersal and dilution, the standard being set to avoid acute, local impacts. The alternative approach is the fixed uniform emission standards

(UESs), which limit all polluting discharges irrespective of existing water quality, the dilution available or the future use of the river. Any discharge of pollutants is considered to be ultimately unacceptable and progress must be made towards their elimination by recovery and recycling. These standards are especially suitable for chronic impacts or for those pollutants with high persistence, such as metals or organochlorines. The UES approach is favoured in the United States and many European countries. In practice a mixed approach of EQSs and UESs is often adopted.

In England and Wales, EQOs have been developed as water quality objectives (WQOs) for individual stretches of river to protect the river ecosystem. The River Ecosystem (RE) classification is based on the concentrations of seven common determinands that have effects on fish populations – dissolved oxygen, BOD, total ammonia, un-ionized ammonia, pH, copper and zinc, i.e. largely pollutants in sewage effluents. A description of the RE classes is given in Table 9.1. A target class is set for a stretch and then regular monitoring assesses compliance with the target. This is a national scheme that reflects conditions over all river types and many lowland rivers do not achieve RE 1 status because of their natural characteristics. In lowland rivers non-compliance is frequently caused by low oxygen but it is uncertain whether this is a natural ecological condition (and hence the targets are too rigidly fixed) or whether it genuinely reflects poor water quality. It is important to know this because stringent controls on effluent discharges could be imposed, at considerable cost, to counteract a natural phenomenon.

The Environment Agency also classifies rivers using the General Quality Assessment (GQA). It provides an annual overview of the quality of rivers and is used to monitor trends over time. Chemical and biological aspects are covered and methods for including nutrients and aesthetics are also being developed. The chemical GQA describes quality in terms of BOD, dissolved oxygen and ammonia. Quality is graded A–F, with the limit set for the determinands coinciding with the ecosystem classes. The biological GQA was described on p. 283.

The standards set for individual sewage and trade effluents are known as *consent conditions*. In Britain the 95 percentile approach to effluent quality is adopted, in

Table 9.1 Description of the River Ecosystem classification

Class	Description of water
RE 1	Very good quality suitable for all fish species
RE 2	Good quality suitable for all fish species
RE 3	Fair quality suitable for good coarse fish populations
RE 4	Fair quality suitable for coarse fish populations
RE 5	Poor quality which is likely to limit coarse fish populations
Unclassified	Bad quality, in which fish are unlikely to be present, or insufficient data available to classify water quality

which recorded values must be less than the stipulated consent for at least 95 per cent of the time and must exceed the consent no more than 5 per cent of the time, an agreed number of samples being taken over a year to assess performance. This probabilistic approach also applies to the quality standards described above, which itself is linked to the frequency of sampling. Some standards are imperative – they must be adhered to – others are considered guideline.

In contrast to rivers, the classification of water quality in lakes, other than in terms of nitrogen and phosphorus loading (p. 155), has received little attention. Moss *et al.* (1997) and Kudelska *et al.* (1997) discuss some recent approaches to classification.

Standards must ensure clear assessments of performances and trends. Water-quality management requires an integrated approach, which includes the sampling of rivers and effluents, the setting of standards for effluent discharges, the determination of river quality targets, the assessment of compliance with standards and the setting of priorities for action.

The European Union has produced directives for many aspects of water quality. The requirements include fixed standards for a wide range of physical, chemical and microbiological determinands that must not be exceeded. Of particular importance is the EU Directive on Urban Waste Water Treatment, controlling discharges associated with the treatment of sewage. Others deal with sewage sludge used in agriculture, bathing waters and shellfish waters, groundwater, surface water abstraction and freshwater fish.

A new EU directive, the Water Framework Directive, enacted in December 2000, pulls together a range of existing European legislation on water matters into one unifying law. It is considered to be the most comprehensive piece of European legislation on the environment so far, in taking a holistic approach to the management of aquatic ecosystems. The directive has two key components:

- a system of managing the aquatic environment based on catchments;
- the introduction of co-ordinated *programmes of measures* to achieve at least *good status* for rivers, lakes and coastal waters by specified deadlines.

The status categories (high, good etc.) will be a measure of the deviation of the status of a particular waterbody from that achieved by a similar water not affected by human activity, i.e. as close as possible to the natural condition. The surface water status is composed of two elements, the *chemical status*, i.e. the compliance with EU environmental quality standards, and the *ecological status*, that is the quality of the structure and functioning of aquatic ecosystems. This consists of a combination of biological (e.g. biodiversity), hydrological (e.g. flow) and physicochemical (e.g. oxygen, temperature) elements. The emphasis of the directive is therefore on ecological targets, not just on water quality targets. Groundwater status is a function of quantity and chemical condition. Quality objectives will then be set for each waterbody and measures to achieve those objectives put in place. Monitoring at regular intervals will ensure that the objectives are met and maintained. The aim is to achieve at least *good status* across the waterbodies of the European Union by 2015. This will certainly require the tackling of diffuse pollution, rather more difficult to control than that from point sources.

The Water Framework Directive envisages river catchment management as encompassing six steps:

1. Establish River Basin Districts by 2003 and ensure that administrative arrangements are in place to achieve the objectives across the whole catchment.
2. Determine the number and types of waterbody within the catchment and determine their reference conditions (by 2004).
3. Identify the environmental stresses (human impact) on waterbodies and carry out an economic analysis of water use to identify the most cost-effective way of achieving the objectives of the directive.
4. Design the measures needed to implement the directive.
5. Monitor the progress and report on the state of implementation (by 2012).
6. Revise the measures, if necessary, based on the monitoring results (by 2015) and every six years thereafter.

Much of this approach is already encompassed in the LEAP procedures in England and Wales (see p. 296). Comprehensive the Water Framework Directive may be, but the policy of subsidiarity within the European Union means that strict common standards are not set, individual member states being allowed to define their own levels of quality. The filthy state of many of Europe's rivers, lakes and coastal waters, despite several decades of apparently stringent standards set by earlier directives, does not bode well for achieving an overall good status of our aquatic ecosystems by 2015.

Pollard and Huxham (1998) and Bloch (1999) provide more details on the Water Framework Directive. To meet its objectives will require considerable expenditure on the part of member states. The cost of meeting more stringent environmental requirements should be targeted at those causing the water quality problem by adopting the *polluter pays principle*; those responsible for pollution contributing towards the cost of monitoring, regulating and controlling its adverse effects.

Where there is reason to expect harmful effects but no conclusive scientific evidence to show a causal link, it is sensible to adopt the precautionary principle (Santillo *et al.*, 1998). This is the avoidance or reduction of risks to the environment before specific environmental hazards are encountered. It can also be used to set stringent environmental standards, where for a variety of reasons causal links have not been proved, to protect a resource that is being damaged, e.g. standards to protect otter (*Lutra lutra*) populations (Mason, 1995).

Risk assessment and the setting of standards are discussed by Landis and Yu (1995), Dobbs and Zabel (1996), Gerba (1996) and in papers in Douben (1998).

RIVER REGULATION AND INTER-RIVER TRANSFER

In most parts of the world the use of water has risen dramatically and there is a water crisis (Cosgrove and Rijsberman, 2000). Only about 1 per cent of global water is readily accessible for use. Of water used, some 73 per cent is for irrigation, 20 per cent is for industry and the remainder for domestic purposes. Some 14 per cent of the world's run-off of freshwater is now impounded behind dams. Irrigation is often hugely inefficient and improving its efficiency would do much to ease the water crisis. In the republics of Uzbekistan and Kazakhstan, large amounts of water have been diverted from rivers draining the inland Aral Sea to irrigate crops of cotton. The Aral Sea was once the world's fourth largest freshwater lake but the surface area has shrunk by 46 per cent, the volume by 69 per cent and the salinity has increased threefold. The fishery has collapsed, as has the shipping industry, while the health of more than one million people is being damaged by polluted water. The result is an ecological calamity (Williams and Aladin, 1991; Pearce, 1995).

Much of the Mediterranean basin is suffering the effects of over-abstraction. The Middle Eastern countries, on average, use 155 per cent of their renewable water supplies each year (Darwish, 1995) and this is an area where wars over water are thought most likely to break out. Shortfalls in water supply have also been regular events in recent years in England and Wales. Over-abstraction has led to the headwaters of many streams running dry, even in winter (p. 232), having a major impact on aquatic communities (Mantle and Mantle, 1992).

Although domestic use makes up a small proportion of the total, it has risen dramatically in developed countries. For example, in the United States it has tripled since 1940, with a per capita consumption of $375\,l\,d^{-1}$. Domestic consumption of water in England and Wales has doubled since 1961, to $160\,l\,d^{-1}$.

Many areas also suffer devastating floods, often with terrible loss of life. Frequently floods are the result of major changes in the watershed, especially deforestation. The increased unpredictability and severity of both floods and droughts are likely to be an early warning of global warming.

An engineering solution to both droughts and floods is river regulation. Almost every river in Europe and North America has been regulated, while there are few major rivers elsewhere that have not been regulated in some way (Petts, 1994). River flows may be regulated by the addition of water from reservoirs, the water later being abstracted for use downstream, or water may be transferred from one river to another. Such schemes may have major impacts on the biota of rivers and may also influence levels of pollution (Petts, 1994). If waters are held back by a dam, for example, the reduced flow in the river downstream may lose much of its capacity to self-purify.

Edwards (1984) summarized a large, interdisciplinary study, extending over five years, into the possible consequences of a river regulation scheme on the ecology of the River Wye in Wales. An enlargement of Craig Goch Reservoir, in the headwaters of the river, was planned in order to allow managed releases of water to

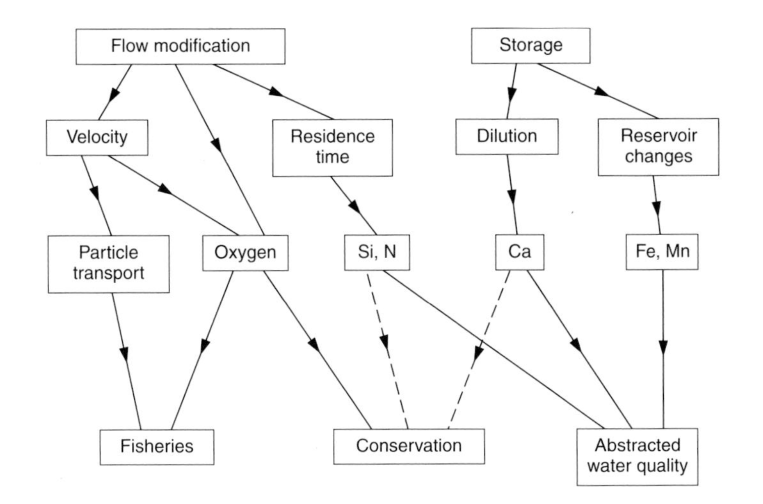

Fig. 9.5 Possible extensive effects of regulation on the aquatic resources of the River Wye; more doubtful effects are shown by dotted lines (from Edwards, 1984).

compensate for abstractions downstream. Some of the impounded water was also to be used to regulate the adjacent catchment of the River Severn.

The predicted general effects of regulation on the River Wye are shown in Fig. 9.5. Over the river as a whole regulation was considered likely to be beneficial to fisheries, because flow augmentation would avoid low oxygen concentrations developing during periods of macrophyte decay, a time at which considerable mortality of Atlantic salmon had occurred in the past. The food supply of salmon should also be increased in summer by enhanced downstream drift of invertebrates, while oxygen-sensitive benthic invertebrates would be protected. The water in the river was likely to become softer and concentrations of iron and manganese were predicted to increase as a result of the release of hypolimnetic water from the reservoir, though it was unlikely to be sufficient to damage water quality. The local effects of regulation were likely to be more varied and more difficult to predict. Increases in water velocity might influence the upstream movement of salmon as local temperature and oxygen regimes were modified. Plankton discharging from the reservoir might increase the food supply of filter-feeding animals and indirectly the growth of salmon. The softening of water might lead to a reduction in the upstream distribution of molluscs and crustaceans, which require calcium.

In the River Ebro catchment in northeastern Spain, over 100 reservoirs have been built to supply water for various uses, but two large dams constructed since 1964, primarily to provide hydroelectric power, have had a major impact downstream (Prat and Ibañez, 1995). Some 95 per cent of sediments carried by the water are retained behind the dam, resulting in a loss of habitat in the delta. A reduction in river flows has caused eutrophication, pollution and seawater penetrating for 30 km up from the mouth of the river, causing great ecological change. Further water

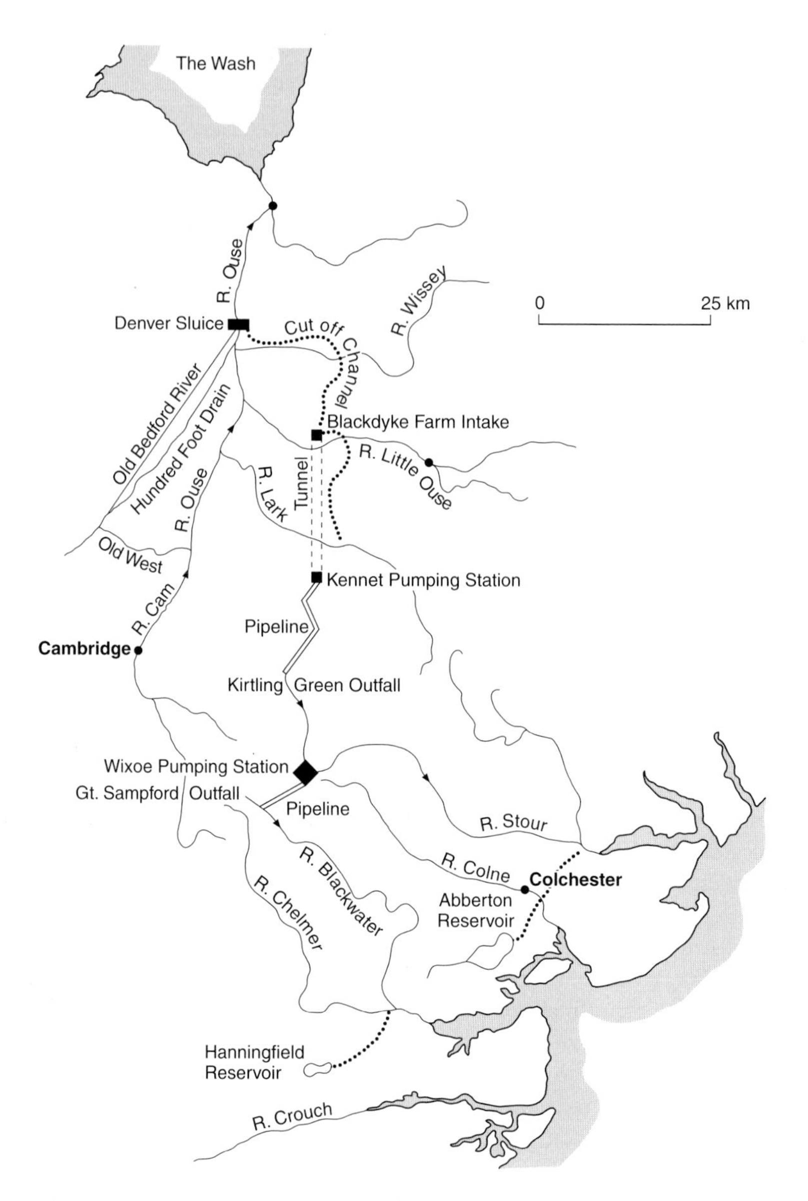

Fig. 9.6 The Ely–Ouse water transfer scheme.

abstractions are planned. It is considered that these will extend the saltwater for a further 18 km upstream, threaten fisheries and aquaculture in the lower river and delta, and lead to salinity problems in cultivated fields in the lower catchment.

Inter-river transfers may import pollutants into catchments where they may do considerable damage. In March 1973, 50 commercial tomato growers in Essex, eastern England, reported severe damage to crops grown in greenhouses, whole crops being destroyed in some instances. The damage was consistent with injury caused by growth regulatory herbicides, eventually identified as 2,3,6-trichlorobenzoic acid (TBA), which was traced to the domestic water supply. Using both chemical analyses of river water at various points, and plant bioassays – applying water to the cotyledons of tomato seedlings – the herbicide was eventually traced to the factory that manufactured it (Williams *et al.*, 1977).

The factory began making TBA in 1971 and the herbicide became a component of its discharge to the River Cam above Cambridge. The effluent met the standards applied and monitoring of the river below the discharge detected only very small and localized effects. Before 1973 the River Cam water was discharged, via the River Ouse, to the sea and none was abstracted for public water supply. In early 1973 the Ely-Ouse to Essex water scheme became operational (Fig. 9.6). Water from the River Ouse is pumped at Denver into a channel which acts as an aqueduct to take water back southwards. From the channel it is pumped through tunnel and pipeline to be discharged into the upper reaches of the Rivers Stour and Blackwater to augment their flows. In the lower reaches of these rivers water is abstracted to Abberton and Hanningfield Reservoirs from where, after treatment, it is put into supply. TBA, the pollutant from the factory near Cambridge, was causing considerable financial losses to tomato growers who obtained their water from different catchments over 200 km away. Despite the dilution of the effluent during the inter-river transfer, the herbicide was still concentrated enough to kill tomatoes, which are sensitive to amounts as low as $0.0005\ mg\,l^{-1}$ (Williams *et al.*, 1977). Charcoal towers were installed at the factory to treat the effluent as soon as it was implicated in the tomato failure and no further problems have been encountered.

As well as transporting pollutants, inter-river transfers may also transport animals and plants to new localities. For example, the predatory zander (*Stizostedion lucioperca*) has used the Ely–Ouse transfer system to colonize the rivers of Essex.

General reviews of the hydrological and ecological impacts of river regulation are provided by Petts (1984, 1994) and Petts *et al.* (1995).

FISHERIES

Interest in recreational fishing has expanded rapidly and it is estimated there are some 2.9 million anglers in England and Wales, whose combined annual spending on their pastime is £3.3–5 billion. It is not known how many jobs are supported but there are at least 12 000 jobs dependent on the sale of fishing tackle alone (Ministry of Agriculture, Fisheries and Food, 2000). On rivers with good runs of salmon and migratory trout the renting of fishing is extremely lucrative to riparian owners. The development and optimization of fisheries resources and their maintenance in relation to other conflicting uses of water requires a high level of expertise. Fisheries biologists must work closely with other disciplines in a multifunctional

Fig. 9.7 Major fish-kill caused by a saltwater intrusion, Norfolk Broads, England (photograph by Jonathan Wortley).

team, including personnel primarily involved in pollution assessment. A single pollution episode can destroy years of work in managing and improving a fishery (Fig. 9.7). In England and Wales during the period 1997/98 there were 1043 reported fish-kills, of which about 25 per cent were related to pollution events. The work carried out by fisheries biologists within the water industry includes:

- assessments of habitat and water quality to support particular species of fish;
- population, growth and production estimates of fish stocks in freshwater habitats;
- investigations of the effects of land drainage and river channel modification works on fisheries;
- culturing of fish for stocking freshwaters;
- studies of populations and movement of migratory fish (especially salmonids);
- emergency investigations following pollution incidents;
- investigations of fish diseases and pathology.

Models are frequently used to predict the effects of management activities on biological resources, including fisheries. One model that is widely used in the United States, and gaining currency elsewhere, is PHABSIM (Physical Habitat Simulation). This examines the effects of changes in discharge (for example caused by increases in resource use) on the habitat preferences of the biota. The underlying principles of PHABSIM are that individual species exhibit habitat preferences within a range of conditions that they can tolerate, that these preferences can be defined for each species and that the area of stream providing these conditions can be quantified as a function of discharge and channel structure (Petts and Maddock, 1996). Changing flow influences both current speed and water depth, primary variables used to predict the impact of altered flows on stream life. The model predicts changes in the physical habitat of the river and how this may influence ecological values. This then allows the biologist to recommend the flows required to maintain the ecological integrity of the site. For species of interest, habitat suitability curves are derived (showing range of conditions tolerated, with optimum conditions) for velocity, width, depth and substratum type. This is done by field survey work, from the literature or by consulting experts. An example of the use of PHABSIM is provided by Petts *et al.* (1995).

As an example of the type of work that may be undertaken, fisheries biologists of the East of England Region of the Environment Agency carry out a rolling programme to estimate the population density and biomass of fish stocks in the region's rivers, a total of over 3000 km. The results of such surveys for the River Waveney are shown in Fig. 9.8. The Waveney is in a popular tourist area, with

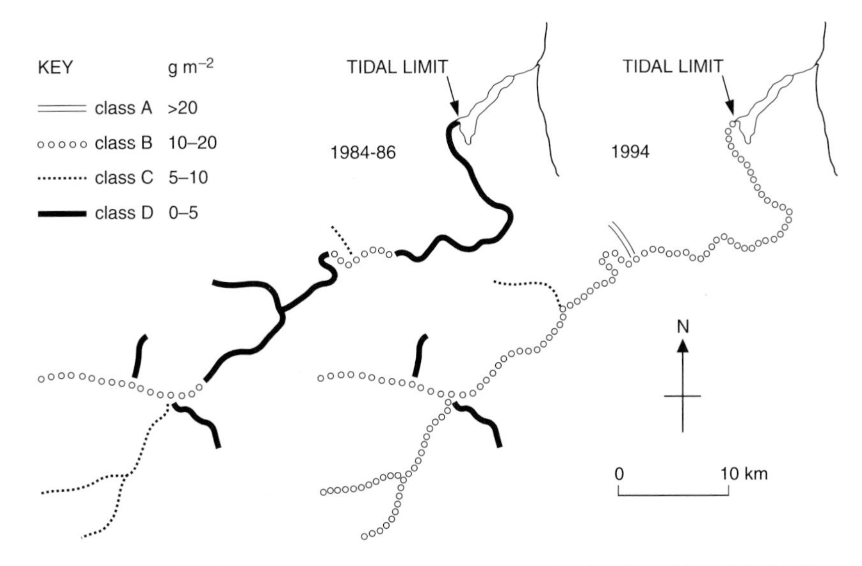

Fig. 9.8 Biomass of fish in the River Waveney, eastern England, 1984–86 and 1994 (from Environment Agency data).

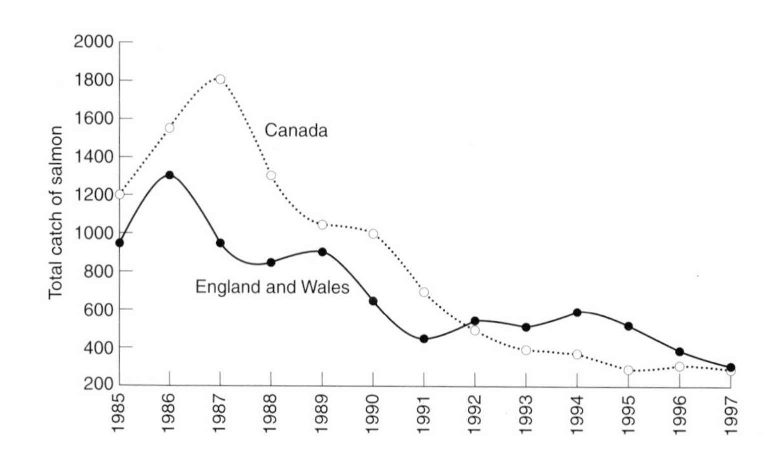

Fig. 9.9 Total Atlantic salmon catches (tonnes of fresh weight), 1985–97, in England and Wales, and Canada.

much recreational angling. The river should support a fishery of at least Class B over much of its length, but in 1984–86 most of it was in Class D, with a biomass of fish of less than 5 g°m^{-2}. The fisheries team then began a detailed investigation, in cooperation with water chemists, and it was found that much of the river regularly had ammonia concentrations in excess of those necessary to maintain coarse fisheries. The source of ammonia was pig units, of which the catchment supports a large number. Efforts to control the discharges of these effluents have been generally successful, with most of the fishery improving to Class B by 1994.

Attention to water quality has led to a general overall improvement in coarse fisheries, though the diversity of fisheries is often limited by habitat degradation (e.g. Garner, 1997). In contrast, fisheries for migratory salmon are in decline. Figure 9.9 shows the deterioration in stocks of Atlantic salmon (*Salmo salar*) in England and Wales, and in Canada, over a ten year period, though individual fisheries may have improved owing to a decrease in pollution. Similar declines have occurred in other Atlantic stocks and in salmon species inhabiting the North Pacific (Knudsen *et al.*, 1999). Salmon put on much of their growth in the oceans, where their preferred temperature is 6 to 9 °C. Increases in surface sea temperatures in the mid-Atlantic and Pacific Oceans are thought to be reducing feeding opportunities and hence survival rates (McFarlane *et al*, 2000; Noakes *et al.*, 2000). However, there are also fishing pressures at sea and in the rivers, while afforestation and agricultural practices in the catchment, acidification, drought and disease may also adversely affect stocks. The relative importance of these factors is unknown and much research is required if the downward trend has any chance of being reversed.

Anglers, spending many hours by rivers, are often the first to detect problems. They know their rivers and so any concerns they may express are worthy of investigation. However, in some activities their role is less than exemplary, in particular with regard to live-baiting and predators. Live-baiting is the use of small fish to catch larger ones. The unfortunate beast is crippled on several hooks and its pathetic movements, together with the release of stress or fear pheromones, attract the larger fish. This cruel method of angling is practised by only a few and condemned by many but neither the angling governing bodies nor the authorities appear willing to take on this minority. It brings the pastime into disrepute.

There is no doubt that predators take fish but a well-managed fishery is little impacted by them. Nevertheless when two anglers get together the conversation invariably turns to cormorants (*Phalacrocorax carbo*), the current *bête noire*. Indeed, cormorants are large and, to some, ugly and, primarily coastal feeders, they have taken to fishing inland over the last two decades or so. Recent research funded by the UK government has suggested that the role of cormorants in the decline of fisheries is a local rather than a general problem and in many cases where cormorants have been blamed for the decline, other factors are at work. Nevertheless, at a regional fisheries committee meeting I attended in 2000 the research was dismissed as an expensive waste of money because its conclusions did not coincide with the prejudices of anglers. The total breeding population of cormorants in Britain and Ireland is less than 12 000 pairs, only 1300 of which breed inland, many of these feeding in estuaries. The wintering population is about 16 000 birds, the majority feeding in the coastal zone. Compare that with the angling population of 2.9 million! Reviews of fisheries management include those of Elliott (1995), Mann (1995) and Cowx (1995).

Angling is not the only recreational use of waterways. The Environment Agency is also responsible for improving access and facilities for boats, which in some areas are a major source of tourist income (Fig. 9.10). British waterways are very underused as a way of moving goods about the country, a situation that the government wishes to change (Department of the Environment, Transport and the Regions, 2000). Access to waterways for casual recreation such as walking and picnicking also requires improvement.

CONSERVATION OF BIODIVERSITY

The Convention on Biological Diversity, signed at Rio de Janeiro in 1992, places a commitment on its signatories to protect biodiversity and to use natural biological resources in a sustainable manner. In the United Kingdom this concept was enshrined within the Environment Act 1995. The water industry, considered in the broadest terms, may impact the wildlife of freshwaters in a number of ways (English Nature, 1997):

- loss of wetland habitats;
- drainage of wetlands;

Fig. 9.10 Many rivers are heavily used for recreation, as here in the Norfolk Broads, eastern England. Traditional wherries (left) cause little environmental damage but motor cruisers (right) are noisy, leave oil films on the water and cause severe bankside erosion, adding to eutrophication problems (photograph by the author).

- increased water abstraction;
- pollution and siltation;
- modification of river habitats, such as channelization;
- introduction of non-native species;
- intensive fisheries management;
- inappropriate development of recreation and navigation.

As the main regulator of the aquatic environment, the Environment Agency has a legal duty, under the Environment Act, to have regard to conserving wildlife in all of its operations, and indeed to promote biodiversity. The basis for this is the EA Action Plan for Conservation (Environment Agency, 2000). Its main objectives are:

- to take full account of conservation before taking policy and operational decisions;
- to give priority to protecting statutory sites;
- to ensure the EA takes full part in implementing the UK Biodiversity Action

Plan by providing a lead in promoting the conservation of key water-related habitats and species;

- to demonstrate, through EA work, the benefits of best environmental practice for conservation of the wider countryside;
- to ensure the EA applies appropriate conservation criteria when considering activities it authorises;
- to influence, at both national and local level, plans for rural and urban development, to the benefit of conservation as a whole.

We can see that wildlife conservation, barely mentioned by the water industry 30 years ago, is now considered central to its operations.

The conservation of threatened or declining species or habitats can be addressed through Biodiversity Action Plans (BAPs). In the UK there are some 400 species action plans and 40 habitat action plans. The EA has taken on responsibility for 39 species and 5 wetland habitats and so is the lead organization in a partnership of government bodies, non-governmental organizations (NGOs) and indeed individuals. Examples of EA lead plans are otter (*Lutra lutra*), vendace (*Coregonus albula*), pearl mussel (*Margaritifera margaritifera*) and eutrophic standing waters. While BAPs may be national or international in design, the actions must be taken locally. An ideal action plan considers the current status of the species (or habitat), the factors causing decline, current action being taken, the action plan and its targets, and proposed action with partners. The problem with many BAPs is that the factors causing the decline are very often not known, so a list of likely factors (habitat loss, pollution etc.) is given. As the budget often does not include money for detailed research, the proposed actions may be ill-conceived and not result in population recovery. For many species there may be no resources to move any of the actions in a plan forward.

The drainage of land for agriculture and channel improvements to aid flood prevention cause great concern to wildlife conservationists. About half of the world's wetlands were destroyed in the twentieth century. For example damming and canalizing the upper part of the middle Danube in 1992 has destroyed some 23 000 ha of wetland, which had an estimated annual value of US$520 million (Balon and Holcik, 1999). Land drainage is not, however, a new phenomenon. In the seventeenth century the Dutch engineer Cornelius Vermuyden constructed two massive parallel channels, 1 km apart and running for 30 km across the fens of eastern England to the sea, enabling 20 000 ha of wetland to be drained and turned into some of the richest arable land in the country. The land between the two channels acts as a safety valve when the incoming tide prevents the escape of water during spate conditions. The immense flood created between the two channels provides habitat for large numbers of waterfowl (Fig. 9.11), with peak winter numbers over 72 000 recorded. The Washes also support good numbers of breeding waterfowl and waders. For miles on either side of the Washes the arable land is an ecological desert of flat, intensively cultivated land and this is an apt description for the end point of most modern drainage schemes.

Fig. 9.11 The Ouse Washes, eastern England. The grazing land between the two channels is flooded each winter (a) to protect arable land and properties outside, which lie below the level of the channels. The floodlands are of international importance for birds, where large populations of wild swans (b) and other waterfowl overwinter (photographs by Sheila Macdonald and the author).

The drainage of wetlands results in the loss of species-rich grasslands, fens and marshes to species-poor, improved grasslands or ultimately to arable monocultures. Since the 1940s, land drainage, followed by ploughing, re-seeding and fertilization has left only 5 per cent of lowland neutral grasslands in England and Wales with any significant wildlife interest. Of plant species in Britain that have become extinct or rare or are rapidly declining, 22 per cent have been adversely affected by land drainage. The attractive snakeshead fritillary (*Fritillaria meleagris*) (Fig. 9.12), once a widespread plant in English lowland meadows, has become restricted to a handful of nature reserves, while the snipe (*Gallinago gallinago*), a wading bird breeding in wetlands, has become extinct over much of lowland England (O'Brien, 1998). A localized drainage scheme may affect the water table over a considerably greater area so that a landowner, interested in preserving a wetland, may find his habitat deteriorating owing to the activities of his neighbours. The management of wet grasslands for birds (Fig. 9.13) is discussed by Ward (1994) and O'Brien (1998).

The costs of modern land drainage schemes can be measured in terms of environmental damage, in terms of the over-production of food (which is bought and stored at the taxpayers' expense) and, in some areas, in a decline in the quality of agricultural land. The benefits are more difficult to determine. Calculations of benefit ignore the intangible environmental losses to the community and normally assume unrealistic yields of crops and a rapid take-up by farmers. The anticipated benefits are calculated on the 'farm gate' prices received by farmers for crops, which are unrealistically high because of the price protection that agriculture enjoys (Bowers, 1995). Individual landowners profit from land drainage – most of the costs, both direct and indirect, are borne by the taxpayer. Once destroyed, the original biodiversity of such wetlands can never be fully re-created so it would seem prudent, in the national interest, to conserve their ecological richness, at least until a need for increased food production can be shown to be essential.

Less obvious than drainage, but equally damaging to wetlands, is over-abstraction of water, leading to a gradual drying out. In England and Wales, some 85 000 ha of wetlands are currently receiving remedial action to counteract water-related problems, at a cost of £100 million over five years.

Another major area of conflict between wildlife conservation and the water industry is river channelization for flood prevention. Some 31 000 km of rivers in England and Wales alone are subjected to regular maintenance. Management may involve the straightening of river beds and the removal of natural obstructions, the destruction of aquatic macrophytes by hand, machines or herbicides and the clearance of bankside vegetation, especially of mature trees (Hey, 1996). The effects of channelization of the river Welland in the English Midlands was to reduce the richness of invertebrate families by 50 per cent and the biomass by 80 per cent (Smith *et al.*, 1990).

Most river banks of the northern hemisphere were once either swamp or forest but many are now devoid of riparian trees. In eastern England up to 70 per cent of bankside trees were removed between 1879 and 1970 (Fig. 9.14), 20 per cent of

Fig. 9.12 Snakeshead fritillary (*Fritillaria meleagris*), once a common plant in the alluvial meadows in England, now restricted to just a few sites (photograph by the author).

Fig. 9.13 Grazing marshes at Strumpshaw Fen, a reserve of the Royal Society for the Protection of Birds. Water levels are manipulated in spring to provide ideal nesting conditions for wading birds and ducks. Once breeding has finished, cattle graze the plant-rich marshes through the summer (photograph by the author).

them in the period 1960 to 1970 (Mason and Macdonald, 1990; Harper *et al.*, 1997). Figure 9.15 contrasts unmanaged and managed stretches. The level of management illustrated in Fig. 9.15b has a severe impact on both riparian and aquatic wildlife (Mason *et al.*, 1984; Mason 1995). Bankside tree cover, for example, is essential to bats (Warren *et al.*, 2000).

The marked decrease in distribution and abundance of the otter, owing to contamination with organochlorines, was described in Chapter 1 (p. 6). Habitat destruction has, however, also had a significant impact. Otters lie up and breed in bankside dens and, in some parts of their range, they favour the eroded root systems of mature trees (Fig. 9.16), especially oak (*Quercus* spp.), ash (*Fraxinus excelsior*) and sycamore (*Acer pseudoplatanus*) (Macdonald and Mason, 1983). The roots of these species dislike waterlogging so, although the trees readily establish and thrive on riverbanks, their major roots tend to grow shallowly and horizontally. In high waters the soil is easily eroded from among such roots, the trees gradually tilt, forming underground cavities, and eventually fall into the river, where they may cause flooding. They were therefore removed by river engineers and such den sites for otters became very scarce. Under pressure from conservationists, however, trees are now more usually pollarded or coppiced rather than ripped out, leaving roots

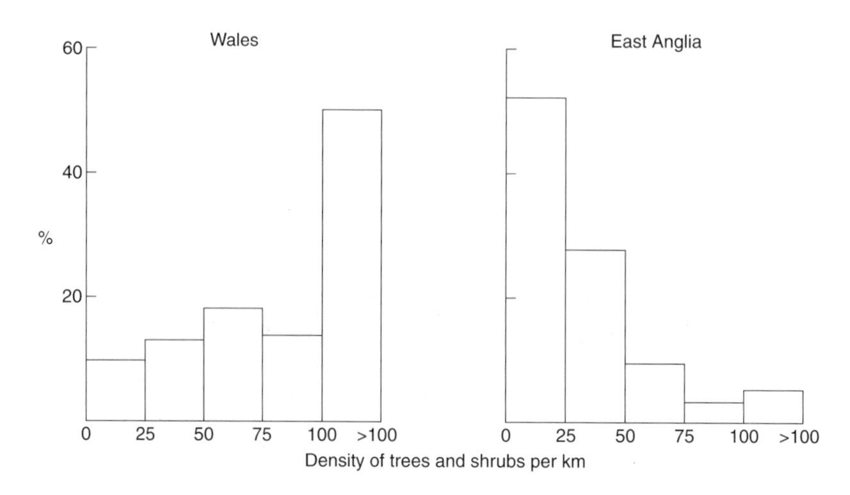

Fig. 9.14 The density of trees and shrubs (number per km) along the banks of rivers in Wales and East Anglia. Notice that heavy management in East Anglia has resulted in a very low density of woody vegetation, while many stretches in Wales have continuous tree-lines. Vegetation along Welsh streams has actually increased in recent years with a change to more ecologically sensitive types of management.

intact and preserving the otters' refuges. The two commonest trees on riverbanks, alder (*Alnus glutinosa*) and willows (*Salix* spp.), have water-seeking roots, the dense fibrous root mat suffering far less erosion, but also not providing cavities for den sites.

Current management practices are expensive even for effective flood prevention. In England and Wales the annual cost of aquatic weed management alone is of the order of £45–75 million and much of this may have become necessary because of the construction of drainage schemes and the excessive removal of bankside trees, providing ideal conditions for macrophyte growth. Dawson (1989) has shown that the cutting of excess plant growth produces only a short-term improvement, but it may stimulate synchronized regrowth, accentuating problems in future years. Macrophyte removal has been shown to cause a dramatic decline in the zooplankton in a river, followed by a rapid decrease in the growth rate of young roach (*Rutilus rutilus*) as they were forced to feed on a less nutritious algal diet (Garner *et al.*, 1996). Replanting riverbanks with trees, to provide a half-shade that reduces, but does not eliminate, macrophytes would prove cost-effective and will benefit wildlife and fisheries. Dawson (1989) reviews options for macrophyte management.

To ensure that the management of rivers does not result in damage to important wildlife habitats, River Habitat Surveys (RHS) are carried out in England and Wales. The RHS provides a classification of rivers based on their habitat quality. The river is considered in sections of 500 m and features are recorded within the channel, banks and river corridor to 50 m on either side of the river. The within-river data will include recording of biotopes (riffle, pool, run etc.) and functional

Fig. 9.15 (a) Low-intensity management on the River Severn, Wales. (b) Typical stream management in eastern England (photographs by the author).

Fig. 9.16 An otter den in the base of a sycamore tree, in Wales (photograph by Sheila Macdonald).

habitats (e.g. gravel, sand, macrophyte types, tree roots) (Newson *et al.*, 1998; Kemp *et al.*, 1999). Such attributes are recorded at 10 equidistant points along the 500 m stretch, while a further checklist records features occurring between the points. From a survey in 1994, 11 broad river types, further divided into 67 subtypes, were identified in England and Wales (Raven *et al.*, 1998, 2000). Field data can be compared with those expected for a largely unmodified, natural example of that class to produce an overall habitat quality assessment (excellent, good, fair, poor, bad). In the mid-1990s, a large number of sites across the UK were surveyed. About 42 per cent of rivers were found to be extensively or heavily modified by human activities.

Using information from RHS or similar data sources, an objective PC-based system has been developed to assess the conservation value of rivers: System for Evaluating Rivers for Conservation or SERCON (Boon *et al.*, 1997; Wilkinson *et al.*, 1998). Sensitive and sensible approaches to river management are described in Brookes (1988), Purseglove (1988), Gore and Petts (1989) and Hey (1995, 1996).

HABITAT RESTORATION AND CREATION

Traditional engineering has then, along with pollution, caused large-scale damage to very many rivers worldwide and these two processes are frequently linked. There is now a growing interest in the restoration of rivers to, as far as is possible, their natural state. Ideally restoration should be viewed at the catchment scale. In particular, floodplains that are currently cropped to the margins of the river should be returned to grazing marshes, wetlands and even riverine woodlands. The values would be immense. Water would be retained for longer in the system, reducing the risk of flooding to life and property, while this long retention time would allow for the replenishment of groundwaters. These new wetlands would retain and transform nutrients and pollutants, and stabilize sediments, reducing both eutrophication and pollution in the river. They would provide habitat for a diverse range of wildlife, both within the floodplain and the river, while the landscape would be greatly improved. Both landscape and wildlife will attract visitors, bringing income to the local economy. These benefits of wetlands have been given an annual value of US\$370 500 ha^{-1} in the United States (Keddy, 2000). Fuglsang (1998) has suggested that, if riparian zones are re-established along an entire watercourse, maintenance within that watercourse (dredging, weed-cutting etc.) will no longer be necessary. So far, floodplain restoration has been on a very small scale (de Waal *et al.*, 1998). However, in the Netherlands, the successful creation of a large wetland area, the Oostvaarderplassen, has resulted in ambitious plans to restore large wetlands (Bijlmakers and de Swart, 1995), including a long-term plan to create a National Ecological Network, linking areas of high conservation value by buying farmland and wetlands to develop corridors of natural habitat (van Rijen, 1998). The restoration of the Everglades in Florida is similarly largescale (Davis *et al.*, 1994).

Gilbert and Anderson (1998) list a number of techniques that are employed in the restoration of river channels:

- putting meanders back into watercourses;
- recreating two-stage or multiple channels;
- creating ponds and backwaters;
- removal of flood embankments to allow flooding;
- lowering floodplain surface;
- creating buffer zones along rivers;
- re-profiling banks to include wet and dry areas;
- raising groundwater levels in floodplains;
- adding diverse substrata, including boulders, to channels;
- creation of riffles, channel deflectors;
- planting emergent macrophytes;
- creation of hiding places for fish;
- removal of concrete channels, walls etc.

Figure 9.17 illustrates how pools and riffles can be created to restore habitat diversity in a regulated river. Hey (1996) and Brookes and Shields (1996) consider the options for restoration in some detail while Holmes (1998) and Crafer *et al.* (2000) provide examples of the rehabilitation of rivers. Artificial riffles and stream concentrators, which make water flow more swiftly, were found to increase the diversity of invertebrate species (Friberg *et al.*, 1994; Ebrahimnezhad and Harper, 1997; Harper *et al.*, 1998, 1999; Gortz, 1998) and colonization takes place quite quickly (Friberg *et al.*, 1998a). Wetlands generally respond rapidly to restoration, as has been shown in the Danube delta in Romania (Edwards, 1997).

In North America, dam-building beavers (*Castor canadensis*) historically had a widespread influence on the landscape of river valleys, acting as keystone species. They were exterminated from many rivers by trappers but they are now being used to rehabilitate streams with a long history of abuse (Naiman *et al.*, 1988).

Water-supply reservoirs in upland areas are normally steep-sided, deep and nutrient-poor, of limited value to wildlife. They often drown important habitat and, where compensation with water below the dam is inadequately provided, the river may dry up to the detriment of aquatic life. In contrast, lowland reservoirs are a definite benefit to wildlife, for they are usually shallow, nutrient-rich and flood farmland. An example is Abberton Reservoir, in eastern England, completed in 1940 and covering 490 ha, with a mean depth of only 5 m. It supports a peak number of over 39 000 waterfowl and, as well as supplying water, is a nature reserve of international importance. Rutland Water, in the English Midlands (see p. 166), was completed in 1976, a nature reserve having been designed at its western end in the early stages of planning and covering 80 ha of water, with 160 ha of adjoining land. The rest of the 1260 ha of water is used for sailing and angling, though still

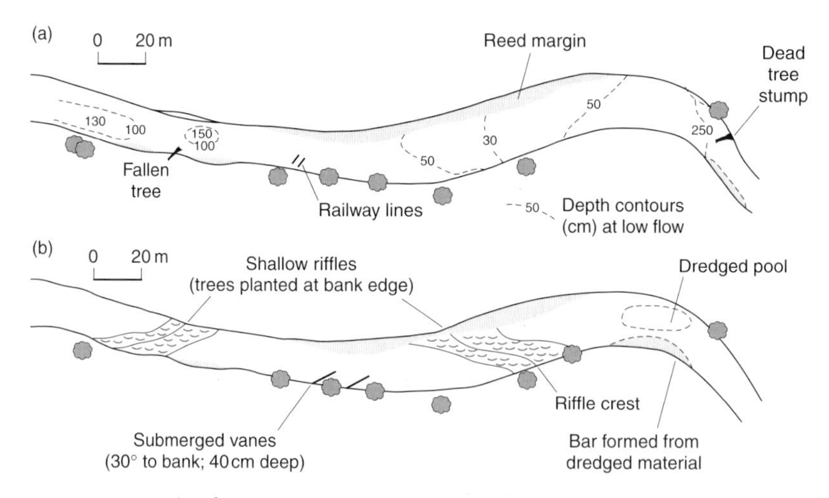

Fig. 9.17 An example of restoration using pools and riffles: (a) original profile, (b) modifications (adapted from Hey, 1996).

supporting wildlife, while the reservoir perimeter has picnic sites, walks and cycle tracks – an excellent example of multiple use of a resource. The wildfowl numbers at Rutland Water quickly became spectacular, with over 40 species recorded and peak totals of over 24 000 birds.

Adjacent to the River Thames in urban west London, a large reservoir became surplus to requirements in the late 1980s. Nine hectares of the site were reclaimed for housing, raising capital for the Wildfowl and Wetlands Trust to convert the remaining 42 ha into a new wetland reserve. Work started in 1995 and the Wetlands Centre opened to the public in 2000. The site has a lake, shallow lagoons, a scrape for wading birds and grazing marshes, together with a state-of-the-art educational and interpretation centre (Branson, 2000). Some 130 species have been recorded annually, even during the construction stage.

The creation of wetlands for wildlife is not a new activity but there is currently much interest in developing new wetlands. Even garden ponds have conservation value, being especially important, for example, for the common frog (*Rana temporaria*) which has largely disappeared from rural habitats. The design of wildlife ponds is described by Williams *et al.* (1997). Wetland creation is discussed by Merritt (1994) and Gilbert and Anderson (1998). Unfortunately very many of the world's key wetlands are threatened by development. Finlayson and Moser (1991) describe the types and values of wetlands throughout the world.

CONCLUSIONS

Many millions of people, and especially children, die from water-related diseases each year in the developing world but the solutions are well-known. Attempts to tackle the problem have been largely unsuccessful because the population is growing at a faster rate than clean water can be provided. The answer to providing clean water lies with politicians rather than with water scientists. It has been more difficult to relate other forms of pollution to adverse effects on human health and welfare. However, many materials, including newly synthesized ones, are constantly added to our waterways as traces in effluents, and the long-term effects of pollutants, acting alone or in combination, remain largely unknown. That populations of some animals have been severely reduced over wide areas by pollutants such as DDT and PCBs, not considered at the time to be environmental contaminants, should warn against complacency.

The water industry will weigh the benefits of pollution control against the costs and the balance may not always tip in favour of the environment. A cut-back in resources for pollution control by the British government in the 1980s led to a clear decline in river water quality, a process that has now been reversed by a greener administration. Politicians may be elected to serve society at large but once in power, they may follow other agendas, as shown by the failure of the international political community to achieve anything concrete towards cutting greenhouse gases (p. 234).

We cannot assume that the environment is safe in the hands of politicians. Neither can we assume that it is entirely safe in the hands of regulatory authorities such as the Environmental Protection Agency or the Environment Agency, for, though notionally independent of government, they are dependent on politicians for their budgets. Citizens must be always vigilant and their voice can be very powerful if expressed through environmental pressure groups such as Friends of the Earth, Greenpeace and the Worldwide Fund for Nature. Biologists have a crucial role, both inside and outside the water industry, in ensuring the sustainability of this most critical of resources – water.

REFERENCES

Abdul-Hussein, M.M. and Mason, C.F. (1988). The phytoplankton community of a eutrophic reservoir. *Hydrobiologia*, **169**, 265–77.

Abel, P.D. (1996). *Water Pollution Biology*. Taylor & Francis, London.

Adrián, M.I. and Delibes, M. (1987). Food habits of the otter (*Lutra lutra*) in the Doñana National Park, S.W. Spain. *J. Zool.*, **212**, 399–406.

Adrian, R. and Deneke, R. (1996). Possible impact of mild winters on zooplankton succession in eutrophic lakes of the Atlantic European area. *Freshwater Biol.*, **36**, 757–70.

Aho, J.M., Gibbons, J.W. and Esch, G.W. (1976). Relationship between thermal loading and parasitism in the mosquitofish. In Esch, G.W. and McFarlane, R.W. (eds) *Thermal Ecology II*, pp. 213–18. Technical Information Service, Springfield, VA.

Aitken, M.N. (1997). Short-term leaf surface adhesion of heavy metals following application of sewage sludge to grassland. *Grass Forage Sci.*, **52**, 73–85.

Ali, M.M., Murphy, K.J. and Abernethy, V.J. (1999). Macrophyte functional variables *versus* species assemblages as predictors of trophic status in flowing waters. *Hydrobiologia*, **415**, 131–8.

Allan, J.D., Erickson, D.L. and Fay, J. (1997). The influence of catchment land use on stream integrity across multiple spatial scales. *Freshwater Biol.*, **37**, 149–61.

Allott, T.E.H., Battarbee, R.W., Curtis, C., Kreiser, A.M., Juggins, S. and Harriman, R. (1995). An empirical model of critical loads for surface waters based on paleolimnological data. In Hornung, M., Sutton, M.A. and Wilson, R.B. (eds) *Mapping and Modelling Critical Loads for Nitrogen – a Workshop Report*, pp. 50–4. ITE, Penicuick.

Altmann, P., Cunningham, J., Dhanesha, U., Ballard, M., Thompson, J. and Marsh, F. (1999). Disturbance of cerebral function in people exposed to drinking water contaminated with aluminium sulphate: a retrospective study of the Camelford water incident. *Brit. Med. J.*, **319**, 807.

Alvo, R., Hussell, D.J.T. and Berrill, M. (1988). The breeding success of common loons (*Gavia immer*) in relation to alkalinity and other lake characteristics in Ontario. *Can. J. Zool.*, **66**, 746–52.

American Public Health Association (1989). *Standard Methods for the Examination of Water and Wastewater*, 17th edn. American Public Health Association, Inc., New York.

Anderson, D.W., Newman, S.H., Kelly, P.R., Herzog, S.K. and Lewis, K.P. (2000). An experimental soft-release of oil-spill rehabilitated American coots

(*Fulica americana*): 1. Lingering effects on survival, condition and behavior. *Environ. Pollut.*, **107**, 285–94.

Andersson, P., Borg, H. and Kärrhage, P. (1995). Mercury in fish muscle in acidified and limed lakes. *Water, Air, Soil Pollut.*, **80**, 889–92.

Andersson, T., Bengtsson, B.-E., Bergqvist, P.-A., Eriksson, T., Larsson, A. and Norrgren, L. (1993). Biochemical and physiological effects in farmed Baltic salmon fed lipids containing xenobiotics extracted from Baltic herring. *J. Aquat. Ecosystem Health*, **2**, 185–96.

Andrén, C., Henrikson, L., Olsson, M. and Nilson, G. (1988). Effects of pH and aluminium on embryonic and early larval stages of Swedish brown frogs *Rana arvalis*, *R. temporaria* and *R. dalmatina*. *Hol. Ecol.*, **11**, 127–35.

Andres, S., Baudrimont, M., Lapaquellerie, Y., Ribeyre, F., Maillet, N., Latouche, C. and Boudou, A. (1999). Field transplantation of the freshwater bivalve *Corbicula fluminea* along a polymetallic gradient (River Lot, France). *Environ. Toxicol. Chem.*, **18**, 2462–71.

Andres, S., Ribeyre, F., Tourencq, J.N. and Boudou, A. (2000). Interspecific comparison of cadmium and zinc contamination in the organs of four fish species along a polymetallic pollution gradient (Lot River, France). *Sci. Total Environ.*, **248**, 11–25.

Annadotter, H., Cronberg, G., Aagren, R., Lundstert, B., Nilsson, P.A. and Strobeck, S. (1999). Multiple techniques for lake restoration. *Hydrobiologia*, **396**, 77–85.

Anonymous (1999). Donaña after the spill. *World Birdwatch*, **21(4)**, 20–2.

Appelberg, M., Degerman, E. and Norrgren, L. (1992). Effects of acidification and liming on fish in Sweden – a review. *Finnish Fish. Res.*, **13**, 77–91.

Armitage, P.D., Moss, D., Wright, J.F. and Furse, M.T. (1983). The performance of a new biological quality score system based on macroinvertebrates over a wide range of unpolluted running-water sites. *Water Res.*, **17**, 333–47.

Armstrong, J., Armstrong, W. and Vanderputten, W.H. (1996a). *Phragmites* dieback–bud and root death, blockages the aeration and vascular systems and the possible role of phytotoxins. *New Phytol.*, **133**, 399–414.

Armstrong, J., Afreenzobayed, F. and Armstrong, W. (1996b). *Phragmites* die-back: sulphide- and acetic acid-induced bud and root death, lignification, and blockages within aeration and vascular systems. *New Phytol.*, **134**, 601–14.

Armstrong, J.D., Braithwaite, V.A. and Fox, M. (1998). The response of wild Atlantic salmon parr to acute reductions in water flow. *J. Anim. Ecol.*, **67**, 292–7.

Arnell, N. (1996). *Global Warming, River Flows and Water Resources*. Wiley, Chichester.

Arthington, A.H. and Mitchell, D.S. (1986). Aquatic invading species. In Groves, R.H. and Burdon, J.J. (eds) *Ecology of Biological Invasions*, pp. 34–53. Cambridge University Press, Cambridge.

Arthur, J.W., Zischke, J.A. and Ericksen, G.L. (1982). Effect of elevated water temperature on macroinvertebrate communities in outdoor experimental channels. *Water Res.*, **16**, 1465–77.

Ashby, J.A., Knapp, E.B. and Ravnborg, H.M. (1998). Involving local organizations in watershed management. In Lutz, E. (ed.) *Agriculture and the Environment*, pp. 118–29. World Bank, Washington, DC.

Aston, R.J. (1973). Tubificids and water quality: a review. *Environ. Pollut.*, **5**, 1–10.

Aston, R.J., Beattie, R.C. and Milner, A.G.P. (1987). Characteristics of spawning sites of the common frog (*Rana temporaria*) with particular reference to acidity. *J. Zool. (Lond.)*, **213**, 233–42.

Atchison, G.J., Henry, M.G. and Sandheinrich, M.B. (1987). Effects of metals on fish behaviour: a review. *Environ. Biol. Fish.*, **18**, 11–25.

Åtland, A. (1998). Behavioural responses of brown trout, *Salmo trutta*, juveniles in concentration gradients of pH and Al – a laboratory study. *Environ. Biol. Fish.*, **53**, 331–45.

Åtland, A. and Barlaup, B.T. (1995). Avoidance of toxic mixing zones by Atlantic salmon (*Salmo salar* L.) and brown trout (*Salmo trutta* L.) in the limed River Audna, southern Norway. *Environ. Pollut.*, **90**, 203–8.

Attrill, M.J. (ed.) (1998). *A Rehabilitated Estuarine Ecosystem. The Environment and Ecology of the Thames Estuary*. Kluwer, Dordrecht.

Atuma, S.S., Linder, C.-E., Wicklund-Glynn, A., Andersson, Ö. and Larsson, L. (1996). Survey of consumption fish from Swedish waters for chlorinated pesticides and polychlorinated biphenyls. *Chemosphere*, **33**, 791–9.

Baldwin, I.G., Harman, M.M.I. and Neville, D.A. (1994). Performance characteristics of a fish monitor for detection of toxic substances. 1. Laboratory trials. *Water Res.*, **28**, 2191–9.

Balls, H., Moss, B. and Irvine, K. (1989). The loss of submerged plants with eutrophication. 1. Experimental design, water chemistry, aquatic plant and phytoplankton biomass in experiments carried out in ponds in Norfolk Broadland. *Freshwater Biol.*, **22**, 71–87.

Balon, E.K. and Holcik, J. (1999). Gabcikova river barrage system: the ecological disaster and economic calamity for the inland delta of the middle Danube. *Environ. Biol. Fish.*, **54**, 1–17.

Barak, N.A.-E. and Mason, C.F. (1989). Heavy metals in water, sediment and invertebrates from rivers in eastern England. *Chemosphere*, **19**, 1709–14.

Barak, N.A.-E. and Mason, C.F. (1990a). Mercury, cadmium and lead in eels and roach: the effects of size, season and locality on metal concentrations in flesh and liver. *Sci. Total Environ.*, **92**, 249–56.

Barak, N.A.-E. and Mason, C.F. (1990b). Mercury, cadmium and lead concentrations in five species of freshwater fish from eastern England. *Sci. Total Environ.*, **92**, 257–63.

Barbour, M.T. and Yoder, C.O. (2000). The multimetric approach to bioassessment, as used in the United States of America. In Wright, J.F., Sutcliffe, D.W. and Furse, M.T. (eds) *Assessing the Biological Quality of Freshwaters*, pp. 281–92. Freshwater Biological Association, Ambleside.

Barko, J.W. and James, W.F. (1998). Effects of submerged aquatic macrophytes in lakes. In Jeppeson, E., Sondergaard, M., Sondergaard, M. and Christofferson, K. (eds) *The Structuring Role of Macrophytes in Lakes*, pp. 197–214. Springer, New York.

Barlaup, B.T., Åtland, A. and Kleiven, E. (1994). Stocking of brown trout (*Salmo trutta* L.) cohorts after liming – effects on survival and growth during five years of reacidification. *Water, Air, Soil, Pollut.*, **72**, 317–30.

Barr, J.F. (1986). Population dynamics of the common loon (*Gavia immer*) associated with mercury-contaminated waters in northwestern Ontario. Canadian Wildlife Service, Occasional Paper no. 56.

Barreto, G.R. and Macdonald, D.W. (2000). The decline and local extinction of a population of water voles, *Arvicola terrestris*, in southern England. *J. Säugertier.*, **65**, 110–20.

Barreto, G.R., Macdonald, D.W. and Strachan, R. (1998). The tightrope hypothesis: an explanation for plummeting water vole numbers in the Thames catchment. In Bailey, R.G., José, P.V. and Sherwood, R.R. (eds) *United Kingdom Floodplains*, pp. 311–27. Westbury, Otley.

Barrett, P.R.F., Littlejohn, J.W. and Curnow, J. (1999). Long-term algal control in a reservoir using barley straw. *Hydrobiologia*, **415**, 309–13.

Battarbee, R.W. (1994a). Diatoms, lake acidification and the Surface Water Acidification Programme (SWAP): a review. *Hydrobiologia*, **274**, 1–7.

Battarbee, R.W. (1994b). Surface water acidification. In Roberts, N. (ed.) *The Changing Global Environment*, pp. 213–41. Blackwell, Oxford.

Battarbee, R.W. and Charles, D.F. (1994). Lake acidification and the role of paleolimnology. In Steinberg, C.F.W. and Wright, R.F. (eds) *Acidification of Freshwater Ecosystems; Implications for the Future*, pp. 51–65. Wiley, Chichester.

Battarbee, R.W., Allott, T.E.H., Kreiser, A.M. and Juggins, S. (1993). Setting critical loads for UK surface waters: the diatom model. In Hornung, M. and Skeffington, R.A. (eds) *Critical Loads: Concept and Applications*. ITE symposium 28, pp. 99–102.

Battini, M., Rocco, V., Lozada, M., Tartarotti, B. and Zagarese, H.E. (2000). Effects of ultraviolet radiation on the eggs of landlocked *Galaxius maculatus* (Galaxiidae, Pisces) in northwestern Patagonia. *Freshwater Biol.*, **44**, 547–52.

Baumann, P.C. and Harshbarger, J.C. (1995). Decline in liver neoplasms in wild brown bullhead catfish after coking plant closes and environmental PAHs plummet. *Environ. Health Perspect.*, **103**, 168–70.

Bayley, M., Nielsen, J.R. and Baatrup, E. (1999). Guppy sexual behaviour as an effect biomarker of estrogen mimics. *Ecotoxicol. Environ. Safety*, **43**, 68–73.

Becker, D.S. and Bigham, G.N. (1995). Distribution of mercury in the aquatic food web of Onondaga Lake, New York. *Water, Air, Soil Pollut.*, **80**, 563–71.

Beebee, T.J.C. (1995). Amphibian breeding and climate. *Nature*, **374**, 219–20.

Beil, S., Timmis, K.N. and Pieper, D.H. (1999). Genetic and biochemical analyses of the tec operon suggest a route for evolution of chlorobenzene degradation genes. *J. Bacteriol.*, **181**, 341–6.

Beklioglu, M. and Moss, B. (1996). Mesocosm experiments on the interaction of sediment influence, fish predation and aquatic plants with the structure of phytoplankton and zooplankton communities. *Freshwater Biol.*, **36**, 315–25.

Beklioglu, M., Carvalho, L. and Moss, B. (1999). Rapid recovery of a shallow hypertrophic lake following sewage effluent diversion: lack of chemical resilience. *Hydrobiologia*, **412**, 5–15.

Béland, P., Deguise, S., Girard, C., Lagacé, A., Martineau, D., Michaud, R., Muir, D.C.G., Norstrom, R.J., Pelletier, E., Ray, S. and Shugart, L.R. (1993). Toxic compounds and health and reproductive effects in St. Lawrence beluga whales. *J. Great Lakes Res.*, **19**, 766–75.

Beltman, D.J., Clements, W.H., Lipton, J. and Cacela, D. (1999). Benthic invertebrate metals exposure, and community-level effects downstream from a hard-rock mine site. *Environ. Toxicol. Chem.*, **18**, 299–307.

Benndorf, J. (1987). Foodweb manipulation without nutrient control: a useful strategy in lake restoration? *Schweiz. Z. Hydrol.*, **49**, 237–48.

Benson, A.J. (2000). Documenting over a century of aquatic introductions in the United States. In Claudi, R. and Leach, J.H. (eds) *Nonindigenous Freshwater Organisms*, pp. 1–31. Lewis, Boca Raton, FL.

Bergman, A. and Olsson, M. (1985). Pathology of Baltic grey seal and ringed seal females with special reference to adrenocortical hyperplasia: is environmental pollution the cause of a widely distributed disease syndrome? *Finn. Game Res.*, **44**, 43–62.

Bergman, E., Hansson, L-A., Persson, A., Strand, J., Romare, P., Enell, M., Granéli, W., Svensson, J.M., Hamrin, S.F., Cronberg, G., Andersson, G. and Bergstrand, E. (1999). Synthesis of theoretical and empirical experiences from nutrient and cyprinid reductions in Lake Ringsjön. *Hydrobiologia*, **404**, 145–56.

Berwick, P.G. (1984). Physical and chemical conditions for microbial oil degradation. *Biotechnol. Bioeng.*, **24**, 1294–305.

Beumer, J.P. and Bacher, G.J. (1982). Species of *Anguilla* as indicators of mercury in the coastal rivers and lakes of Victoria, Australia. *J. Fish Biol.*, **21**, 87–94.

Beurskens, J.E.M. and Stortelder, P.B.M. (1995). Microbial transformations of PCBs in sediments: what can we learn to solve practical problems? *Water Sci. Technol.*, **31**, 99–107.

Bignert, A., Göthberg, A., Jensen, S., Litzén, K., Odsjö, T., Olsson, M. and Reutergårdh, L. (1993). The need for adequate biological sampling in ecotoxicological investigations: a retrospective study of twenty years pollution monitoring. *Sci. Total Environ.*, **128**, 121–39.

Bijlmakers, L.L. and de Swart, E.O.A.M. (1995). Large-scale wetland-restoration of the Ronde Venen, The Netherlands. *Water Sci. Technol.*, **31**, 197–205.

Birkhead, M. and Perrins, C.M. (1986). *The Mute Swan*. Croom Helm, London.

Bishop, C.A., Ng, P., Pettit, K.E., Kennedy, S.W., Stegeman, J.J., Norstrom, R.J. and Brooks, R.J. (1998). Environmental contamination and developmental abnormalities in eggs and hatchlings of the common snapping turtle (*Chelydra serpentina serpentina*) from the Great Lakes–St Lawrence River basin (1989–91). *Environ. Pollut.*, **101**, 143–56.

Bitton, G. (1999). *Wastewater Microbiology*. 2nd edn. Wiley, New York.

Björklund, I., Borg, H. and Johansson, K. (1984). Mercury in Swedish lakes – its regional distribution and causes. *Ambio*, **13**, 118–21.

Black, J.J. and Baumann, P.C. (1991). Carcinogens and cancers in freshwater fishes. *Environ. Health Perspect.*, **90**, 27–33.

Black, P.E. (1987). *Conservation of Water and Related Land Resources*. Rownan and Littlefield, Totowa, NJ.

Blajeski, A., Duffy, L.K. and Bowyer, R.T. (1996). Differences in faecal profiles of porphyrins among river otters exposed to the *Exxon-Valdez* oil-spill. *Biomarkers*, **1**, 262–6.

Blaylock, B.G. (1965). Chromosomal aberrations in a natural population of *Chironomus tentans* exposed to chronic low-level radiation. *Evolution*, **13**, 421–8.

Blewett, D.A., Wright, S.E., Casemore, D.P., Booth, N.E. and Jones, C.E. (1993). Infective dose size studies on *Cryptosporidium parvum* using gnotobiotic lambs. *Water Sci. Technol.*, **27**, 61–4.

Bloch, H. (1999). The European Union Water Framework Directive, taking water policy into the next millenium. *Water Sci. Technol.*, **40**, 67–71.

Blockwell, S.J., Taylor, E.J., Phillips, D.R., Turner, M. and Pascoe, D. (1996a). A scanning electron microscope investigation of the effects of pollutants on the hepatopancreatic ceca of *Gammarus pulex* (L.). *Ecotoxicol. Environ. Safety*, **35**, 209–21.

Blockwell, S.J., Pascoe, D. and Taylor, E.J. (1996b). Effects of lindane on the growth of the freshwater amphipod *Gammarus pulex* (L.). *Chemosphere*, **32**, 1795–803.

Blockwell, S.J., Maund, S.J. and Pascoe, D. (1998). The acute toxicity of lindane to *Hyalella azteca* and the development of a sublethal bioassay based on precopulatory guarding behaviour. *Arch. Environ. Contam. Toxicol.*, **35**, 432–40.

Boar, R.R., Crook, C.E. and Moss, B. (1989). Regression of *Phragmites australis* reedswamps and recent changes of water chemistry in Norfolk Broadland, England. *Aquat. Bot.* **35**, 41–55.

Boar, R.R., Lister, D.H. and Clough, W.T. (1995). Phosphorus loads in a small groundwater-fed river during the 1989–92 East Anglian drought. *Water Res.*, **29**, 2167–73.

Boening, D.W. (2000). Ecological effects, transport, and fate of mercury: a general review. *Chemosphere*, **40**, 1335–52.

Boers, P., Ballegooijen, L. van and Uunk, J. (1991). Changes in phosphorus cycling in a shallow lake due to food web manipulations. *Freshwater Biol.*, **25**, 9–20.

Boers, P., Does, J. van der, Quaak, M. and Vlugt, J. van der (1994). Phosphorus fixation with iron (III) chloride: a new method to combat internal phosphorus loading in shallow lakes? *Arch. Hydrobiol.*, **129**, 339–51.

Boon, P.J., Holmes, N.T.H., Maitland, P.S., Rowell, T.A. and Davies, J. (1997). A system for evaluating rivers for conservation (SERCON): development, structure and function. In Boon, P.J. and Howell, D.L. (eds) *Freshwater Quality: Defining the Indefinable?*, pp. 299–326. The Stationary Office, Edinburgh.

Boorman, L.A. and Fuller, R.M. (1981). The changing status of reedswamps in the Norfolk Broads. *J. Appl. Ecol.*, **18**, 241–69.

Bootsma, M.A., Barendregt, A. and van Alphen, J.C.A. (1999). Effectiveness of reducing external nutrient load entering a eutrophicated shallow lake ecosystem in the Naardermeer nature reserve, The Netherlands. *Biol. Conserv.*, **90**, 193–201.

Boryslawskyj, M., Garrood, A.C., Pearson, J.T. and Woodhead, D. (1987). Rates of accumulation of dieldrin by a freshwater filter feeder: *Sphaerium corneum*. *Environ. Pollut.*, **43**, 3–13.

Bouché, M.-L., Habets, F., Biagianti-Risbourg, S. and Vernrt, G. (2000). Toxic effects and bioaccumulation of cadmium in the aquatic oligochaete *Tubifex tubifex*. *Ecotoxicol. Environ. Safety*, **46**, 246–51.

Bowerman, W.W., Best, D.A., Grubb, T.G., Sikarskie, J.G. and Giesy, J.P. (2000). Assessment of environmental endocrine disruptors in bald eagles of the Great Lakes. *Chemosphere*, **41**, 1569–74.

Bowers, J. (1995).The interface between ecology and economics in catchment management. In Harper, D.M. and Ferguson, A.J.D. (eds) *The Ecological Basis for River Management*, pp. 515–23. Wiley, Chichester.

Brakke, D.F., Baker, J.P., Bohmer, J. *et al.* (1994). Group report: physiological and ecological effects of acidification on aquatic biota. In Steinberg, C.E.W. and Wright, R.F. (eds) *Acidification of Freshwaters: Implications for the Future*, pp. 275–312. Wiley, Chichester.

Branson, A. (2000). The Wetland Centre, London. *British Wildlife*, **11**, 274–7.

Brett, M.T. (1989). Zooplankton communities and acidification processes (a review). *Water, Air, Soil Pollut.*, **44**, 387–414.

Brettum, P. (1996). Changes in the volume and composition of phytoplankton after experimental acidification of a humic lake. *Environ. Int.*, **22**, 619–28.

Breukelaar, A.W., Lammens, E.H.R.R., Klein Breteler, J.G.P. and Tatrai, I. (1994). Effects of benthivorous bream (*Abramis brama*) and carp (*Cyprinus carpio*) on sediment resuspension and concentrations of nutrients and chlorophyll *a*. *Freshwater Biol.*, **32**, 113–21.

Brevik, E.M., Grande, M., Knutzen, J., Polder, A. and Skaare, J.U. (1996). DDT contamination of fish and sediments from Lake Orsjoen, southern Norway: comparison of data from 1975 and 1994. *Chemosphere*, **33**, 2189–200.

Brix, H. (1994). Functions of macrophytes in constructed wetlands. *Water Sci. Technol.*, **29**, 76–8.

Brix, H. and Schierup, H.-H. (1989). The use of aquatic macrophytes in water pollution control. *Ambio*, **18**, 100–7.

Brook, A.J. (1994). Algae. In Maitland, P.S., Boon, P.J. and McLusky, D.S. (eds) *The Fresh Waters of Scotland*, pp. 131–46. Wiley, Chichester.

Brooker, M.P. and Edwards, R.W. (1975). Aquatic herbicides and the control of water weeds. *Water Res.*, **9**, 1–15.

Brookes, A. (1988). *Channelized Rivers: Perspectives for Environmental Management*. Wiley, Chichester.

Brookes, A. and Shields, F.D. (1996). *River Channel Restoration: Guiding Principles for Sustainable Projects*. Wiley, Chichester.

Brooks, A.W., Maltby, L., Saul, A.J. and Calow, P. (1996). A simple indoor artificial stream system designed to study the effects of toxicant pulses on aquatic organisms. *Water Res.*, **30**, 285–90.

Brown, A.F. and Pascoe, D. (1989). Parasitism and host sensitivity to cadmium: an acanthacephalan infection of the freshwater amphipod *Gammarus pulex*. *J. Appl. Ecol.*, **26**, 473–87.

Brown, D.J.A. and Sadler, K. (1989). Fish survival in acid waters. In Morris, R., Taylor, E.W., Brown, D.J.A. and Brown, J.A. (eds) *Acid Toxicity and Aquatic Animals*, pp. 31–44. Cambridge University Press, Cambridge.

Brown, L.R. (1987). Oil-degrading microorganisms. *Chem. Eng. Progr.*, **42**, 35–40.

Brown, M. (1989a). Biodegradation of oil in freshwaters. In Green, J. and Trett, M. (eds) *The Fate and Effects of Oil in Freshwater*, pp. 197–213. Applied Science Publishers, London.

Brown, M. (1989b). Clean-up technology. In Green, J. and Trett, M. (eds) *The Fate and Effects of Oil in Freshwater*, pp. 215–26. Applied Science Publishers, London.

Brown, M.J., Linton, E. and Rees, E.C. (1992). Causes of mortality among wild swans in Britain. *Wildfowl*, **43**, 70–9.

Brown, V.M., Jordan, D.H.M. and Tiller, B.A. (1967). The effect of temperature on the acute toxicity of phenol to rainbow trout in hard water. *Water Res.*, **1**, 587–94.

Brunberg, A.-K. and Boström, B. (1992). Coupling between benthic biomass of *Microcystis* and phosphorus release from the sediments of a highly eutrophic lake. *Hydrobiologia*, **235/236**, 375–85.

Bruns, I., Friese, K., Markert, B. and Krauss, G.-J. (1997). The use of *Fontinalis antipyretica* L. ex Hedw. as a bioindicator for heavy metals. *Sci. Total Environ.*, **204**, 161–76.

Brusle, J. (1991). The eel (*Anguilla* sp.) and organic chemical pollutants. *Sci. Total Environ.*, **102**, 1–19.

Buckton, S.T., Brewin, P.A., Lewis, A., Stevens, P. and Ormerod, S.J. (1998). The distribution of dippers, *Cinclus cinclus* (L.), in the acid-sensitive region of Wales, 1984–95. *Freshwater Biol.*, **39**, 387–96.

Bull, K.R. (1991). The critical load/levels approach to gaseous pollutant emission control. *Environ. Pollut.*, **69**, 105–23.

Burton, S.M., Rundle, S.D. and Jones, M.B. (2001). The relationship between trace metal contamination and stream meiofauna. *Environ. Pollut.*, **111**, 159–67.

Burton, T.M. and Allan, J.W. (1986). Influence of pH, aluminium, and organic matter on stream invertebrates. *Can. J. Fish Aquat. Sci.*, **43**, 1285–9.

Burton, T.M., Stanford, R.M. and Allan, J.W. (1985). Acidification effects on stream biota and organic matter processing. *Can. J. Fish. Aquat. Sci.*, **42**, 669–75.

Bury, R.B. (1972). The effects of diesel fuel on a stream fauna. *Calif. Fish. Game*, **58**, 291–5.

Bush, M.B. (1997). *Ecology of a Changing Planet*. Prentice Hall, Englewood Cliffs, NJ.

Cabelli, V.J., Dufour, A.P., McCabe, L.J. and Levin, M.A. (1983). A marine recreational water quality criterion consistent with indicator concepts and risk analysis. *J. Water Pollut. Control Fed.*, **55** 1306–14.

Cairncross, S. and Feacham, R.G. (1983). *Environmental Health Engineering in the Tropics*. Wiley, Chichester.

Cairns, J. and Atkinson, R.B. (1994). Constructing ecosystems and determining their connectivity to the larger ecological landscape. In Hester, R.E. and Harrison, R.M. (eds) *Mining and its Environmental Impact*, pp. 111–19. Royal Society of Chemistry, Cambridge.

Cairns, J. and Cherry, D.S. (1993). Freshwater multi-species test systems. In Calow, P. (ed.) *Handbook of Ecotoxicology*, vol.1, pp. 101–16. Blackwell, Oxford.

Calamari, D. and Marchetti, R. (1973). The toxicity of mixtures of metals and surfactants to rainbow trout (*Salmo gairdneri* Rich.). *Water Res.*, **7**, 1453–64.

Campbell, C. and Ogden, M. (1999). *Constructed Wetlands in the Sustainable Landscape*. Wiley, Chichester.

Campbell, P.G.C. and Stokes, P.M. (1985). Acidification and toxicity of metals to aquatic biota. *Can. J. Fish Aquat. Sci.*, **42**, 2034–3049.

Campbell, R.N.B., Maitland, P.S. and Lyle, A.A. (1986). Brown trout deformities: an association with acidification? *Ambio*, **15**, 244–5.

Camplin, W.C. and Aarkrog, A. (1989). Radioactivity in north European waters: report of Working Group 2 of CEC Project MARINA. *Fisheries Research Data Report No. 20*. Ministry of Agriculture, Fisheries and Food, Lowestoft.

Canton, S.P. and Chadwick, J.W. (1988). Variability in benthic invertebrate density estimates from stream samples. *J. Freshwater Ecol.*, **4**, 291–7.

Cao, Y., Bark, A.W. and Williams, W.P. (1997). Analysing benthic macroinvertebrate community changes along a pollution gradient: a framework for the development of biotic indices. *Water Res.*, **31**, 884–92.

Caraco, N.F., Cole, J.J., Findlay, S.E.G., Fischer, D.T., Lampman, G.G., Pace, M.L. and Strayer, D.L. (2000). Dissolved oxygen declines in the Hudson River associated with the invasion of the zebra mussel (*Dreissena polymorpha*). *Environ. Sci. Technol.*, **34**, 1204–10.

Carpenter, D. (1998). Human health effects of environmental pollutants: new insights. *Environ. Monit. Assess.*, **53**, 245–58.

Carpenter, S.R. and Kitchell, J.F. (1993). *The Trophic Cascade in Lakes*. Cambridge University Press, Cambridge.

Carpenter, S.R., Fisher, S.G., Grimm, N.B. and Kitchell, J.F. (1992). Global change and freshwater ecosystems. *Annu. Rev. Ecol. Syst.*, **23**, 119–39.

Carson, R. (1962). *Silent Spring*. Houghton-Mifflin, Boston, MA.

Cartwright, R.Y. (1997). Viruses and drinking water. In Sutcliffe, D.W. (ed.) *The Microbiological Quality of Water*, pp. 43–7. The Freshwater Biological Association, Ambleside.

Cave, S. (1991). A green revolution down at the sewer ponds. *Our Planet*, **3(5)**, 10–11.

CEFAS (2000). Monitoring and surveillance of non-radioactive contaminants in the aquatic environment and activities regulating the disposal of wastes at sea, 1997. *Aquatic Monitoring Report No. 52*. CEFAS, Lowestoft.

Chambers, B.J., Garwood, T.W.D. and Unwin, R.J. (2000). Controlling soil water erosion and phosphorus losses from arable land in England and Wales. *J. Environ. Quality*, **29**, 145–50.

Chamier, A.-C. (1987). Effect of pH on microbial degradation of leaf litter in seven streams of the English Lake District. *Oecologia*, **71**, 491–500.

Champ, M.A. and Seligman, P.F. (1996). *Organotin: Environmental Fate and Effects*. Chapman and Hall, London.

Chevreuil, M., Carru, A.-M., Chesterikoff, A., Boet, P., Tales, E. and Allardi, J. (1995). Contamination of fish from different areas of the River Seine (France) by organic (PCB and pesticides) and metallic (Cd, Cr, Fe, Mn, Pb and Zn) micropollutants. *Sci. Total Environ.*, **162**, 31–42.

Chorus, I. and Bartrum, J. (1999). *Toxic Cyanobacteria in Water.* E. & F. N. Spon, London.

Claudi, R. and Leach, J.H. (eds) (2000). *Nonindigenous Freshwater Organisms: Vectors, Biology, and Impacts.* Lewis, Boca Raton, FL.

Claudi, R. and Mackie, G.L. (1994). *Practical Manual for Zebra Mussel Monitoring and Control.* Lewis, Boca Raton, FL.

Cleary, D. and Thornton, I. (1994). The environmental impact of gold mining in the Brazilian Amazon. In Hester, R.E. and Harrison, R.M. (eds) *Mining and its Environmental Impact*, pp. 17–29. Royal Society of Chemistry, Cambridge.

Coghlan, A. (1997). Lamb's liver with cadmium garnish. *New Scientist*, 22 March, 4.

Colborn, T. and Clement, C. (eds) (1992). *Chemically-Induced Alterations in Sexual and Functional Development: The Wildlife/Human Connection.* Princeton Sci. Publ. Co., Princeton, NJ.

Colborn, T., Dumanoski, D. and Myers, J.P. (1996). *Our Stolen Future.* Dutton, New York.

Cole, S. (1998). The emergence of treatment wetlands. *Environ. Sci. Technol.*, **33**, 218–23.

Coles, P. (1996). Cholera cure? You're wearing it. *The Independent*, 16 December.

Connell, D.W. and Miller, G.J. (1984). *The Chemistry and Ecotoxicology of Pollution.* Wiley, New York.

Connell, D.W., Lam, P., Richardson, B. and Wu, R. (1999). *Ecotoxicology – An Introduction.* Blackwell Science, Oxford.

Connor, S. (2000). Chernobyl's contamination zone: home to red deer, beavers, sea eagles and carp. *The Independent*, 6 June, 17.

Constanza, R., D'Arge, R., de Groot, R., Farber, S., Grasso, M., Hannon, B., Limburg, K., Naeem, S., O'Neill, R.V., Paruelo, J., Raskin, R.G., Sutton, P. and van den Belt, M. (1997). The value of the world's ecosystem and natural capital. *Nature*, **387**, 253–60.

Cooke, G.D. (1993). *Restoration and Management of Lakes and Reservoirs.* Lewis, Boca Raton, FL.

Cooney, J.J., Silver, S.A. and Beck, E.A. (1985). Factors influencing hydrocarbon degradation in three freshwater lakes. *Microb. Ecol.*, **11**, 127–37.

Cooper, P.F., Hobson, J.A. and Jones, S. (1989). Sewage treatment by reed bed systems. *J. Inst. Water Environ. Manage.*, **3**, 60–74.

Correa, M. (1987). Physiological effects of metal toxicity on the tropical freshwater shrimp *Macrobrachium carcinus* (L., 1758). *Environ. Pollut.*, **45**, 149–55.

Cosgrove, W.J. and Rijsberman, F.R. (2000). *World Water Vision.* Earthscan, London.

Coutant, C.C., Cox, D.K. and Moored, K.W. (1976). Further studies of cold-shock effects on susceptibility of young channel catfish to predation. In Esch, G.W. and McFarlane, R.W. (eds) *Thermal Ecology II*, pp. 154–8. Technical Information Service, Springfield, VA.

Cover, E.C. and Harrel, R.C. (1978). Sequences of colonization, diversity, biomass and productivity of macroinvertebrates on artificial substrates in a freshwater canal. *Hydrobiologia*, **59**, 81–95.

Cowx, I.G. (1995). Fish stock assessment – a biological basis for sound ecological management. In Harper, D.M. and Ferguson, A.J.D. (eds) *The Ecological Basis for River Management*, pp. 375–88. Wiley, Chichester.

Crafer, M., De Garis, Y., Lutt, N., Newsome, A. and Tagg, A. (2000). River rehabilitation: riding a flow of consents. *Ecos*, **21(2)**, 59–64.

Craig, F. and Craig, P. (1989). *Britain's Poisoned Water.* Penguin, London.

Crane, M., Johnson, I. and Maltby, L. (1996). *In situ* assays for monitoring the toxic impacts of waste in rivers. In Tapp, J.F., Wharfe, J.R. and Hunt, S.M. (eds) *Toxic Impacts of Wastes on the Aquatic Environment*, pp. 116–24. Royal Society of Chemistry, Cambridge.

Cresser, M.S. (2000). The critical loads concept: milestone or millstone for the new millennium? *Sci. Total Environ.*, **249**, 51–62.

Cronberg, G., Annadotter, H. and Lawton, L.A. (1999). The occurrence of toxic blue-green algae in Lake Ringsjön, southern Sweden, despite nutrient reduction and fish biomanipulation. *Hydrobiologia*, **404**, 123–9.

Crossland, M.R. (2000). Direct and indirect effects of the introduced toad *Bufo marinus* (Anura: Bufonidae) on populations of native anuran larvae in Australia. *Ecography*, **23**, 283–90.

Crozier, W.W. (2000). Escaped farmed salmon, *Salmo salar* L., in the Glenarm River, Northern Ireland: genetic status of the wild population 7 years on. *Fish. Manage. Ecol.*, **7**, 437–46.

Cullen, P. and Forsberg, C. (1988). Experiences with reducing point sources of phosphorus to lakes. *Hydrobiologia*, **170**, 321–36.

Curtis, E.J.C. and Curds, C.R. (1971). Sewage fungus in rivers in the United Kingdom: the slime community and its constituent organisms. *Water Res.*, **5**, 1147–59.

Curtis, E.J.C., Delves-Broughton, J. and Harrington, D.W. (1971). Sewage fungus: studies of *Sphaerotilus* slimes using laboratory recirculating channels. *Water Res.*, **5**, 267–79.

Czarnezki, J.M. (1985). Accumulation of lead in fish from Missouri streams impacted by lead mining. *Bull. Environ. Contam. Toxicol.*, **34**, 736–45.

Daldorph, P.G.W. (1999). A reservoir in management-induced transition between ecological states. *Hydrobiologia*, **396**, 325–33.

Daldorph, P. and Price, R. (1994). Long term phosphorus control at three eutrophic reservoirs in south-eastern England. *Arch. Hydrobiol.*, **40**, 231–43.

Dallinger, R., Prosi, F., Segner, H. and Back, H. (1987). Contaminated food and uptake of heavy metals by fish: a review and proposal for further research. *Oecologia*, **73**, 91–8.

Daly, H.B. (1993). Laboratory rat experiments show consumption of Lake Ontario salmon causes behavioural changes: support for wildlife and human research results. *J. Great Lakes Res.*, **19**, 784–8.

Danilov, R.A. and Ekelund, N.G.A. (2000). The use of epiphyton and epilithon data as a base for calculating ecological indices in monitoring of eutrophication in lakes in central Sweden. *Sci. Total Environ.*, **248**, 63–70.

Darwish, A. (1995). Arid waters. *Our Planet*, **7(3)**, 26–7.

Davies, I.J. (1984). Sampling aquatic insect emergence. In Downing, J.A. and Rigler, F.H. (eds) *A Manual on Methods for the Assessment of Secondary Productivity in Fresh Waters*, pp. 161–227. Blackwell, Oxford.

Davies, P.E. (2000). Development of a national river bioassessment system (AUSRIVAS) in Australia. In Wright, J.F., Sutcliffe, D.W.and Furse, M.T. (eds) *Assessing the Biological Quality of Fresh Waters*, pp. 113–24. Freshwater Biological Association, Ambleside.

Davies, P.H. and Woodling, J.D. (1980). Importance of laboratory-derived metal toxicity results in predicting in-stream response of resident salmonids. In Eaton, J.G., Parish, P.R. and Hendricks, D.C. (eds) *Aquatic Toxicology*, pp. 281–99. American Society for Testing and Materials, Philadelphia, PA.

Davis, S.M., Ogden, J.C. and Park, W.A. (eds). (1994). *Everglades – The Ecosystem and its Restoration.* St. Lucie Press, Delray Beach, FL.

Dawson, F.H. (1989). Ecology and management of water plants in lowland streams. *Freshwater Biol. Assoc. Ann. Rep.*, **57**, 43–60.

Dawson, F.H. (1994). Spread of *Crassula helmsii* in Britain. In de Waal, L., Child, L.E., Wade, P.M. and Brock, J.H. (eds) *Ecology and Management of Invasive Riverside Plants*, pp. 1–14. Wiley, Chichester.

Dawson, F.H. and Holland, D. (1999). The distribution in bankside habitats of three alien invasive plants in the UK in relation to the development of control strategies. *Hydrobiologia*, **415**, 193–201.

De Bernardi, R. (1989). Biomanipulation of aquatic food chains to improve water quality in eutrophic lakes. In Ravera, O. (ed.) *Ecological Assessment of Environmental Degradation, Pollution and Recovery*, pp. 195–215. Elsevier, Amsterdam.

De Bisthoven, L.J., Huymans, C., Vannevel, R., Goemans, G. and Ollevier, F. (1997). Field and experimental morphological response of *Chironomus* larvae (Diptera, Nematocera) to xylene and toluene. *Neth. J. Zool.*, **47**, 227–39.

De Bisthoven, L.J., Vermeulen, A. and Ollevier, F. (1998). Experimental induction of morphological deformities in *Chironomus riparius* larvae by chronic exposure to copper and lead. *Arch. Environ. Contam. Toxicol.*, **35**, 249–56.

De Boer, J. and Hagel, P. (1994). Spacial differences and temporal trends of chlorobiphenyls in yellow eel (*Anguilla anguilla*) from inland waters of the Netherlands. *Sci. Total Environ.*, **14**, 155–74.

De Lafontaine, Y., Gagne, F., Blaise, C., Costan, G., Gagnon, P. and Chan, H.M. (2000). Biomarkers in zebra mussels (*Dreissena polymorpha*) for the assessment and monitoring of water quality of the St Lawrence River (Canada). *Aquat. Toxicol.*, **50**, 51–71.

Department of the Environment, Transport and the Regions (2000). *Waterways for Tomorrow*. DETR, London.

Department of Trade and Industry (1999). *UK Energy in Brief.* DTI, London.

Depledge, M.H. (1986). Global implications of Chernobyl. *Mar. Pollut. Bull.*, **17**, 281–2.

Depledge, M. (1989). The rational basis for detection of the early effects of marine pollutants using physiological indicators. *Ambio*, **18**, 301–2.

De Waal, L., Child, L.E. and Wade, P.M. (1995). The management of three alien invasive riparian plants: *Impatiens glandulifera* (Himalayan balsam), *Heracleum mantegazzianum* (giant hogweed) and *Fallopia japonica* (Japanese knotweed). In Harper, D.M. and Ferguson, A.J.D. (eds) *The Ecological Basis for River Management*, pp. 315–21. Wiley, Chichester.

De Waal, L.C., Child, L.E., Wade, P.M. and Brock, J.H. (eds) (1994). *Ecology and Management of Invasive Riverside Plants*. Wiley, Chichester.

De Waal, L.C., Large, A.R.G. and Wade, P.M. (eds). (1998). *Rehabilitation of Rivers: Principles and Implementation.* Wiley, Chichester.

Dheer, J.M.S. (1988). Haematological, haematopoitic and biochemical responses to thermal stress in an air-breathing freshwater fish, *Channa punctatus* Bloch. *J. Fish. Biol.*, **32**, 197–206.

Diamond, J., Collins, M. and Gruber, D. (1988). An overview of automated biomonitoring – past developments and future needs. In Gruber, D. and Diamond, J. (eds) *Automated Biomonitoring: Living Sensors as Environmental Monitors*, pp. 23–39. Ellis Horwood, Chichester.

Dickman, M. and Rygiel, G. (1996). Chironomid larval deformity frequencies, mortality, and diversity in heavy-metal contaminated sediments of a Canadian riverine wetland. *Environ. Internat.*, **22**, 693–703.

Diggins, T.P. and Stewart, K.M. (1993). Deformities of aquatic larval midges (Chironomidae: Diptera) in the sediments of the Buffalo River, New York. *J. Great Lakes Res.*, **19**, 648–59.

Dillon, P.J. and Rigler, F.H. (1975). A simple method for predicting the capacity of a lake for development based upon lake trophic status. *J. Fish. Res. Bd Can.*, **32**, 1519–31.

Dobbs, , A.J. and Zabel, T.F. (1996). Water-quality control. In Petts, G. and Calow, P. (eds) *River Restoration*, pp. 44–59. Blackwell, Oxford.

Dodds, W.K. and Gudder, D.A. (1992). The ecology of *Cladophora*. *J. Phycol.*, **28**, 415–27.

Douben, P.E.T. (1989). Uptake and elimination of waterborne cadmium by the fish *Noemacheilus barbatulus* L. (Stone loach). *Arch. Environ. Contam. Toxicol.*, **18**, 576–86.

Douben, P.E.T. (ed.) (1998). *Pollution Risk Assessment and Management.* Wiley, Chichester.

Downey, D.M., French, C.R. and Odom, M. (1994). Low cost limestone treatment of acid sensitive trout streams in the Appalachian Mountains of Virginia. *Water, Air, Soil Pollut.*, **77**, 49–77.

Downing, J.A. (1984). Sampling the benthos of standing waters. In Downing, J.A. and Rigler, F.H. (eds) *A Manual on Methods for the Assessment of Secondary Productivity in Fresh Waters*, pp. 87–130. Blackwell, Oxford.

Duffy, L.K., Bowyer, R.T., Testa, J.W. and Faro, J.B. (1994). Chronic effects of the *Exxon Valdez* oil spill on blood and enzyme chemistry of river otters. *Environ. Toxicol. Chem.*, **13**, 643–7.

Dunstone, N. (1993). *The Mink.* Poyser, London.

Ebrahimnezhad, M. and Harper, D.M. (1997). The biological effectiveness of artificial riffles in river rehabilitation. *Aquat. Conserv.*, **7**, 187–97.

Edmondson, W.T. (1969). Eutrophication in North America. In *Eutrophication: Causes, Consequences, Correctives*, pp. 124–49. National Academy of Sciences, Washington, DC.

Edmondson, W.T. (1970). Phosphorus, nitrogen and algae in Lake Washington after diversion of sewage. *Science*, **169**, 690–1.

Edmondson, W.T. (1972). Lake Washington. In Goldman, C.R. (ed.) *Environmental Quality and Water Development*, pp. 281–98. Freeman and Co., New York.

Edmondson, W.T. (1991). *The Uses of Ecology: Lake Washington and Beyond.* University of Washington Press, Seattle, WA.

Edwards, R. (1997). Return of the pelican. *New Scientist*, 27 March, 33–5.

Edwards, R.W. (1975). A strategy for the prediction and detection of effects of pollution on natural communities. *Schweiz. Z. Hydrol.*, **37**, 135–43.

Edwards, R.W. (1984). Predicting the environmental impact of a major reservoir development. In Roberts, R.D. and Roberts, T.M. (eds) *Planning and Ecology*, pp. 55–79. Chapman and Hall, London.

Edwards, R.W. (1989). Ecological assessment of degradation and recovery of rivers from pollution. In Ravera, O. (ed.) *Ecological Assessment of Environmental Degradation, Pollution and Recovery*, pp. 159–94. Elsevier, Amsterdam.

Edwards, R.W., Hughes, B.D. and Read, M.W. (1975). Biological survey in the detection and assessment of pollution. In Chadwick, M.J. and Goodman, G.T. (eds) *The Ecology of Resource Degradation and Renewal*, pp. 139–56. Blackwell, Oxford.

Edwards, R.W., Ormerod, S.J. and Turner, C. (1991). Field experiments to assess biological effects of pollution episodes in streams. *Verh. Internat. Verein. Limnol.*, **24**, 1734–7.

Egglishaw, H., Gardiner, R. and Foster, J. (1986). Salmon catch decline and forestry in Scotland. *Scott. Geogr. Mag.*, **102**, 57–61.

Eisler, R. (1995). Ecological and toxicological aspects of the partial meltdown of the Chernobyl nuclear power plant reactor. In Hoffman, D.J., Rattner, B.A., Burton, G.A. and Cairns, J. (eds) *Handbook of Ecotoxicology*, pp. 549–64. Lewis, Boca Raton, FL.

Eklund, M.W. and Dowell, V.R. (1987). *Avian Botulism, an International Perspective*. Charles C. Thomas, Springfield, IL.

Elliott, J.M. (1981). Some aspects of thermal stress in teleosts. In Pickering, A.D. (ed.) *Stress and Fish*, pp. 209–45. Academic Press, London.

Elliott, J.M. (1995). The ecological basis for management of fish stocks in rivers. In Harper, D.M. and Ferguson, A.J.D. (eds) *The Ecological Basis for River Management*, pp. 323–37. Wiley, Chichester.

Elliott, J.M. (2000). Pools as refugia for brown trout during two summer droughts: trout responses to thermal and oxygen stress. *J. Fish Biol.*, **56**, 938–48.

Elliott, J.M. and Drake, C.M. (1981). A comparative study of seven grabs for sampling macroinvertebrates in rivers. *Freshwater Biol.*, **11**, 99–120.

Ellis, J.B., Revitt, D.M., Shutes, R.B.E. and Langley, J.M. (1994). The performance of vegetated biofilters for highway runoff control. *Sci. Total Environ.*, **146/147**, 543–50.

Ellis, J.C. and Hunt, D.T.E. (1986). *Surface Water Acidification: An Assessment of Historic Water Quality Records*. Report TR 240. Water Research Centre, Medmenham.

Ellis, K.V. (1989). *Surface Water Pollution and its Control*. Macmillan, London.

Elser, J.J. (1999). The pathway to noxious cyanobacteria blooms in lakes: the food web as the final turn. *Freshwater Biol.*, **42**, 537–43.

Elton, C.S. (1958). *The Ecology of Invasions by Animals and Plants*. Methuen, London.

Englemann, C.J. and McDiffet, W.F. (1996). Accumulation of aluminium and iron by bryophytes affected by acid-mine drainage. *Environ. Pollut.*, **94**, 76–84.

English Nature (1997). *Wildlife and Fresh Water*. English Nature, Peterborough.

Environment Agency (1998). *Endocrine-Disrupting Substances in the Environment: What Should be Done?* Environment Agency, Bristol.

Environment Agency (2000). *Focus on Biodiversity*. Environment Agency, Bristol.

Environmental Data Services (1998). Climate change 'may be' behind decline in river quality. *ENDS Rep.*, **283**, 10.

Evans, G.P., Johnson, D. and Withell, C. (1986). *Development of the WRC Mk III Fish Monitor: Description of the System and its Response to Some Commonly Encountered Pollutants*. Report TR 233. Water Research Centre, Medmenham.

Everall, N.C. and Lees, D.R. (1996). The use of barley-straw to control general and blue-green algal growth in a Derbyshire reservoir. *Water Res.*, **30**, 269–76.

Everall, N.C. and Lees, D.R (1997). The identification and significance of chemicals released from decomposing barley straw during reservoir algal control. *Water Res.*, **31**, 614–20.

Everard, M. (1996). The importance of periodic droughts for maintaining diversity in freshwater environments. *Freshwater Forum*, **7**, 33–50.

Everard, M. (2000). Aquatic ecology, economy and society: the place of aquatic ecology in the sustainability agenda. *Freshwater Forum*, **13**, 31–46.

Extence, C.A., Bates, A.J., Forbes, W.J. and Barham, P.J. (1987). Biologically based water quality management. *Environ. Pollut.*, **45**, 221–36.

Fallowfield, H.J., Svoboda, I.F. and Martin, N.J. (1992). Aerobic and photosynthetic treatment of animal slurries. In Fry, J.C., Gadd, G.M., Herbert, R.A., Jones, C.W. and Watson-Craik, I.A. (eds) *Microbial Control of Pollution*, pp. 171–97. Cambridge University Press, Cambridge.

Fayer, R. (1997). *Cryptosporidium and Cryptosporidiosis*. CRC Press, Boca Raton, FL.

Fennessy, M.S. and Mitsch, W.J. (1989). Design and use of wetlands for renovation of drainage from coal mines. In Mitsch, W.J. and Jørgensen, S.E. (eds) *Ecological Engineering*, pp. 231–53. Wiley, New York.

Findlay, D.L., Hecky, R.E., Hendzel, L.L., Stainton, M.P. and Regehr, G.W. (1994). Relationship between N_2-fixation and heterocyst abundance and its relevance to the nitrogen budget of Lake 227. *Can. J. Fish. Aquat. Sci.*, **51**, 2254–66.

Finlayson, M. and Moser, M. (1991). *Wetlands*. Facts on File, Oxford.

Flannagan, J.F. (1970). Efficiencies of various grabs and corers in sampling freshwater benthos. *J. Fish. Res. Bd Can.*, **27**, 1691–700.

Flannagan, J.F. and Rosenberg, D.M. (1982). Types of artificial substrates used for sampling freshwater benthic macro-invertebrates. In Cairns, J. (ed.) *Artificial Substrates*, pp. 237–66. Ann Arbor Science Publishers, Ann Arbor.

Fleming, W.J., Clark, D.R. and Henny, C.J. (1983). Organochlorine pesticides and PCBs: a continuing problem for the 1980s. *Trans. North Amer. Wildlife Res. Conf.*, **48**, 186–99.

Flower, R.J. and Battarbee, R.W. (1983). Diatom evidence for recent acidification of two Scottish lochs. *Nature*, **305**, 130–3.

Flower, R.J., Battarbee, R.W. and Appleby, P.G. (1987). The recent palaeolimnology of acid lakes in Galloway, south-west Scotland: diatom analysis, pH trends, and the role of afforestation. *J. Ecol.*, **75**, 797–824.

Focht, D.D. (1995). Strategies for the improvement of aerobic metabolism of polychlorinated biphenyls. *Curr. Opin. Biotechnol.*, **6**, 341–6.

Forbes, V.E. and Forbes, T.L. (1994). *Ecotoxicology in theory and practice*. Chapman and Hall, London.

Forsyth, D.J., Martin, P.A., de Smet, K.D. and Rishe, M.E. (1994). Organochlorine contaminants and eggshell thinning in grebes from Prairie Canada. *Environ. Pollut.*, **85**, 51–8.

Foster, R.B. and Bates, J.M. (1978). Use of mussels to monitor point source industrial discharges. *Environ. Sci. Technol.*, **12**, 958–62.

Fox, G.A. (1993). What have biomarkers told us about the effects of contaminants on the health of fish-eating birds in the Great Lakes? The theory and a literature review. *J. Great Lakes Res.*, **19**, 722–36.

Fox, G.A., Kennedy, S.W., Norstrom, R.J. and Wigfield, D.C. (1988). Porphyria in herring gulls: a biochemical response to chemical contamination of Great Lakes food chains. *Environ. Toxicol. Chem.*, **7**, 831–9.

Fox, G.A., Collins, B., Hayakawa, H., Weseloh, D.V., Ludwig, J.P., Kubiak, T.J. and Erdman, T.C. (1991). Reproductive outcomes in colonial fish-eating birds: a biomarker for development toxicants in Great Lakes food chains. II. Spatial variation in the occurrence and prevalence of bill defects in young double-crested cormorants in the Great Lakes, 1979–87. *J. Great Lakes Res.*, **17**, 158–67.

Foy, R. and Lennox, S. (2000). Contributions of diffuse and point sources to the phosphorus loads in the River Main over a 22-year period. *Boreal Environ. Res.*, **5**, 27–37.

France, R.L. (1996). Biomass and production of amphipods in low alkalinity lakes affected by acid precipitation. *Environ. Pollut.*, **94**, 189–93.

France, R.L. and Welbourn, P.M. (1992). Influence of lake pH and macrograzers on the distribution and abundance of nuisance metaphytic algae in Ontario, Canada. *Can. J. Fish Aquat. Sci.*, **49**, 185–95.

Freda, J. (1986). The influence of acidic pond water on amphibians: a review. *Water, Air, Soil Pollut.*, **30**, 439–50.

Freeman, C., Cresswell, R., Guasch, H., Hudson, J., Lock, M.A., Reynolds, B., Sabater, F. and Sabater, S. (1994). The role of drought in the impact of climate change on the microbiota of peatland streams. *Freshwater Biol.*, **32**, 223–30.

Freitas, R.J. and Burr, M.D. (1996). Animal wastes. In Pepper, I.L., Gerba, C.P. and Brusseau, M.L. (eds) *Pollution Science*, pp. 237–51. Academic Press, San Diego, CA.

Friberg, N., Kronvang, B., Svendsen, L.M., Hansen, H.O. and Nielsen, M.B. (1994). Restoration of a channelized reach of the River Gelsa, Denmark: effects on the macroinvertebrate community. *Aquat. Conserv.*, **4**, 289–96.

Friberg, N., Kronvang, B. Hansen, H.O. and Svendsen, L.M. (1998). Long-term, habitat-specific response of a macroinvertebrate community to river restoration. *Aquat. Conserv.*, **8**, 87–99.

Frick, K.G. and Herrmann, J. (1990). Aluminium and pH effects on sodium ion regulation in mayflies. In Mason, B.J. (ed.) *The Surface Waters Acidification Programme*, pp. 409–12. Cambridge University Press, Cambridge.

Friedrich, G. (1990). A revision of the saprobien system. *Z. Wass. Abwass. Forsch.*, **23**, 141–52.

Friedrich, G. and Müller, D. (1984). Rhine. In Whitton, B.A. (ed.) *Ecology of European Rivers*, pp. 265–315. Blackwell, Oxford.

Fryer, G. (1980). Acidity and species diversity in freshwater crustacean faunas. *Freshwater Biol.*, **10**, 41–5.

Fuglsang, A. (1998). Rehabilitation of rivers by using wet meadows as nutrient filters. In de Waal, L., Large, A.R.G. and Wade, P.M. (eds) *Rehabilitation of Rivers: Principles and Implementation*, pp. 97–111. Wiley, Chichester.

Furse, M.T., Wright, J.F., Armitage, P.D. and Moss, D. (1981). An appraisal of pond-net samples for biological monitoring of lotic macro-invertebrates. *Water Res.*, **6**, 79–89.

Garcia-Criado, F., Tome, A., Vega, F.J. and Antolin, C. (1999). Performance of some diversity and biotic indices in rivers affected by coal mining in northwestern Spain. *Hydrobiologia*, **394**, 209–17.

Gardiner, J. (1996). The use of EIA in delivering sustainable development through integrated water management. *European Water Pollut. Control*, **6**, 50–9.

Gariboldi, J.C., Jagoe, C.H. and Bryan, A.L. (1998). Dietary exposure to mercury in nestling wood storks (*Mycteria americana*) in Georgia. *Arch. Environ. Contam. Toxicol.*, **34**, 398–405.

Garner, P. (1997). Habitat use by 0+ cyprinid fish in the River Great Ouse, East Anglia. *Freshwater Forum*, **8**, 2–27.

Garner, P., Bass, J.A.B. and Collett, G.D. (1996). The effects of weed cutting upon the biota of a large regulated river. *Aquat. Conserv.*, **6**, 21–9.

Garrison, P.A., Tullis, K., Aarts, J.J.M.J.G., Brouwer, A., Giesy, J.P. and Denison, M.S. (1996). Species-specific recombinant cell lines as bioassay systems for the detection of 2,3,7,8-tetrachlorobenzo-*p*-dioxine-like chemicals. *Fund. Appl. Toxicol.*, **30**, 194–203.

Gee, A.S. and Stoner, J.H. (1988). The effects of afforestation and acid deposition on the water quality of upland Wales. In Usher, M.B. and Thompson, D.B.A. (eds) *Ecological Change in the Uplands*, pp. 273–87. Blackwell, Oxford.

George, M. (1992). *The Land Use, Ecology and Conservation of Broadland*. Packard, Chichester.

Gerba, C.P. (1996). Risk assessment. In Pepper, I.L., Gerba, C.P. and Brusseau, M.L. (eds) *Pollution Science*. Academic Press, San Diego.

Gerba, C.P., Walter, R. and Farrah, S.R. (1995). *Water Contamination by Viruses: Occurrence, Detection, Treatment*. Lewis Publishers, Boca Raton, FL.

Gerdeaux, D. (1998). Fluctuations in lake fisheries and global warming. In Jones, J.G., Puncochar, P., Reynolds, C.S. and Sutcliffe, D.W. (eds) *Management of Lakes and Reservoirs During Global Climatic Change*, pp. 263–72. Kluwer, Dordrecht.

Gerhardt, A. (1993). Review of impact of heavy metals on stream invertebrates with special emphasis on acid conditions. *Water, Air, Soil Pollut.*, **66**, 289–314.

Giesy, J.P., Ludwig, J.P and Tillitt, D.E. (1994). Deformities in birds of the Great Lakes region: assigning causality. *Environ. Sci. Technol.*, **28**, 128A–135A.

Gilbert, O.L. and Anderson, P. (1998). *Habitat Creation and Repair*. Oxford University Press, Oxford.

Giller, P.S. and Malmqvist, B. (1998). *The Biology of Streams and Rivers*. Oxford University Press, Oxford.

Girling, A.E., Pascoe, D., Janssen, C.R. *et al.* (2000). Development of methods for evaluating toxicity to freshwater ecosystems. *Ecotoxicol. Environ. Safety*, **45**, 148–76.

Goldschmidt, T., Witte, F. and Wanink, J. (1993). Cascading effects of the introduced Nile perch on the detritivorous/phytoplanktivorous species of the sublittoral areas of Lake Victoria. *Conserv. Biol.*, **7**, 686–700.

Gomot, A. (1998). Toxic effects of cadmium on reproduction, development, and hatching in the freshwater snail *Lymnaea stagnalis* for water quality monitoring. *Ecotoxicol. Environ. Safety*, **41**, 288–97.

Goncalves, E.P.R., Boaventura, R.A.R. and Mouvet, C. (1992). Sediments and aquatic mosses as pollution indicators for heavy metals in the Ave river basin (Portugal). *Sci. Total Environ.*, **114**, 7–24.

Goodyear, K.L. and McNeill, S. (1999). Bioaccumulation of heavy metals by aquatic macroinvertebrates of different feeding guilds: a review. *Sci. Total Environ.*, **229**, 1–19.

Gophen, M., Ochumba, P.B.O. and Kaufman, L.S. (1995). Some aspects of perturbation in the structure and biodiversity of the ecosystem of Lake Victoria (East Africa). *Aquat. Living Resources*, **8**, 27–41.

Gore, J.A. (1997). Water quality in the USA: evolving perpectives and public perception. In Boon, P.J. and Howell, D.L. (eds) *Freshwater Quality: Defining the Indefinable?*, pp. 69–85. The Stationary Office, Edinburgh.

Gore, J.A. and Petts, G.E. (1989). *Alternatives in Regulated River Management*. CRC Press, Boca Raton, FL.

Gortz, P. (1998). Effects of stream restoration on the macroinvertebrate community in the River Esrom, Denmark. *Aquat. Conserv.*, **8**, 115–30.

Gosling, L.M. and Baker, S.J. (1991). Coypu *Myocastor coypus*. In Corbet, G.B. and Harris, S. (eds) *The Handbook of British Mammals*, pp. 267–75. Blackwell, Oxford.

Gower, A.M., Myers, G., Kent, M. and Foulkes, M.E. (1994). Relationships between macroinvertebrate communities and environmental variables in metal-contaminated streams in south-west England. *Freshwater Biol.*, **32**, 199–221.

Grahn, O. (1986). Vegetation structure and primary production in acidified lakes in south-western Sweden. *Experentia*, **42**, 465–70.

Grasman, K.A., Scanlon, P.F. and Fox, G.A. (1998). Reproductive and physiological effects of environmental contaminants in fish-eating birds of the Great Lakes: a review of historical trends. *Environ. Monit. Assess.*, **53**, 117–45.

Graves, J. and Reavey, D. (1996). *Global Environmental Change: Plants, Animals and Communities*. Longman, Harlow.

Gray, N.F. (1985). Heterotrophic slimes in flowing waters. *Biol. Rev.*, **60**, 499–548.

Gray, N.F. (1994). *Drinking Water Quality*. Wiley, Chichester.

Green, D.W.J., Williams, K.A., Hughes, D.R.L., Shaik, G.A.R. and Pascoe, D. (1988). Toxicity of phenol to *Asellus aquaticus* (L.) – effects of temperature and episodic exposure. *Water Res.*, **22**, 225–31.

Green, J. and Trett, M.W. (eds) (1989). *The Fate and Effects of Oil in Freshwater*. Applied Science Publishers, London.

Green, R.H. (1979). *Sampling Design and Statistical Methods for Environmental Biologists*. Wiley, New York.

Grimalt, J.O., Ferrer, M. and MacPherson, E. (1999). The mine tailing accident at Aznalcollar. *Sci. Total Environ.*, **242**, 3–11.

Gruber, D., Frago, C.H. and Rasnake, W.J. (1994). Automated biomonitors – first line of defence. *J. Aquat. Ecosystem Health*, **3**, 87–92.

Güde, H. and Gries, T. (1998). Phosphorus fluxes in Lake Constance. *Arch. Hydrobiol.*, **53**, 505–44.

Guilhermino, L., Diamantino, T., Silva, M.C. and Soares, A.M.V.M. (2000). Acute toxicity test with *Daphnia magna*: an alternative to mammals in the prescreening of chemical toxicity? *Ecotoxicol. Environ. Safety*, **46**, 357–62.

Guillette, L.J., Brock, J.W., Rooney, A.A. and Woodward, A.R. (1999). Serum concentrations of various environmental contaminants and their relationship to sex steroid concentrations and phallus size in juvenile American alligators. *Arch. Environ. Contam. Toxicol.*, **36**, 447–55.

Guiney, P.D., Sykora, J.L. and Keleti, G. (1987). Environmental impact of an aviation kerosene spill on stream water quality in Cambria County, Pennsylvania. *Environ. Toxicol. Chem.*, **6**, 977–88.

Gulati, R.D. (1989). Concept of stress and recovery in aquatic ecosystems. In Ravera, O. (ed.) *Ecological Assessment of Environmental Degradation, Pollution and Recovery*, pp. 81–119. Elsevier, Amsterdam.

Gullbring, P., Hammar, T., Helgée, A., Troedsson, B., Hansson, K. and Hansson, F. (1998). Remediation of PCB-contaminated sediments in Lake Järnsjön: investigations, considerations and remedial actions. *Ambio*, **27**, 374–84.

Haines, T.A. (1981). Acidic precipitation and its consequences for aquatic ecosystems: a review. *Trans. Amer. Fish. Soc.*, **110**, 669–707.

Haines, T.A. and Baker, J.P. (1986). Evidence of fish population responses to acidification in the eastern United States. *Water, Air, Soil Pollut.*, **31**, 605–29.

Håkanson, L., Andersson, T. and Nilsson, A. (1989). Caesium-137 in perch in Swedish lakes after Chernobyl – present situation, relationships and trends. *Environ. Pollut.*, **58**, 195–212.

Hakkari, L. (1992). Effects of pulp and paper mill effluents on fish populations in Finland. *Finn. Fish. Res.*, **13**, 93–106.

Häkkinen, I. and Häsänen, E. (1980). Mercury in eggs and nestlings of the osprey (*Pandion haliaetus*) in Finland and bioaccumulation from fish. *Ann. Zool. Fenn.*, **17**, 131–9.

Hall, R.J., Likens, G.E., Fiance, S.B. and Hendrey, G.R. (1980). Experimental acidification of a stream in the Hubbard Brook experimental forest, New Hampshire. *Ecology*, **61**, 976–89.

Hall, R.J., Driscoll, C.T. and Likens, G.E. (1987). Importance of hydrogen ions and aluminium in regulating the structure and function of stream ecosystems: an experimental test. *Freshwater Biol.*, **18**, 17–43.

Ham, L., Quinn, R. and Pascoe, D. (1995). Effects of cadmium on the predator–prey interaction between the turbellarian *Dendrocoelum lacteum* (Müller, 1774) and the isopod crustacean *Asellus aquaticus* (L.). *Arch. Environ. Contam. Toxicol.*, **29**, 358–65.

Hamilton, D.J. and Ankney, C.D. (1994). Consumption of zebra mussels *Dreissena polymorpha* by diving ducks in Lakes Erie and St. Clair. *Wildfowl*, **45**, 159–66.

Hamilton, R.S. and Harrison, R.M. (1991). *Highway Pollution*. Elsevier, London.

Hancock, S.J. and Buddhavarapu, L. (1993). Control of algae using duckweed (*Lemna*) systems. In Moshiri, G.A. (ed.) *Constructed Wetlands for Water Quality Improvement*, pp. 399–406. Lewis, Boca Raton, FL.

Handy, R.D. and Depledge, M.H. (1999). Physiological responses: their measurement and use as environmental biomarkers in ecotoxicology. *Ecotoxicology*, **8**, 329–49.

Hann, B.J. and Turner, M.A. (2000). Littoral microcrustacea in Lake 302S in the Experimental Lakes Area of Canada: acidification and recovery. *Freshwater Biol.*, **43**, 133–46.

Hansson, L.A., Annadotter, H., Bergman, E., Hamrin, S.F., Jeppeson, E., Kairesalo, T., Luokkanen, E., Nilsson, P.A., Sondergaard, M. and Strand, J. (1998). Biomanipulation as an application of food-chain theory: constraints, synthesis and recommendations for temperate lakes. *Ecosystems*, **1**, 558–74.

Hardman, D.J., McEldowney, S. and Waite, S. (1993). *Pollution: Ecology and Biotreatment*. Longman, Harlow.

Hare, L., Saouter, E., Campbell, P.G.C., Tessier, A., Ribeyre, F. and Boudou, A. (1991). Dynamics of cadmium, lead and zinc exchange between nymphs of the burrowing mayfly *Hexagenia rigida* (Ephemeroptera) and the environment. *Can. J. Fish. Aquat. Sci*, **48**, 39–47.

Hargreaves, J.W., Lloyd, E.J.H. and Whitton, B.A. (1975). Chemistry and vegetation of highly acidic streams. *Freshwater Biol.*, **5**, 563–76.

Harper, D. (1992). *Eutrophication of Freshwaters*. Chapman and Hall, London.

Harper, D., Witkowski, F., Kemp-McCarthy, D. and Crabb, J. (1997). The distribution and abundance of riparian trees in English lowland floodplains. *Global Ecol. Biogeog. Lett.*, **6**, 297–306.

Harper, D., Ebrahimnezhad, M. and Cot, F.C. (1998). Artificial riffles in river rehabilitation: setting the goals and measuring the successes. *Aquat. Conserv.*, **8**, 5–16.

Harper, D.M., Ebrahimnezhad, M., Taylor, E., Dickinson, S., Decamp, O., Verniers, G. and Balbi, T. (1999). A catchment-scale approach to the physical restoration of lowland UK rivers. *Aquat. Conserv.*, **9**, 141–57.

Harrad, S.J., Sewart, A.P., Alcock, R., Boumphrey, R., Burnett, V., Duarte-Davidson, R., Halsall, C., Sanders, G., Waterhouse, K., Wild, S.R. and Jones, K.C. (1994). Polychlorinated biphenyls (PCBs) in the British environment: sinks, sources and temporal trends. *Environ. Pollut.*, **85**, 131–46.

Harriman, R. and Wells, D.E. (1985). Causes and effects of surface water acidification in Scotland. *Water Pollut. Control*, **84**, 215–24.

Harriman, R., Morrison, B.R.S., Caines, L.A., Cullen, P. and Watt, A.W. (1987). Long-term changes in fish populations of acid streams and lochs in Galloway, south west Scotland. *Water, Air, Soil Pollut.*, **32**, 89–112.

Harrison, A.D. (1984). The acidophilic thiobacilli and other acidophilic bacteria that share their habitat. *Annu. Rev. Microbiol.*, **38**, 265–92.

Harrison, R.M. (2001). Chemistry and climate change in the troposphere. In Harrison, R.M. (ed.) *Pollution: Causes, Effects and Control*, 4th edn, pp. 194–219. Royal Society of Chemistry, Cambridge.

Hartmann, J. (1977). Fischereiliche Veranderungen in kulturbedingt eutrophierenden Seen. *Schweiz. Z. Hydrol.*, **39**, 243–54.
Haslam, S.M. (1987). *River Plants of Western Europe.* Cambridge University Press, Cambridge.
Havas, M. and Likens, G.E. (1985). Changes in ^{22}Na influx in *Daphnia magna* (Straus) as a function of elevated Al concentrations in soft water at low pH. *Proc. Natl Acad. Sci. USA*, **82**, 7345–9.
Havas, M., Hutchinson, T.C. and Likens, G.E. (1984). Red herrings in acid rain research. *Environ. Sci. Technol.*, **18**, 176A–186A.
Havens, K. and DeCosta, J. (1985). The effect of acidification in enclosures on the biomass and population size structure of *Bosmina longirostris*. *Hydrobiologia*, **122**, 153–8.
Hawkes, H.A. (1975). River zonation and classification. In Whitton, B. (ed.) *River Ecology*, pp. 312–74. Blackwell, Oxford.
Hawkes, H.A. (1997). Origin and development of the Biological Monitoring Working Party score system. *Water Res.*, **32**, 964–8.
Hawkes, H.A. and Davies, L.J. (1971). Some effects of organic enrichment on benthic invertebrate communities in stream riffles. In Duffey, E.A. and Watt, A.S. (eds) *The Scientific Management of Animal and Plant Communities for Conservation*, pp. 271–93. Blackwell, Oxford.
Haycock, N.E., Pinnay, G. and Walker, C. (1993). Nitrogen retention in river corridors: European perspective. *Ambio*, **22**, 340–1.
Hayes, C.R., Clark, R.G., Stent, R.F. and Redshaw, C.J. (1984). The control of algae by chemical treatment in a eutrophic water supply reservoir. *J. Water Eng. Sci.*, **38**, 149–62.
Hayes, J.T. (1991). Global climate change and water resources. In Wyman, R.L. (ed.) *Global Climate Change and Life on Earth*, pp. 18–42. Routledge, Chapman and Hall, New York.
Heath, R., Steynberg, R., Guglielmi, A. and Maritz, A. (1998). The implications of point source phosphorus management to potable water treatment. *Water Sci. Technol.*, **37**, 343–50.
Heaney, S.I., Correy, J.E. and Lishman, J.P. (1992). Changes of water quality and sediment phosphorus of a small productive lake following decreased phosphorus loading. In Sutcliffe, D.W. and Jones, J.G. (eds) *Eutrophication: Research and Application to Water Supply*, pp. 119–31. Freshwater Biological Association, Ambleside.
Hedtke, S.F. and Puglisi, F.A. (1982). Short-term toxicity of five oils to four freshwater species. *Arch. Environ. Contam. Toxicol.*, **11**, 425–30.
Hellawell, J.M. (1986). *Biological Indicators of Freshwater Pollution and Environmental Management*. Applied Science Publishers, London.
Hemsley-Flint, B. (2000). Classification of the biological quality of rivers in England and Wales. In Wright, J.F., Sutcliffe, D.W. and Furse, M.T. (eds) *Assessing the Biological Quality of Fresh Waters*, pp. 55–69. Freshwater Biological Association, Ambleside.
Henderson-Sellers, B. and Markland, H.R. (1987). *Decaying Lakes*. Wiley, Chichester.
Hendrey, G.R. and Vertucci, F. (1980). Benthic plant communities in acid Lake Colden, New York: *Sphagnum* and the algal mat. In Drablos, D. and Tolan, A. (eds) *Ecological Impact of Acid Precipitation*, pp. 314–15. SNFC Project, Oslo.
Henriksen, A. (1989). Air pollution effects on aquatic ecosystems and their restoration. In Ravera, O. (ed.) *Ecological Assessment of Environmental Degradation, Pollution and Recovery*, pp. 291–312. Elsevier, Amsterdam.
Henriksen, A., Skogheim, O.K. and Rosseland, B.O. (1984). Episodic changes in pH and aluminium-speciation kill fish in a Norwegian salmon river. *Vatten*, **40**, 255–60.

Henriksen, A., Lien, L., Rosseland, B.O., Traaen, T.S. and Seveldrud, I.S. (1989). Lake acidification in Norway: present and predicted fish status. *Ambio*, **18**, 314–21.

Henriksen, A., Kämäri, J., Posch, M. and Wilander, A. (1992). Critical loads of acidity: Nordic surface waters. *Ambio*, **21**, 356–63.

Henrikson, L., Nyman, H.G., Oscarson, H.G. and Stenson, J.A.E. (1980). Trophic changes without changes in the external nutrient loading. *Hydrobiologia*, **68**, 257–63.

Henry, M.G. and Atchison, G.J. (1986). Behavioural changes in social groups of bluegills exposed to copper. *Trans. Amer. Fish. Soc.*, **115**, 590–5.

Herbert, D.W.M. (1961). Freshwater fisheries and pollution control. *Proc. Soc. Water Treat. J.*, **10**, 135–56.

Hernández, L.M., Gomara, B., Fernandez, M., Jiménez, B., Gonzalez, M.J., Baos, R., Hiraldo, F., Ferrer, M., Benito, B., Suñer, M.A., Devesa, V., Muñoz, O. and Montoro, R. (1999). Accumulation of heavy metals and As in wetland birds in the area around Doñana National Park affected by the Aznalcollar toxic spill. *Sci. Total Environ.*, **242**, 293–308.

Hessen, D.O. and Lydersen, E. (1996). The zooplankton story of humic Lake Skjervatjern during whole catchment acidification. *Environ. Int.*, **22**, 643–52.

Hewitt, C.N. (2001). Radioactivity in the environment. In Harrison, R.M. (ed.) *Pollution: Causes, Effects and Control*, 4th edn, pp. 474–99, Royal Society of Chemistry, Cambridge.

Hey, R. (1995). River processes and management. In O'Riordan, T. (ed.) *Environmental Science for Environmental Management*, pp. 131–50. Longman, Harlow.

Hey, R. (1996). Environmentally sensitive river engineering. In Petts, G. and Calow, P. (eds) *River Restoration*, pp. 80–105. Blackwell, Oxford.

Heyman, U. and Lundgren, A. (1988). Phytoplankton biomass and production in relation to phosphorus. *Hydrobiologia*, **170**, 211–27.

Higler, L.W.G. and van Liere, L. (1997). Freshwater quality in Europe: tales from the continent. In Boon, P.J. and Howell, D.L. (eds) *Freshwater Quality: Defining the Indefinable?*, pp. 59–68. The Stationary Office, Edinburgh.

Hill, I.R., Heimbach, F., Leeuwangh, P. and Mathiessen, P. (eds) (1994). *Freshwater Field Tests for Hazard Assessment of Chemicals*. Lewis, Boca Raton, FL.

Hoffman, D.J. (1979). Embryotoxic and teratogenic effects of crude oil on mallard embryos on day one of development. *Bull. Environ. Contam. Toxicol.*, **22**, 632–7.

Hogg, I.D. and Williams, D.D. (1996). Response of stream invertebrates to a global-warming thermal regime; an ecosystem-level manipulation. *Ecology*, **77**, 395–407.

Holdgate, M.W. (1979). *A Perspective of Environmental Pollution*. Cambridge University Press, Cambridge.

Holdich, D. (1991). The native crayfish and threats to its existence. *Brit. Wildlife*, **2**, 141–51.

Holland, D.G. and Harding, J.P.C. (1984). Mersey. In Whitton, B. (ed.) *Ecology of European Rivers*, pp. 113–44. Blackwell, Oxford.

Holmes, N.T.H. (1998). The River Restoration Project and its demonstration sites. In de Waal, L., Large, A.R.G. and Wade, P.M. (eds) *Rehabilitation of Rivers: Principles and Implementation*, pp. 133–48. Wiley, Chichester.

Holmes, N.T.H. (1999a). Recovery of headwater stream flora following the 1989–1992 groundwater drought. *Hydrol. Processes*, **13**, 341–54.

Holmes, N.T.H. (1999b). British river macrophytes – perceptions and uses in the 20th century. *Aquat. Conserv.*, **9**, 535–9.

Holmes, N.T.H., Boon, P.J. and Rowell, T.A. (1998). A revised classification for British rivers based on their aquatic plant communities. *Aquat. Conserv.*, **8**, 555–78.

Horan, N.J. (1998). *Biological Wastewater Treatment Systems*, 2nd edn. Wiley, Chichester.

Hosper, S.H. (1989). Biomanipulation, new perspectives for restoration of shallow, eutrophic lakes in the Netherlands. *Hydrobiol. Bull.*, **23**, 5–10.

Hosper, S.H. (1998). Stable states, buffers and switches: an ecosystem approach to the restoration and management of shallow lakes in the Netherlands. *Water Sci. Technol.*, **37**, 151–64.

Hough, R.A., Fornwall, M.D., Negele, B.J., Thompson, R.L. and Putt, D.A. (1989). Plant community dynamics in a chain of lakes: principal factors in the decline of rooted macrophytes with eutrophication. *Hydrobiologia*, **173**, 199–217.

Houghton, J.T. (1997). *Global Warming: The Complete Briefing*. Cambridge University Press, Cambridge.

Howells, G. and Dalziel, T.R.K. (1995). A decade of studies at Loch Fleet, Galloway (Scotland): a catchment liming project and restoration of a brown trout fishery. *Freshwater Forum*, **5**, 4–38.

Howells, G., Dalziel, T.R.K. and Turnpenny, A.W.H. (1992). Loch Fleet: liming to restore a brown trout fishery. *Environ. Pollut*, **78**, 131–9.

Howells, G.D. and Gammon, K.M. (1984). Role of research in meeting environmental assessment needs for power station siting. In Roberts R.D. and Roberts, T.M. (eds) *Planning and Ecology*, pp. 310–30. Chapman and Hall, London.

Hughes, B.D. (1975). A comparison of four samplers for benthic macro-invertebrates inhabiting coarse river deposits. *Water Res.*, **9**, 61–9.

Hughes, B.D. (1978). The influence of factors other than pollution on the value of Shannon's diversity index for benthic macro-invertebrates in streams. *Water Res.*, **12**, 357–64.

Hughes-Clarke, S.A. and Mason, C.F. (1992). Ecological development of field corner tree plantations on arable land. *Landscape Urban Plann.*, **22**, 59–72.

Hugla, J.L. and Thomé, J.P. (1999). Effects of polychlorinated biphenyls on liver ultrastructure, hepatic monooxygenases, and reproductive success in barbel. *Ecotoxicol. Environ. Safety*, **42**, 265–73.

Hugla, J.L., Philippart, J.C., Kremers, P., Goffinet, G. and Thomé, J.P. (1995). PCB contamination of the common barbel, *Barbus barbus* (Pisces, Cyprinidae), in the River Meuse in relation to hepatic monooxygenase activity and ultrastructural liver change. *Neth. J. Aquat. Ecol.*, **29**, 135–45.

Hultberg, H. and Skeffington, R. (1998). *Experimental Reversal of Acid Rain Effects. The Gårdsjön Roof Project.* Wiley, Chichester.

Hunn, J.B., Cleveland, L. and Little, E.E. (1987). Influence of pH and aluminium on developing brook trout in low calcium water. *Environ. Pollut.*, **43**, 63–73.

Hunt, E.G. and Bischoff, A.I. (1960). Inimical effects on wildlife of DDD application to Clear Lake. *Calif. Fish. Game*, **46**, 91–106.

Hunter, P.R. (1997). *Waterborne Diseases: Epidemiology and Ecology*. Wiley, Chichester.

Hutchinson, N.J., Neary, B.P. and Dillon, P.J. (1991). Validation and use of Ontario's Trophic Status Model for establishing lake development guidelines. *Lake Reserv. Manage.*, **7**, 13–23.

Hynes, H.B.N. (1960). *The Biology of Polluted Waters*. Liverpool University Press, Liverpool.

Ip, H.M.H. and Phillips, D.J.H. (1989). Organochlorine chemicals in human breast milk in Hong Kong. *Arch. Environ. Contam. Toxicol.*, **18**, 490–4.

Irvine, K., Moss, B. and Balls, H. (1989). The loss of submerged plants with eutrophication

II. Relationships between fish and zooplankton in a set of experimental ponds, and conclusions. *Freshwater Biol.*, **22**, 89–107.

Irvine, K., Stansfield, J. and Moss, B. (1991). The use of enclosures to demonstrate the enhancement of *Daphnia* populations when isolated from fish predation in a shallow eutrophic lake. *Mem. Ist. ital. Idrobiol.*, **48**, 325–44.

Jackson, D. (1992). Environmental impact of the United Kingdom nuclear fuel reprocessing industry. In Drake, J.A.G. (ed.) *The Chemical Industry – Friend to the Environment?*, pp. 126–48. Royal Society of Chemistry, London.

Jackson, S.T. and Charles, D.F. (1988). Aquatic macrophytes in Adirondack (New York) lakes: patterns of species composition in relation to environment. *Can. J. Bot.*, **66**, 1449–60.

Jacobsen, D. (1998). The effect of organic pollution on the macroinvertebrate fauna of Ecuadorian highland streams. *Arch. Hydrobiol.*, **143**, 179–95.

Jacobsen, L. and Perrow, M.R. (1998). Predation risk from piscivorous fish influencing the diel use of macrophytes by planktivorous fish in experimental ponds. *Ecol. Freshwater Fish*, **7**, 78–86.

Jacobson, J.L. and Jacobson, S.W. (1993). A 4-year follow-up study of children born to consumers of Lake Michigan fish. *J. Great Lakes Res.*, **19**, 776–83.

Jaeger, D. (1994). Effects of hypolimnetic water aeration and iron-phosphate precipitation on the trophic level of Lake Krupunder. *Hydrobiologia*, **275/276**, 433–44.

Jagoe, C.H., Dallas, C.E., Chesser, R.K., Smith, M.H., Lingenfelser, S.K., Lingenfelser, J.T., Holloman, K. and Lomakin, M. (1998). Contamination near Chernobyl: radiocaesium, lead and mercury in fish and sediment radiocaesium from waters within the 10 km zone. *Ecotoxicology*, **7**, 201–9.

Janzen, F. (1994). Climate change and temperature-dependent sex determination in reptiles. *Proc. Natl Acad. Sci.*, **91**, 7487–90.

Jarvis, P.J. (2000). *Ecological Principles and Environmental Issues.* Prentice Hall, Harlow.

Jenkins, A., Boorman, D. and Renshaw, M. (1996). The U.K. Acid Waters Monitoring Network; an assessment of chemistry data, 1988–93. *Freshwater Biol.*, **36**, 169–78.

Jeppeson, E., Kristensen, P., Jensen, J.P., Søndergaard, M., Mortensen, E. and Lauridsen, T. (1991). Recovery resilience following a reduction in external phosphorus loading of shallow, eutrophic Danish lakes: duration, regulating factors and methods for overcoming resilience. *Mem. Ist. ital. Idrobiol.*, **48**, 127–48.

Jepson, P.D., Bennett, P.M., Allchin, C.R., Law, R.J., Kuiken, T., Baker, J.R., Rogan, E. and Kirkwood, J.K. (1999). Investigating potential associations between chronic exposure to polychlorinated biphenyls and infectious disease mortality in harbour porpoises from England and Wales. *Sci. Total Environ*, **243/244**, 339–48.

Jobling, S., Nolan, M., Tyler, C.R., Brighty, G. and Sumpter, J.P. (1998). Widespread sexual disruption in wild fish. *Environ. Sci. Technol.*, **32**, 2498–506.

Johnes, P.J. and Hodgkinson, R.A (1998). Phosphorus loss from agricultural catchments: pathways and implications for management. *Soil Use Manage.*, **14**, 175–85.

Johnes, P.J., Moss, B. and Phillips, G.L. (1996). The determination of total nitrogen and total phosphorus concentrations in freshwaters from land use, stock headage and population data: testing of a model for use in conservation and water quality management. *Freshwater Biol.*, **36**, 451–73.

Johnson, R.K. and Goedkoop, W. (2000). The 1995 national survey of Swedish lakes and streams: assessment of ecological status using macroinvertebrates. In Wright, J.F.,

Sutcliffe, D.W. and Furse, M.T. (eds) *Assessing the Biological Quality of Fresh Waters*, pp. 229–40. Freshwater Biological Association, Ambleside.

Jones, J.C. and Reynolds, J.D. (1997). Effects of pollution on reproductive behaviour of fishes. *Rev. Fish Biol. Fisheries*, **7**, 463–91.

Jones, K. and Telford, D. (1991). On the trail of a seasonal microbe. *New Scientist*, 6 April, 36–9.

Jones, K.C., Sanders, G., Wild, S.R., Burnett, V.B. and Johnston, A.E. (1992). Evidence for the decline in PCBs and PAHs in rural vegetation and air. *Nature*, **356**, 137–40.

Jonsson, B., Forseth, T. and Ugedal, O. (1999). Chernobyl radioactivity persists in fish. *Nature*, **400**, 417.

Jørgensen, S.E. (1980). *Lake Management*. Pergamon, Oxford.

José, P. (1989). Long-term nitrate trends in the River Trent and four major tributaries. *Regul. Rivers*, **4**, 43–57.

Joyce, C. (1990). Lead poisoning lasts beyond childhood. *New Scientist*, 13 January, 26.

Kadlec, R.H. and Knight, R.L. (1996). *Treatment wetlands*. Lewis, Boca Raton, FL.

Karås, P., Neuman, E. and Sandström, O. (1991). Effects of a pulp mill effluent on the population dynamics of perch *Perca fluviatilis*. *Can. J. Fish. Aquat. Sci.*, **48**, 28–34.

Karim, M.M. (2000). Arsenic in groundwater and health problems in Bangladesh. *Water Res.*, **34**, 304–10.

Karr, J.R. (1991). Biological integrity: a long-neglected aspect of water resource management. *Ecol. Appl.*, **1**, 66–84.

Karr, J.R. (1999). Defining and measuring river health. *Freshwater Biol.*, **41**, 221–34.

Karr, J.R. and Chu, E.W. (1999). *Restoring Life in Running Waters: Better Biological Monitoring*. Island Press, Washington, DC.

Karr, J.R. and Dudley, D.R. (1981). Ecological perspectives on water quality goals. *Environ. Manage.*, **5**, 55–68.

Kazan, J., Sinnott, D. and Kazan, E.D.O. (1987). The toxicity of pyrene in the fish *Pimephales promelas*: synergism by piperonyl butoxide and by ultraviolet light. *Chemosphere*, **16**, 10–12.

Keddy, P.A. (2000). *Wetland Ecology: Principles and Conservation*. Cambridge University Press, Cambridge.

Kedwards, T.J., Blockwell, S.J., Taylor, E.J. and Pascoe, D. (1996). Design of an electronically operated flow-through respirometer and its use to investigate the effects of copper on the respiration rate of the amphipod *Gammarus pulex* (L.). *Bull. Environ. Contam. Toxicol.*, **57**, 610–16.

Kelce, W.R., Stone, C.R., Laws, S.C., Gray, L.E., Kemppainen, J.A. and Wilson, E.M. (1995). Persistent DDT metabolite p,p′-DDE is a potent androgen receptor antagonist. *Nature*, **375**, 581–5.

Kelly, L.A. and Smith, S. (1996). The nutrient budget of a small eutrophic loch and the effectiveness of straw bales in controlling algal blooms. *Freshwater Biol.*, **36**, 411–18.

Kelly, M. (1988). *Mining and the Freshwater Environment*. Elsevier, London.

Kelly, M.G. (1998). The use of the trophic diatom index to monitor eutrophication in rivers. *Water Res.*, **32**, 236–42.

Kelly, M.G. and Whitton, B.A. (1995). The Trophic Diatom Index: a new index for monitoring eutrophication in rivers. *J. Appl. Phycol.*, **7**, 433–44.

Kelly, M.G. and Whitton, B.A. (1998). Biological monitoring of eutrophication in rivers. *Hydrobiologia*, **384**, 55–67.

Kelly, M.G., Girton, C. and Whitton, B.A. (1987). Use of moss-bags for monitoring heavy metals in rivers. *Water Res.*, **21**, 1429–35.

Kemp, J.L., Harper, D.M. and Crosa, G.A. (1999). Use of 'functional habitats' to link ecology with morphology and hydrology in river rehabilitation. *Aquat. Conserv.*, **9**, 159–78.

Khan, R.A. (1999). Study of pearl dace (*Margariscus margarita*) inhabiting a stillwater pond contaminated with diesel fuel. *Bull. Environ. Contam. Toxicol.*, **62**, 638–45.

Khatami, S.H., Pascoe, D. and Learner, M.A. (1998). The acute toxicity of phenol and unionized ammonia, separately and together, to the ephemeropteran *Baetis rhodani* (Pictet). *Environ. Pollut.*, **99**, 379–87.

Kihlström, J.E., Olsson, M., Jensen, S., Johansson, J., Ahlbom, J. and Bergman, A. (1992). Effects of PCB and different fractions of PCB on the reproduction of the mink (*Mustela vison*). *Ambio*, **21**, 563–9.

Kime, D.E. (1995). The effects of pollution on reproduction in fish. *Rev. Fish Biol. Fisheries*, **5**, 52–96.

Kime, D.E. (1999). A strategy for assessing the effects of xenobiotics on fish reproduction. *Sci. Total Environ.*, **225**, 3–11.

King, A. and Clifford, S. (2000). *The River's Voice.* Green Books, Totnes.

King, J.M. and Coley, K.S. (1985). Toxicity of aqueous extracts of natural and synthetic oils to three species of *Lemna*. In Bahner, R.C. and Hansen, D.J. (eds) *Aquatic Toxicology and Hazard Assessment.* American Society for Testing and Materials, Philadelphia, PA.

Klaine, S.J. and Lewis, M.A. (1995). Algal and plant toxicity testing. In Hoffman, D.J., Rattner, B.A., Burton, G.A. and Cairns, J. (eds) *Handbook of Ecotoxicology*, pp. 163–84. Lewis, Boca Raton, FL.

Knudsen, E.E., Steward, C.R., Macdonald, D.D., Williams, J.E. and Reiser, D.W. (1999). *Sustainable Fisheries Management: Pacific Salmon.* Lewis, Boca Raton, FL.

Köck, G., Noggler, M. and Hofer, R. (1996). Pb in otoliths and opercula of Arctic char (*Salvelinus alpinus*) from oligotrophic lakes. *Water Res.*, **30**, 1919–23.

Koivisto, S. (1995). Is *Daphnia magna* an ecologically representative zooplankton species in toxicity tests? *Environ. Pollut.*, **90**, 263–7.

Kraak, M.H.S., Scholten, M.C.T., Peters, W.H.M. and de Kock, W.C. (1991). Biomonitoring of heavy metals in the western European Rivers Rhine and Meuse using the freshwater mussel *Dreissena polymorpha*. *Environ. Pollution*, **74**, 101–14.

Kraak, M.H.S., Kuipers, F., Schoon, H., de Groot, C.J. and Admiraal, W. (1994). The filtration rate of the zebra mussel *Dreissena polymorpha* used for water quality assessment in Dutch rivers. *Hydrobiologia*, **294**, 13–16.

Kratz, K.W., Cooper, S.D. and Melack, J.M. (1994). Effects of single and repeated experimental acid pulses on invertebrates in a high altitude Sierra Nevada stream. *Freshwater Biol.*, **32**, 161–83.

Kristensen, P. (1994). Sensitivity of embryos and larvae in relation to other stages in the life cycle of fish: a literature review. In Müller, R. and Lloyd, R. (eds) *Sublethal and Chronic Effects of Pollutants on Freshwater Fish*, pp. 155–66. Fishing News Books, Oxford.

Kubiak, T.J., Harris, H.J., Smith, L.M., Stalling, D.L., Schwartz, T.R., Trick, J.A., Sileo, L., Docherty, D.E. and Erdman, T.C. (1989). Microcontaminants and reproductive success of the Forster's Tern on Green Bay, Lake Michigan – 1983. *Arch. Environ. Contam. Toxicol.*, **18**, 706–27.

Kudelska, D., Soszka, H. and Cydzik, D. (1997). Polish practice in lake quality assessment. In Boon, P.J. and Howell, D.L. (eds) *Freshwater Quality: Defining the Indefinable?*, pp. 149–54. The Stationary Office, Edinburgh.

Kurmayer, R. and Wanzenböck, J. (1996). Top-down effects of underyearling fish on a phytoplankton community. *Freshwater Biol.*, **36**, 599–609.

Laliberté, G., Proulx, D., de Pauw, N. and de la Noüe, J. (1994). Algal technology in wastewater treatment. *Ergebn. Limnol.*, 42, 283–302.

Lammens, E.H.R.R. (1989). Causes and consequences of the success of bream in Dutch eutrophic lakes. *Hydrobiol. Bull.*, **23**, 11–18.

Lammens, E.H.R.R. (1999). The central role of fish in lake restoration and management. *Hydrobiologia*, **396**, 191–8.

Lampo, M. and de Leo, G.A. (1998). The invasion ecology of the toad *Bufo marinus*: from South America to Australia. *Ecol. Appl.*, **8**, 288–396.

Landis, W.G. and Yu, M-H. (1995). *Introduction to Environmental Toxicology*. Lewis, Boca Raton, FL.

Langford, T.E. (1970). The temperature of a British river upstream and downstream of a heated discharge from a power station. *Hydrobiologia*, **35**, 353–75.

Langford, T.E. (1975). The emergence of insects from a British river, warmed by power station cooling water. Part II. The emergence patterns of some species of Ephemeroptera, Trichoptera and Megaloptera in relation to water temperature and river flow, upstream and downstream of the cooling-water outfalls. *Hydrobiologia*, **47**, 91–133.

Langford, T.E. (1983). *Electricity Generation and the Ecology of Natural Waters*. Liverpool University Press, Liverpool.

Larsson, A., Haux, C. and Sjobeck, M.-L. (1985). Fish physiology and metal pollution: results and experiences from laboratory and field studies. *Ecotoxicol. Environ. Safety*, **9**, 250–81.

Larsson, P. (1984a). Uptake of sediment-released PCBs by the eel *Anguilla anguilla* in static model systems. *Ecol. Bull.*, **36**, 62–7.

Larsson, P. (1984b). Transport of PCBs from aquatic to terrestrial environments by emerging chironomids. *Environ. Pollut.(A)*, **34**, 283–9.

Larsson, P. (1986). Zooplankton and fish accumulate chlorinated hydrocarbons from contaminated sediments. *Can. J. Fish Aquat. Sci.*, **43**, 1463–6.

Latif, M.A., Bodaly, R.A., Johnston, T.A. and Fudge, R.J.P. (2001). Effects of environmental and maternally derived methylmercury on the embryonic and larval stages of walleye (*Stizostedion vitreum*). *Environ. Pollut.*, **111**, 139–48.

Laurén, D.J. and McDonald, D.G. (1987a). Acclimation to copper by rainbow trout, *Salmo gairdneri*: physiology. *Can J. Fish. Aquat. Sci.*, **44**, 99–104.

Laurén, D.J. and McDonald, D.G. (1987b). Acclimation to copper by rainbow trout, *Salmo gairdneri*: biochemistry. *Can. J. Fish. Aquat. Sci.*, **44**, 105–11.

Lauridsen, T.L., Jeppesen, E. and Søndergaard, M. (1994). Colonization and succession of submerged macrophytes in shallow Lake Vaeng during the first five years following fish manipulation. *Hydrobiologia*, **275/276**, 233–42.

Laws, E.A. (1993). *Aquatic Pollution*. Wiley, New York.

Lawson, T. (1996). Brent duck. *Ecos*, **17(2)**, 27–35.

Lawson, T. (1998). Ruddy duck – NGOs close in. *Ecos*, **19(3/4)**, 76–7.

Lawton, L.A. and Codd, G.A. (1991). Cyanobacterial (blue-green algal) toxins and their significance in UK and European waters. *J. Inst. Water Environ. Manage.*, **5**, 460–5.

Layton, A.C., Gregory, B., Schultz, T.W. and Sayler, G.S. (1999). Validation of genetically engineered bioluminescent surfactant resistant bacteria as toxicity assessment tools. *Ecotoxicol. Environ. Safety*, **43**, 222–8.

Leach, J. and Dawson, H. (1999). *Crassula helmsii* in the British Isles – an unwelcome invader. *Brit. Wildlife*, **10**, 234–9.

Lean, G., Hinrichsen, D. and Markham, A. (1990). *Atlas of the Environment*. Hutchinson, London.

Lefcort, H., Meguire, R.A., Wilson, L.H. and Ettinger, W.F. (1998). Heavy metals alter the survival, growth, metamorphosis, and antipredatory behavior of Columbia spotted frog (*Rana luteiventris*) tadpoles. *Arch. Environ. Contam. Toxicol.*, **35**, 447–56.

Lefcort, H., Ammann, E. and Eiger, S.M. (2000). Antipredatory behaviour as an index of heavy-metal pollution? A test using snails and caddisflies. *Arch. Environ. Contam. Toxicol.*, **38**, 311–16.

Lehmann, A. and Lachavanne, J-B. (1999). Changes in the water quality of Lake Geneva indicated by submerged macrophytes. *Freshwater Biol.*, **42**, 457–66.

Leivestad, H., Jensen, E., Kjartansson, H. and Xingfu, L. (1987). Aqueous speciation of aluminium and toxic effects on Atlantic salmon. *Ann. Soc. R. Zool. Belg.*, **117**, 387–98.

Lelek, A. and Kohler, C. (1990). Restoration of fish communities of the Rhine River two years after a heavy pollution wave. *Regul. Rivers*, **5**, 57–66.

Lemonick, M.D. (1988). Nightmare on the Monongahela. *Time*, 18 January, 34–5.

Lester, J.N. and Edge, D. (2001). Sewage and sewage sludge treatment. In Harrison, R.M. (ed.) *Pollution: Causes, Effects and Control*, 4th edn, pp. 113–44. Royal Society of Chemistry, Cambridge.

Levin, R. (1987). Reducing lead in drinking water: a benefit analysis. Office of Policy Planning and Evaluation, US Environmental Protection Agency, Report no. EPA–23–09–86–019, Washington, DC.

Lewis, M.A. (1993). Freshwater primary producers. In Calow, P. (ed.) *Handbook of Ecotoxicology*, vol. 1, pp. 28–50. Blackwell, Oxford.

Leynen, M., van den Berckt, T., Aerts, J.M., Castelein, B., Berckmans, D. and Ollevier, F. (1999). The use of Tubificidae in a biological early warning system. *Environ. Pollution*, **105**, 151–4.

Liere, L. van, Parma, S. and Gulati, R.D. (1992). Working group Water Quality Research Loosdrecht Lakes: its history, structure, research programme, and some results. *Hydrobiologia*, **233**, 1–9.

Likens, G.E., Bormann, F.H., Johnson, N.M., Fisher, D.W. and Pierce, R.S. (1970). Effects of forest cutting and herbicide treatment on nutrient budgets in the Hubbard Brook Watershed ecosystem. *Ecol. Monogr.*, **40**, 23–47.

Lilius, H., Isomaa, B. and Holmström, T. (1994). A comparison of the toxicity of 50 reference chemicals to freshly isolated rainbow trout hepatocytes and *Daphnia magna*. *Aquat. Toxicol.*, **30**, 47–60.

Lisk, D.J. (1991). Environmental effects of landfills. *Sci. Total Environ.*, **100**, 415–68.

Litten, S., Mead, B. and Hassett, J. (1993). Application of passive samplers (PISCES) to locating a source of PCBs in the Black River, New York. *Environ. Toxicol. Chem.*, **12**, 639–47.

Litvak, M.K. and Mandrak, N.E. (2000). Baitfish trade as a vector of aquatic introductions. In Claudi, R. and Leach, J.H. (eds) *Nonindigenous Freshwater Organisms*, pp. 162–80. Lewis, Boca Raton, FL.

Lloyd, R. (1960). The toxicity of zinc sulphate to rainbow trout. *Ann. Appl. Biol.*, **48**, 84–94.

Lloyd, R. (1992). *Pollution and Freshwater Fish*. Fishing News Books, Oxford.

Lockhart, W.L., Wagemann, R., Tracey, B., Sutherland, D. and Thomas, D.J. (1992). Presence and implications of chemical contaminants in the freshwaters of the Canadian Arctic. *Sci. Total Environ.*, **122**, 165–243.

Lodenius, M., Seppänen, A. and Herranen, M. (1983). Accumulation of mercury in fish and man from reservoirs in northern Finland. *Water, Air, Soil Pollut.*, **19**, 237–46.

Loganathan, B.G., Tanabe, S., Hidaka, Y., Kawano, M., Hidaka, H. and Tatsukawa, R. (1993). Temporal trends of persistent organochlorine residues in human adipose tissue from Japan, 1928–85. *Environ. Pollut.*, **81**, 31–9.

Logie, J.W. (1995). Effects of stream acidity on non-breeding dippers *Cinclus cinclus* in the south-central highlands of Scotland. *Aquat. Conserv.*, **5**, 25–35.

Long, S.P. and Mason, C.F. (1983). *Saltmarsh Ecology*. Blackie, Glasgow.

Lonky, E., Reihman, J., Darvill, T., Mather, J. and Daly, H. (1996). Neonatal behavioral assessment scale performance in humans influenced by maternal consumption of environmentally contaminated Lake Ontario fish. *J. Great Lakes Res.*, **22**, 198–212.

Lopez, J., Vazquez, M.D. and Carballeira, A. (1994). Stress responses and metal exchange kinetics following transplant of the aquatic moss *Fontinalis antipyretica*. *Freshwater Biol.*, **32**, 185–98.

Lovett Doust, L., Lovett Doust, J. and Schmidt, M. (1993). In praise of plants as biomonitors – send in the clones. *Functional Ecol.*, **7**, 754–8.

Lovett Doust, J., Schmidt, M. and Lovett Doust, L. (1994). Biological assessment of aquatic pollution; a review, with emphasis on plants as biomonitors. *Biol. Rev.*, **69**, 147–86.

Lowe-McConnell, R.H. (1994). The changing ecosystem of Lake Victoria, East Africa. *Freshwater Forum*, **4**, 76–89.

Ludwig, J.P., Giesy, J.P., Summer, C.L., Bowerman, W., Aulerich, R., Bursian, S., Auman, H.J., Jones, P.D., Williams, L.L., Tillitt, D.E. and Gilbertson, M. (1993). A comparison of water quality criteria for the Great Lakes based on human and wildlife health. *J. Great Lakes Res.*, **19**, 789–807.

Lugo, A.E. (1997). Maintaining an open mind on exotic species. In Meffe, G.K. and Carroll (eds), *Principles of Conservation Biology*, pp. 245–7. Sinauer, Sunderland, MA.

Lükewille, A. (1994). Billion dollar problem, billion dollar solution? Transboundary air pollution calls for transboundary solutions. In Steinberg, C.E.W. and Wright, R.F. (eds) *Acidification of Freshwater Ecosystems: Implications for the Future*, pp. 17–31. Wiley, Chichester.

Lund, J.W.G. (1978). Experiments with lake phytoplankton in large enclosures. *46th Annual Report of the Freshwater Biological Association*, pp. 32–9.

Lundgren, L. (1993). Alternative approaches to evaluating the impact of radionuclide release events – Chernobyl from the Swedish perspective. *Ambio*, **22**, 369–77.

Lynch, M. and Shapiro, J. (1980). Predation, enrichment and phytoplankton community structure. *Limnol. Oceanogr.*, **26**, 86–102.

Mac, M.J., Schwartz, T.R., Edsall, C.C. and Frank, A.M. (1993). Polychlorinated biphenyls in Great Lakes trout and their eggs: relations to survival and congener composition 1979–1988. *J. Great Lakes Res.*, **19**, 752–65.

Macdonald, S.M. and Mason, C.F. (1983). Some factors influencing the distribution of otters (*Lutra lutra*). *Mammal Rev.*, **13**, 1–10.

Macdonald, S.M. and Mason, C.F. (1994). *Status and conservation needs of the otter (Lutra lutra) in the western Palearctic*. Council of Europe, Nature and Environment no. 67, pp. 1–54, Strasbourg.

Madigan, M.T., Martinko, J.M. and Parker, J. (2000). *Biology of Microorganisms*, 9th edn. Prentice Hall, Upper Saddle River, NJ.

Magnuson, J.J., Webster K.E., Assel, R.A. *et al.* (1997). Potential effects of climate changes on aquatic systems: Laurentian Great Lakes and Precambrian Shield region. *Hydrol. Processes*, **11**, 825–71.

Magurran, A.E. (1988). *Ecological Diversity and its Measurement*. Croom Helm, London.

Mahaney, P.A. (1994). Effects of freshwater petroleum contamination on amphibian hatching and metamorphosis. *Environ. Toxicol. Chem.*, **13**, 259–65.

Mainstone, C., Parr, W. and Day, M. (2000). *Phosphorus and River Ecology – Tackling Sewage Inputs*. Environment Agency, Peterborough.

Maitland, P.S. (1995). Ecological impact of angling. In Harper, D.M. and Ferguson, A.J.D. (eds) *The Ecological Basis for River Management*, pp. 443–52. Wiley, Chichester.

Malle, K.-G. (1994). Accidental spills – frequency, importance, control and countermeasures. *Water Sci. Technol.*, **29**, 149–63.

Malmqvist, B. and Hoffsten, P.O. (1999). Influence of drainage fron old mine deposits on benthic macroinvertebrate communities in central Swedish streams. *Water Res.*, **33**, 2415–23.

Maltby, L. (1992). The use of physiological energetics of *Gammarus pulex* to assess toxicity: a study using artificial streams. *Environ. Toxicol. Chem.*, **11**, 79–85.

Maltby, L., Naylor, C. and Calow, P. (1990a). Effect of stress on a freshwater benthic detritivore: scope for growth in *Gammarus pulex*. *Ecotoxicol. Environ. Safety*, **19**, 285–91.

Maltby, L., Naylor, C. and Calow, P. (1990b). Field deployment of a scope for growth assay involving *Gammarus pulex*, a freshwater benthic invertebrate. *Ecotoxicol. Environ. Safety*, **19**, 292–300.

Maltby, L., Forrow, D.M., Boxall, A.B.A., Calow, P. and Betton, C.I. (1995). The effects of motorway run-off on freshwater ecosystems. 1. Field-study. *Environ. Toxicol. Chem.*, **14**, 1079–92.

Mandrak, N.E. (1989). Potential invasion of the Great Lakes by fish species associated with global warming. *J. Great Lakes Res.*, **15**, 306–16.

Manly, B.F.J. (1994). *Multivariate Statistical Methods: A Primer*, 2nd edn. Chapman and Hall, London.

Mann, R.H.K. (1995). Natural factors influencing recruitment success in coarse fish populations. In Harper, D.M. and Ferguson, A.J.D. (eds) *The Ecological Basis for River Management*, pp. 339–48. Wiley, Chichester.

Mantle, A. and Mantle, G. (1992). Impact of low flows on chalk streams and water meadows. *Brit. Wildlife*, **4**, 4–14.

Marsden, M.W., Smith, M.R. and Sargent, R.J. (1997). Trophic status of rivers in the Forth catchment, Scotland. *Aquat. Cons.*, **7**, 211–21.

Marshall, B.E. and Junor, F.J.R. (1981). The decline of *Salvinia molesta* on Lake Kariba. *Hydrobiologia*, **83**, 477–84.

Mason, C.F. (1977a). Populations and production of benthic animals in two contrasting shallow lakes in Norfolk. *J. Anim. Ecol.*, **46**, 147–72.

Mason, C.F. (1977b). The performance of a diversity index in describing the zoobenthos of two lakes. *J. Appl. Ecol.*, **14**, 363–7.

Mason, C.F. (1978). Artificial oases in a lacustrine desert. *Oecologia*, **36**, 93–102.

Mason, C.F. (1986). Invertebrate populations and biomass over four years in a shallow, coastal lagoon. *Hydrobiologia*, **133**, 21–9.

Mason, C.F. (1987). A survey of mercury, lead and cadmium in muscle of British freshwater fish. *Chemosphere*, **16**, 901–6.

Mason, C.F. (1995). River management and mammal populations. In Harper, D.M. and Ferguson, A.J.D. (eds) *The Ecological Basis for River Management*, pp. 289–305. Wiley, Chichester.

Mason, C.F. (1998). Decline of PCB levels in otters (*Lutra lutra*). *Chemosphere*, **36**, 169–71.

Mason, C.F. and Abdul-Hussein, M.M. (1991). Population dynamics and production of *Daphnia hyalina* and *Bosmina longirostris* in a shallow, eutrophic reservoir. *Freshwater Biol.*, **25**, 243–60.

Mason, C.F. and Barak, N.A.-E. (1990). A catchment survey for heavy metals using the eel (*Anguilla anguilla*). *Chemosphere*, **21**, 695–9.

Mason, C.F. and Bryant, R.J. (1975). Changes in the ecology of the Norfolk Broads. *Freshwater Biol.*, **5**, 257–70.

Mason, C.F. and Macdonald, S.M. (1986). *Otters: Ecology and Conservation*. Cambridge University Press, Cambridge.

Mason, C.F. and Macdonald, S.M. (1987). Acidification and otter (*Lutra lutra*) distribution on a British river. *Mammalia*, **51**, 81–7.

Mason, C.F. and Macdonald, S.M. (1988). Radioactivity in otter scats in Britain following the Chernobyl accident. *Water, Air, Soil Pollut.*, **37**, 131–7.

Mason, C.F. and Macdonald, S.M. (1989). Acidification and otter (*Lutra lutra*) distribution in Scotland. *Water, Air, Soil Pollut.*, **43**, 365–74.

Mason, C.F. and Macdonald, S.M. (1990). The riparian woody plant community of regulated rivers in eastern England. *Regul. Rivers*, **5**, 159–66.

Mason, C.F. and Madsen, A.B. (1993). Organochlorine pesticide residues and PCBs in Danish otters (*Lutra lutra*). *Sci. Total Environ.*, **133**, 73–81.

Mason, C.F. and Wren, C.D. (2001). Carnivora. In Shore, R. and Rattner, B. (eds), *Ecotoxicology of wild Mammals*, pp. 315–70. Wiley, Chichester.

Mason, C.F., Macdonald, S.M. and Hussey, A. (1984). Structure, management and conservation value of the riparian woody plant community. *Biol. Conserv.*, **29**, 201–16.

Mason, R.P., Laporte, J.M. and Andres, S. (2000). Factors controlling the bioaccumulation of mercury, methylmercury, arsenic, selenium, and cadmium by freshwater invertebrates and fish. *Arch. Environ. Contam. Toxicol.*, **38**, 283–97.

Matthiessen, P. (2000). Is endocrine disruption a significant ecological issue? *Ecotoxicology*, **9**, 21–4.

Maund, S.J., Taylor, E.J. and Pascoe, D. (1992). Population responses of the freshwater amphipod crustacean *Gammarus pulex* (L.) to copper. *Freshwater Biol.*, **28**, 29–36.

Mayer, F.L. and Ellersieck, M.R. (1988). Experiences with single-species tests for acute toxic effects on freshwater animals. *Ambio*, **17**, 367–75.

McCahon, C.P. and Pascoe, D. (1988a). Culture techniques for three freshwater macroinvertebrate species and their use in toxicity tests. *Chemosphere*, **17**, 2471–80.

McCahon, C.P. and Pascoe, D. (1988b). Cadmium toxicity to the freshwater amphipod *Gammarus pulex* (L.) during the moult cycle. *Freshwater Biol.*, **19**, 197–203.

McCahon, C.P. and Pascoe, D. (1988c). Use of *Gammarus pulex* (L.) in safety evaluation tests: culture and selection of sensitive life stages. *Ecotoxicol. Environ. Safety*, **15**, 245–52.

McCahon, C.P. and Pascoe, D. (1990). Episodic pollution: causes, toxicological effects and ecological significance. *Functional Ecol.*, **4**, 375–83.

McCahon, C.P. and Poulton, M.J. (1991). Lethal and sub-lethal effects of acid, aluminium and lime on *Gammarus pulex* during repeated simulated episodes in a Welsh stream. *Freshwater Biol.*, **25**, 169–78.

McCahon, C.P., Whiles, A.J. and Pascoe, D. (1989a). The toxicity of cadmium to different larval instars of the trichopteran larvae *Agapetus fuscipes* Curtis and the importance of life cycle information to the design of toxicity tests. *Hydrobiologia*, **185**, 153–62.

McCahon, C.P., Brown, A.F., Poulton, M.J. and Pascoe, D. (1989b). Effects of acid, aluminium and lime additions on fish and invertebrates in a chronically acidic Welsh stream. *Water, Air, Soil Pollut.*, **45**, 345–59.

McCarthy, J.F. and Shugart, L.R. (1990). *Biomarkers of Environmental Contamination*. Lewis, Boca Raton, FL.

McCarthy, I.D. and Houlihan, D.F. (1997). The effects of temperature on protein metabolism in fish: the possible consequences for wild Atlantic salmon (*Salmo salar* L.) stocks in Europe as a result of global warming. In Wood, C.M. and McDonald, D.G. (eds) *Global Warming: Implications for Freshwater and Marine Fish*, pp. 51–7. Cambridge University Press, Cambridge.

McFarlane, G.A., King, J.R. and Beamish, R.J. (2000). Have there been recent changes in climate? Ask the fish. *Prog. Oceanogr.*, **47**, 147–69.

McMahon, B.R. and Stuart, S.A. (1989). The physiological problems of crayfish in acid waters. In Morris, R., Taylor, E.W., Brown, D.J.A. and Brown, J.A. (eds) *Acid Toxicity and Aquatic Animals*, pp. 171–99. Cambridge University Press, Cambridge.

Meffe, G.K. (1991). Life history changes in eastern mosquitofish (*Gambusia holbrooki*) induced by thermal elevation. *Can. J. Fish. Aquat. Sci.*, **48**, 60–6.

Meharg, A.A., Osborn, D., Pain, D.J., Sanchez, A. and Naveso, M.A. (1999). Contamination of Doñana food-chains after the Aznalcollar mine disaster. *Environ. Pollut.*, **105**, 387–90.

Meier, P.G., Penrose, D.L. and Polak, L. (1979). The rate of colonization by macro-invertebrates on artificial substrate samplers. *Freshwater Biol.*, **9**, 381–92.

Meijer, M.L., Raat, A.J.P. and Doef, R.W. (1989). Restoration by biomanipulation of Lake Bleiswijkse Zoom (the Netherlands): first results. *Hydrobiol. Bull.*, **23**, 49–57.

Meijer, M.L., Jeppesen, E., van Donk, E. *et al.* (1994). Long-term responses to fish-stock reduction in small, shallow lakes: interpretation of five-year results of four biomanipulation cases in the Netherlands and Denmark. *Hydrobiologia*, **275/276**, 457–66.

Meijer, M.L., De Boois, I., Scheffer, M., Portielje, R. and Hosper, H. (1999). Biomanipulation in shallow lakes in the Netherlands: an evaluation of 18 case studies. *Hydrobiologia*, **409**, 13–30.

Melloul, A.A. and Hassani, L. (1999). *Salmonella* infection in children from the wastewater-spreading zone of Marrakesh city (Morocco). *J. Appl. Microbiol.*, **87**, 536–9.

Meregalli, G., Pluymers, L. and Ollevier, F. (2001). Induction of mouthpart deformities in *Chironomus riparius* larvae exposed to 4-*n*-nonylphenol. *Environ. Pollut.*, **111**, 241–6.

Merritt, A. (1994). *Wetlands, Industry and Wildlife: A Manual of Principles and Practices*. The Wildfowl & Wetlands Trust, Slimbridge.

Mes, J. (1990). Trends in the levels of some chlorinated hydrocarbon residues in adipose tissue of Canadians. *Environ. Pollut.*, **65**, 269–78.

Metcalfe, C.D. (ed.) (1994). Chemical contaminants and fish tumours. *Sci. Total Environ.*, **154**, 1–167.

Metcalfe, J.L. (1989). Biological water quality assessment of running waters based on macroinvertebrate communities: history and present status in Europe. *Environ. Pollut.*, **60**, 101–39.

Metcalfe-Smith, J.L. (1996). Biological water-quality assessment of rivers: use of macroinvertebrate communities. In Petts, G. and Calow, P. (eds) *River Restoration*, pp. 17–43. Blackwell, Oxford.

Meyers-Schöne, L., Shugart, L.R., Beauchamp, J.J. and Walton, B.T. (1993). Comparison of two freshwater turtle species as monitors of radionuclide and chemical contamination – DNA-damage and residue analysis. *Environ. Toxicol. Chem.*, **12**, 1477–96.

Middleboe, A.L. and Markager, S. (1997). Depth limits and minimum light requirements of aquatic macrophytes. *Freshwater Biol.*, **37**, 553–68.

Mierle, G., Clark, K. and France, R. (1986). The impact of acidification on aquatic biota in North America: a comparison of field and laboratory results. *Water, Air, Soil Pollut.*, **31**, 593–604.

Milleman, R.E., Birge, W.J., Black, J.A., Cushman, R.M., Daniels, K.L., Franco, P.J., Giddings, J.M., McCarthy, J.F. and Stewart, A.J. (1984). Comparative acute toxicity to aquatic organisms of components of coal derived synthetic fuels. *Trans. Amer. Fish. Soc.*, **113**, 74–85.

Miller, K.J. (1997). Cryptosporidiosis: a waterborne disease. In Sutcliffe, D.W. (ed.) *The Microbiological Quality of Water*, pp. 139–41. Freshwater Biological Association, Ambleside.

Mills, E.L., Leach, J.H., Carlton, J.T. and Secor, C.L. (1994). Exotic species and the integrity of the great Lakes. *BioScience*, **44**, 666–75.

Ministry of Agriculture, Fisheries and Food (1994). *Radionuclides in Foods*. HMSO, London.

Ministry of Agriculture, Fisheries and Food (2000). *Salmon and Freshwater Fisheries Review*. MAFF, London.

Mitsch, W.J. (1994). *Global Wetlands: Old World and New*. Elsevier, Amsterdam.

Molen, D.T. van der and Boers C.P.M. (1994). Influence of internal loading on phosphorus concentration in shallow lakes before and after reduction of the external loading. *Hydrobiologia*, **275/276**, 379–89.

Moore, M.J. and Myers, M.S. (1994). Pathobiology of chemical-associated neoplasia in fish. In Malins, D.C. and Ostrander, G.K. (eds) *Aquatic Toxicology*, pp. 327–86. Lewis, Boca Raton, FL.

Moore, M.M. (1986). Lead in humans. In Lansdown, R. and Yule, W. (eds) *The Lead Debate: The Environment, Toxicology and Child Health*, pp. 54–95. Croom Helm, London.

Morgan, W.S.G. and Kuhn, P.C. (1988). Effluent discharge control at a South African industrial site utilizing continuous automatic biological surveillance techniques. In Gruber, D. and Diamond, J. (eds) *Automated Biomonitoring: Living Sensors as Environmental Monitors*, pp. 91–103. Ellis Horwood, Chichester.

Morse, G.K., Brett, S.W., Guy, J.A. and Lester, J.N. (1998). Review: phosphorus removal and recovery technologies. *Sci. Total Environ.*, **212**, 69–81.

Moss, B. (1972). Studies on Gull Lake, Michigan. II. Eutrophication – evidence and prognosis. *Freshwater Biol.*, **2**, 309–20.

Moss, B. (1980). Further studies on the palaeolimnology and changes in the phosphorus budget of Barton Broad, Norfolk. *Freshwater Biol.*, **10**, 261–79.

Moss, B. (1983). The Norfolk Broadland: experiments in the restoration of a complex wetland. *Biol. Rev.*, **58**, 521–61.

Moss, B. (1992). The scope for biomanipulation for improving water quality. In Sutcliffe, D.W. and Jones, J.G. (eds) *Eutrophication: Research and Application to Water Supply*, pp. 73–81. Freshwater Biological Association, Ambleside.

Moss, B. and Leah, R.T. (1982). Changes in the ecosystem of a guanotrophic and brackish shallow lake in eastern England: potential problems in its restoration. *Int. Rev. ges. Hydrobiol.*, **67**, 625–39.

Moss, B., Balls, H., Irvine, K. and Stansfield, J. (1986). Restoration of two lowland lakes by isolation from nutrient-rich water sources with and without removal of sediment. *J. Appl. Ecol.*, **23**, 391–414.

Moss, B., Madgwick, J. and Phillips, G. (1996). *A Guide to the Restoration of Nutrient-Rich Shallow Lakes*. Broads Authority/Environment Agency, Norwich.

Moss, B., Johnes, P. and Phillips, G.L. (1997). New approaches to monitoring and classifying standing waters. In Boon, P.J. and Howell, D.L. (eds) *Freshwater Quality: Defining the Indefinable?*, pp. 118–33. The Stationary Office, Edinburgh.

Moss, D., Wright, J.F., Furse, M.T. and Clarke, R.T. (1999). A comparison of alternative techniques for prediction of the fauna of running-water sites in Great Britain. *Freshwater Biol.*, **41**, 167–81.

Motluk, A. (1995). Deadlier than the harpoon? *New Scientist*, 1 July, 12–13.

Mouvet, C. (1985). The use of aquatic bryophytes to monitor heavy metals pollution of freshwaters as illustrated by case studies. *Verh. int. verein. Limnol.*, **22**, 2420–5.

Mouvet, C., Morhain, E., Sutter, C. and Couturieux, N. (1993). Aquatic mosses for the detection and follow-up of accidental discharges in surface waters. *Water, Air, Soil Pollut.*, **66**, 333–48.

Moyle, P.B. and Light, T. (1996). Biological invasions of fresh water: empirical rules and assembly theory. *Biol. Conserv.*, **78**, 149–61.

Moysich, K. and Michalek, A. (2000). Health effects of the Chernobyl nuclear power plant accident – symposium overview. *Epidemiology*, **11**, 753.

Müller, H. (1987). Hydrocarbons in the freshwater environment; a literature review. *Arch. Hydrobiol.*, **24**, 1–69.

Muniz, J.P. (1991). Freshwater acidification: its effects on species and communities of freshwater microbes, plants and animals. *Proc. R. Soc. Edinburgh*, **97B**, 227–54.

Murdoch, M.H. and Hebert, P.D.N. (1994). Mitochondrial DNA diversity of brown bullhead from contaminated and relatively pristine sites in the Great Lakes. *Environ. Toxicol. Chem.*, **13**, 1281–9.

Murk, A.J., Leonards, P.E.G., van Hattum, B., Luit, R., van der Weiden, M.E.J. and Smit, M. (1998). Application of biomarkers for exposure and effect of polyhalogenated aromatic hydrocarbons in naturally exposed European otters (*Lutra lutra*). *Environ. Toxicol. Pharmacol.*, **6**, 91–102.

Murphy, P.M. (1978). The temporal variability in biotic indices. *Environ. Pollut.*, **17**, 227–36.

Myers, M.S., Stehr, C.S., Olson, O.P., Johnson, L.L., McCain, B.B., Chan, S.-L. and Varanasi, U. (1994). Relationships between toxicopathic hepatic lesions and exposure to chemical contaminants in English sole (*Pleuronectes vetulus*), starry flounder (*Platichthys stellatus*) and white croaker (*Genyohemus lineatus*) from selected marine sites on the Pacific coast, U.S.A. *Env. Health Perspective*, **102**, 200–15.

Naiman, R.J., Johnston, C.A. and Kelly, J.C. (1988). Alteration of North American streams by beaver. *BioScience*, **38**, 753–62.

Nakai, S., Inoue, Y., Hosomi, M. and Murakami, A. (2000). *Myriophyllum spicatum*-released allelopathic polyphenols inhibiting growth of blue-green algae *Microcystis aeruginosa*. *Water Res.*, **34**, 3026–32.

Nalepa, T.F., Fahnenstiel, G.L. and Johengen, T.H. (2000). Impacts of the zebra mussel (*Dreissena polymorpha*) on water quality: a case study in Saginaw Bay, Lake Huron. In Claudi, R. and Leach, J.H. (eds) *Nonindigenous Freshwater Organisms*, pp. 255–72. Lewis, Boca Raton, FL.

Nalewajko, C. and Dunstall, T.G. (1994). Miscellaneous pollutants: thermal effluents, halogens, organochlorines, radionuclides. *Ergeb. Limnol.*, **42**, 235–65.

Nasu, Y., Kugimoto, M., Tanaka, O. and Takimoto, A. (1984). *Lemna* as an indicator of water pollution and the absorption of heavy metals by *Lemna*. In Pascoe, D. and Edwards, R.W. (eds) *Freshwater Biological Monitoring*, pp. 113–20. Pergamon, Oxford.

National Research Council (1987). Committee on biological markers. *Environ. Health Perspective*, **74**, 3–9.

Naylor, C., Pindar, L. and Calow, P. (1990). Inter- and intraspecific variation in sensitivity to toxins: the effects of acidity and zinc on the freshwater crustaceans *Asellus aquaticus* (L.) and *Gammarus pulex* (L.). *Water Res.*, **24**, 757–62.

Neal, C., Reynolds, B., Wilkinson, J., Hill, T., Neal, M., Hill, S. and Harrow, M. (1998). The impacts of conifer harvesting on run-off water quality; a regional survey for Wales. *Hydrol. Earth System Sci.*, **2**, 323–44.

Needleman, H.L., Schell, A., Bellinger, D., Leviton, A. and Allred, E.N. (1990). The long-term effects of exposure to low doses of lead in childhood: an 11-year follow-up report. *New Engl. J. Med.*, **322**, 83–8.

Newbold, C. (1975). Herbicides in aquatic systems. *Biol. Conserv.*, **7**, 97–118.

Newman, J.R. and Barrett, P.R.F. (1993). Control of *Microcystis aeruginosa* by decomposing barley straw. *J. Aquat. Plant Manage.*, **31**, 203–6.

Newson, M.D., Harper, D.M., Padmore, C.L., Kemp, J.L. and Vogel, B. (1998). A cost effective approach for linking habitats, flow types and species requirements. *Aquat. Conserv.*, **8**, 431–46.

Noakes, D.J., Beamish, R.J. and Kent, M.L. (2000). On the decline of Pacific salmon and speculative links to salmon farming in British Columbia. *Aquaculture*, **183**, 363–86.

Nriagu, J.O. (1988). A silent epidemic of environmental metal poisoning? *Environ. Pollut.*, **50**, 139–61.

Nuttall, P.M. (1999). *Constructed Wetland Technology*. Wiley, Chichester.

Nyberg, P. (1998). Biotic effects in planktonic crustacean communities in acidified Swedish forest lakes after liming. *Water, Air, Soil Pollut.*, **101**, 257–88.

O'Brien, M. (1998). Distribution and conservation of breeding waders on floodplain grasslands in the British Isles. In Bailey, R.G., José, P.V. and Sherwood, B.R. (eds) *United Kingdom Floodplains*, pp. 301–10. Westbury, Otley.

O'Donoghue, P.J. (1995). *Cryptosporidium* and cryptosporidiosis. *Int. J. Parasitol.*, **25**, 139–95.

Odum, H.T. (2000). *Heavy Metals in the Environment. Using Wetlands for their Removal.* Lewis, Boca Raton, FL.

O'Halloran, J., Meyers, A.A. and Duggan, P.F. (1989). Some sub-lethal effects of lead on mute swan *Cygnus olor. J. Zool. Lond.*, **218**, 627–32.

Økland, J. and Økland, K.A. (1980). pH level and food organisms for fish: studies in 1000 lakes in Norway. In Drablos, D. and Tollan, A. (eds) *Ecological Impact of Acid Precipitation*, pp. 326–7. SMSF, Oslo.

Olaveson, M.M. and Nalewajko, C. (1994). Acid rain and freshwater algae. *Ergebn. Limnol.*, **42**, 99–123.

Olsson, M. and Reutergardh, L. (1986). DDT and PCB pollution trends in the Swedish aquatic environment. *Ambio*, **15**, 103–9.

Openshaw, S. (1992). Radiation and the environment; types, sources, impacts and management. In Newson, M. (ed.) *Managing the Human Impact on the Natural Environment: Patterns and Processes*, pp. 213–31. Belhaven Press, London.

Organization for Economic Cooperation and Development (1982). *Eutrophication of Waters: Monitoring, Assessment and Control.* OECD, Paris.

Oris, J.T. and Giesy, J.P. (1987). The photo-induced toxicity of polycyclic aromatic hydrocarbons to larvae of the fathead minnow (*Pimephales promelas*). *Chemosphere*, **16**, 1395–404.

Ormerod, S.J. and Edwards, R.W. (1985). Stream acidity in some areas of Wales in relation to historical trends in afforestation and the usage of agricultural limestone. *J. Environ. Manage.*, **20**, 189–97.

Ormerod, S.J. and Tyler, S.J. (1993). Birds as indicators of change in water quality. In Furness, R.W. and Greenwood, J.J.D. (eds) *Birds as Monitors of Environmental Change*, pp. 179–216. Chapman and Hall, London.

Ormerod, S.J. and Wade, K.R. (1990). The role of acidity in the ecology of Welsh lakes and streams. In Edwards, R.W., Gee, A.S. and Stoner, J.H. (eds) *Acid Waters in Wales*, pp. 93–119. Kluwer, Dordrecht.

Ormerod, S.J., Allinson, N., Hudson, D. and Tyler, S.J. (1986). The distribution of breeding dippers (*Cinclus cinclus* (L.); Aves) in relation to stream acidity in upland Wales. *Freshwater Biol.*, **16**, 501–7.

Ormerod, S.J., Wade, K.R. and Gee, A.S. (1987a). Macro-floral assemblages in upland Welsh streams in relation to acidity, and their importance to invertebrates. *Freshwater Biol.*, **18**, 545–57.

Ormerod, S.J., Boole, P., McCahon, C.P., Weatherley, N.S., Pascoe, D. and Edwards, R.W. (1987b). Short-term experimental acidification of a Welsh stream: comparing the biological effects of hydrogen ions and aluminium. *Freshwater Biol.*, **17**, 341–56.

Ormerod, S.J., Donald, A.P. and Brown, S.J. (1989). The influence of plantation forestry on the pH and aluminium concentration of upland Welsh streams: a re-examination. *Environ. Pollut.*, **62**, 47–62.

Ormerod, S.J., Weatherley, N.S. and Gee, A.S. (1990). Modelling the ecological impact of changing acidity in Welsh streams. In Edwards, R.W., Gee, A.S. and Stoner, J.H. (eds) *Acid Waters in Wales*, pp. 279–98. Kluwer, Dordrecht.

Ormerod, S.J., Tyler, S.J. and Jüttner, I. (2000). Effects of point-source PCB contamination on breeding performance and post-fledging survival in the dipper *Cinclus cinclus*. *Environ. Pollut.*, **109**, 505–13.

Örn, S., Andersson, P.L., Förlin, L., Tysklind, M. and Norrgren, L. (1998). The impact on reproduction of an orally administered mixture of selected PCBs in zebrafish (*Danio rerio*). *Arch. Environ. Contam. Toxicol.*, **35**, 52–7.

Osborne, P.L. (1981). Phosphorus and nitrogen budgets of Barton Broad and predicted effects of a reduction in nutrient loading on phytoplankton biomass in Barton, Sutton and Stalham Broads, Norfolk, United Kingdom. *Int. Rev. ges. Hydrobiol.*, **66**, 171–202.

Oskam, G. and Breeman, L. van (1992). Management of Biesbosch Reservoirs for quality control with special reference to eutrophication. In Sutcliffe, D.W. and Jones, J.G. (eds) *Eutrophication: Research and Application to Water Supply*, pp. 197–213. Freshwater Biological Association, Ambleside.

Ostendorp, W. (1989). Die-back of reeds in Europe – a critical review of literature. *Aquat. Bot.*, **35**, 5–26.

Pain, D.J., Sanchez, A. and Meharg, A.A. (1998). The Doñana ecological disaster: contamination of a world heritage estuarine marsh ecosystem with acidified pyrite mine waste. *Sci. Total Environ.*, **222**, 45–54.

Palmateer, G.A., Dutka, B.J., Janzen, E.M., Meissner, S.M. and Sakellaris, M.G. (1991). Coliphage and bacteriophage as indicators of recreational water quality. *Water Res.*, **25**, 355–7.

Palmer, M.A., Bell, S.L. and Butterfield, I. (1992). A botanical classification of standing waters in Britain: applications for conservation and monitoring. *Aquat. Conserv.*, **2**, 125–43.

Panter, G.H., Thompson, R.S. and Sumpter, J.P. (1998). Adverse reproductive effects in male fathead minnows (*Pimephales promelas*) exposed to environmentally relevant concentrations of the natural oestrogens, oestradiol and oestrone. *Aquat. Toxicol.*, **42**, 243–53.

Pantle, R. and Buck, H. (1955). Die biologische Uberwachung der Gewasser und die Darstellung der Ergebnisse. *Gas-u Wasserfach*, **96**, 604.

Parnell, J.F., Shields, M.A. and Frierson, D. (1985). Hatching success of brown pelicans (*Pelicanus occidentalis*) eggs after contamination with oil. *Colon. Waterbirds*, **7**, 22–4.

Paschal, D.C., Burt, V., Caudill, S.P., Gunter, E.W., Pirkle, J.L., Sampson, E.J., Miller, D.T. and Jackson, R.J. (2000). Exposure of the US population aged 6 years and older to cadmium: 1988–94. *Arch. Environ. Contam. Toxicol.*, **38**, 377–83.

Pascoe, D. and Beattie, J.H. (1979). Resistance to cadmium by pretreated rainbow trout alevins. *J. Fish Biol.*, **14**, 303–8.

Pascoe, D., Williams, K.A. and Green, D.W.J. (1989). Chronic toxicity of cadmium to *Chironomus riparius* Meigen – effects upon larval development and adult emergence. *Hydrobiologia*, **175**, 109–15.

Pascoe, D., Gower, D.E., McCahon, C.P., Poulton, M.J., Whiles, A.J. and Wulfhorst, J. (1991). Behavioural responses to pollutants – application in freshwater bioassays. In Jeffrey, D.W and Madden, B. (eds) *Bioindicators and Environmental Management*, pp. 245–54. Academic Press, London.

Payne, A.G. (1975). Responses of the three test algae of the Algal Assay Procedure: Bottle Test. *Water Res.*, **9**, 437–45.

Peakall, D. (1992). *Animal Biomarkers as Pollution Indicators*. Chapman and Hall, London.

Peakall, D. and Fairbrother, A. (1998). Biomarkers for monitoring and measuring effects. In Douben, P.E.T. (ed.) *Pollution Risk Assessment and Management*, pp. 351–76. Wiley, Chichester.

Pearce, F. (1995). Poisoned waters. *New Scientist*, 21 October, 29–33.

Pearce, F. (1996). A heavy responsibility. *New Scientist*, 27 July, 12–13.

Pearce, F. (1997a). Why is the apparently pristine Arctic full of toxic chemicals that started off thousands of kilometres away? *New Scientist*, 31 May, 24–7.

Pearce, F. (1997b). Sheep dips poison river life. *New Scientist*, 11 January, 4.

Peckarsky, B.L. (1984). Sampling the stream benthos. In Downing, J.A. and Rigler, F.H. (eds) *A Manual on Methods for the Assessment of Secondary Productivity in Freshwaters*, pp. 131–60. Blackwell, Oxford.

Pelley, J. (1999). Nearly 100,000 lakes in danger despite acid rain cuts. *Environ. Sci. Technol.*, **33**, 352A–353A.

Pepper, I.L., Gerba, C.P. and Brendecke, J.W. (1995). *Environmental Microbiology. A Laboratory Manual.* Academic Press, London.

Perdikaki, K. and Mason, C.F. (1999). Impact of road run-off on receiving streams in eastern England. *Water Res.*, **33**, 1627–33.

Perrins, C.M. and Sears, J. (1991). Collisions with overhead wires as a cause of mortality in Mute Swans *Cygnus olor*. *Wildfowl*, **42**, 5–11.

Perrow, M.R., Moss, B. and Stansfield, J. (1994). Trophic interactions in a shallow lake following a reduction in nutrient loading: a long-term study. *Hydrobiologia*, **275/276**, 43–52.

Perrow, M.R., Schutten, J.H., Howes, J.R., Holzer, T., Madgwick, F.J. and Jowitt, A.J.D.

(1997). Interactions between coot (*Fulica atra*) and submerged macrophytes; the role of birds in the restoration process. *Hydrobiologia*, **342**, 241–55.

Perrow, M.R., Jowitt, A.J.D., Stansfield, J.H. and Phillips, G.L. (1999). The practical importance of interactions between fish, zooplankton and macrophytes in shallow lake restoration. *Hydrobiologia*, **396**, 199–210.

Perry, J.A., Troelstrup, N.H., Newsom, N. and Shelley, B. (1987). Results of a recent whole ecosystem manipulation: the search for generality. *Water Sci. Technol.*, **19**, 55–72.

Petersen, R.C., Landner, L. and Blanck, H. (1986). Assessment of the impact of the Chernobyl reactor accident on the biota of Swedish streams and lakes. *Ambio*, **15**, 327–31.

Petts, G.E. (1984). *Impounded Rivers: Perspectives for Ecological Management*. Wiley, Chichester.

Petts, G.E. (1994). Large-scale river regulation. In Roberts, N. (ed.) *The Changing Global Environment*, pp. 262–84. Blackwell, Oxford.

Petts, G.E. and Maddock, I. (1996). Flow allocation for in-river needs. In Petts, G.E. and Calow, P. (eds) *River Restoration*, pp. 60–79. Blackwell, Oxford.

Petts, G., Maddock, I., Bickerton, M. and Ferguson, A.J.D. (1995). Linking hydrology and ecology: the scientific basis for river management. In Harper, D.M. and Ferguson, A.J.D. (eds) *The Ecological Basis for River Management*, pp. 1–16. Wiley, Chichester.

Phillips, D.J.H. (1993). Bioaccumulation. In Calow, P. (ed.) *Handbook of Ecotoxicology*, vol. 1, pp. 378–96. Blackwell, Oxford.

Phillips, D.J.H. and Rainbow, P.S. (1993). *Biomonitoring of Trace Aquatic Contaminants*. Elsevier, London.

Phillips, G.L. (1992). A case study in restoration: shallow eutrophic lakes in the Norfolk Broads. In Harper, D. *Eutrophication of Freshwaters*, pp. 251–78. Chapman and Hall, London.

Phillips, G.L. and Kerrison, P. (1991). The restoration of the Norfolk Broads: the role of biomanipulation. *Mem. Ist. ital. Idrobiol*, **48**, 75–97.

Phillips, G.L., Jackson, R., Bennett, C. and Chilvers, A. (1994). The importance of sediment phosphorus release in the restoration of very shallow lakes (the Norfolk Broads, England) and implications for biomanipulation. *Hydrobiologia*, **275/276**, 445–56.

Phillips, G.L., Perrow, M.R. and Stansfield, J. (1996). Manipulating the fish–zooplankton interaction in shallow lakes: a tool for restoration. In Greenstreet, S.P.R. and Tasker, M.L. (eds) *Aquatic Predators and their Prey*, pp. 174–83. Fishing News Books, Oxford.

Phillips, G.L., Bramwell, A., Pitt, J., Stansfield, J. and Perrow, M. (1999). Practical application of 25 years research into the management of shallow lakes. *Hydrobiologia*, **396**, 61–76.

Pieters, H. and Hagel, P. (1992). Biomonitoring of mercury in European eel (*Anguilla anguilla* L.) in the Netherlands, compared with pike-perch (*Stizostedion lucioperca*): statistical analysis. In Vernet, J.P. (ed.) *Impact of Heavy Metals on the Environment*, pp. 203–17. Elsevier, Amsterdam.

Pillay, T.V.T. (1992). *Aquaculture and the Environment*. Fishing News Books, Oxford.

Pinder, L.C.V. (1989). Biological surveillance of chalk-streams. *Freshwater Biological Association Annual Report*, **1989**, 81–92.

Pinder, L.C.V. and Farr, I.S. (1987a). Biological surveillance of water quality. 2. Temporal and spatial variation in the macroinvertebrate fauna of the River Frome, a Dorset chalk stream. *Arch. Hydrobiol.*, **109**, 321–31.

Pinder, L.C.V. and Farr, I.S. (1987b). Biological surveillance of water quality. 3. The influence of organic enrichment on the macroinvertebrate fauna of small chalk streams. *Arch. Hydrobiol.*, **109**, 619–37.

Pinder, L.C.V., Ladle, M., Gledhill, T., Bass, J.A.B. and Matthews, A. (1987). Biological surveillance of water quality. 1. A comparison of macroinvertebrate surveillance methods in relation to assessment of water quality in a chalk stream. *Arch. Hydrobiol.*, **109**, 207–26.

Poleo, A.B.S., Ostbye, K., Oxnevad, S.A., Andersen, R.A., Heibo, E. and Vollestad, L.A. (1997). Toxicity of acid aluminium-rich water to seven freshwater fish species; a comparative laboratory study. *Environ. Pollut.*, **96**, 129–39.

Pollard, P. and Huxham, M. (1998). The European Water Framework Directive: a new era in the management of aquatic ecosystem health? *Aquat. Conserv.*, **8**, 773–92.

Polloth, C. and Mangelsdorf, I. (1997). Commentary on the application of (Q)SAR to the toxicological evaluation of existing chemicals. *Chemosphere*, **35**, 2525–42.

Polman, H.J.G. and de Zwart, D. (1994). The toxicity of organic concentrates to *Photobacterium phosphorium* of River Meuse water in the stretch between Remilly (France) and Keizersveer (the Netherlands). *Water Sci. Tech.*, **29**, 253–6.

Polprasert, C. (1989). *Organic Waste Recycling*. Wiley, Chichester.

Poulton, M.J. and Pascoe, D. (1990). Disruption of pre-copula in *Gammarus pulex* (L.): development of a behavioural bioassay for evaluating pollutant and parasite induced stress. *Chemosphere*, **20**, 403–15.

Power, M., Attrill, M.J. and Thomas, R.M. (1999). Heavy metal concentration trends in the Thames estuary. *Water Res.*, **33**, 1672–80.

Poysa, H., Rask, M. and Nummi, P. (1994). Acidification and ecological interactions at higher trophic levels in small forest lakes – perch and the common goldeneye. *Ann. Zool. Fenn.*, **31**, 397–404.

Prat, N. and Ibañez, C. (1995). Effects of water transfers projected in the Spanish National Hydrological Plan on the ecology of the lower River Ebro (N.E. Spain) and its delta. *Water Sci. Technol*, **31**, 79–86.

Preston, A. (1974). Application of critical path analysis techniques to the assessment of environmental capacity and the control of environmental waste disposal. In *Comparative Studies of Food and Environmental Contamination*, pp. 573–83. International Atomic Energy Agency, Vienna.

Pretty, J. (1998). *The Living Land*. Earthscan, London.

Pretty, J.N., Brett, C., Gee, D., Hine, R.E., Mason, C.F., Morison, J.I.L., Raven, H., Rayment, M. and van der Bijl, G. (2000). An assessment of the external costs of agriculture. *Agric. Systems*, **65**, 113–36.

Price, M-A., Jurd, R.D. and Mason, C.F. (1997). A field investigation into the effect of sewage effluent and general water quality in selected immunological indicators in carp (*Cyprinus carpio* L.). *Fish Shellfish Immunol.*, **7**, 193–207.

Prince, R.C. (1992). Bioremediation of oil spills with particular reference to the spill from the *Exxon Valdez*. In Fry, J.C., Gadd, G.M., Herbert, R.A., Jones, C.W. and Watson-Craik, I.A. (eds) *Microbial Control of Pollution*, pp. 19–34. Cambridge University Press, Cambridge.

Purdom, C.E., Hardiman, P.A., Bye, V.J., Eno, N.C., Tyler, C.R. and Sumpter, J.P. (1994). Estrogenic effects of effluents from sewage treatment works. *Chem. Ecol.*, **8**, 275–85.

Purseglove, J. (1988). *Taming the Flood*. Oxford University Press, Oxford.

Pysek, P. and Prach, K. (1995). Invasion dynamics of *Impatiens glandulifera* – a century of spreading reconstructed. *Biol. Conserv.*, **74**, 41–8.

Randall, S., Harper, D. and Brierley, B. (1999). Ecological and ecophysiological impacts of ferric dosing in reservoirs. *Hydrobiologia*, **396**, 355–64.

Rapp, P. and Timmis, K.N. (1999). Degradation of chlorobenzenes at nanomolar concentrations by *Burkholderia* sp strain PS14 in liquid cultures and in soil. *Appl. Environ. Microbiol.*, **65**, 2547–52.

Rask, M. (1992). Effects of acidification and liming on fish populations in Finland. *Finnish Fish. Res.*, **13**, 107–17.

Rast, W. and Holland, M. (1988). Eutrophication of lakes and reservoirs: a framework for making management decisions. *Ambio*, **17**, 2–12.

Raven, P.J. and George, J.J. (1989). Recovery by riffle macroinvertebrates in a river after a major accidental spillage of chlorpyrifos. *Environ. Pollut.*, **59**, 55–70.

Raven, P.J., Holmes, N.T.H., Dawson, F.H. and Everard, M. (1998). Quality assessment using River Habitat Survey data. *Aquat. Conserv.*, **8**, 477–99.

Raven, P.J., Holmes, N.T.H., Naura, M. and Dawson, F.H. (2000). Using river habitat survey for environmental assessment and catchment planning in the UK. *Hydrobiologia*, **422**, 359–67.

Ravera, O. (1989). Lake ecosystem degradation and recovery studied by the enclosure method. In Ravera, O. (ed.) *Ecological Assessment of Environmental Degradation, Pollution and Recovery*, pp. 217–43. Elsevier, Amsterdam.

Read, M. (1989). Arrows v. arrogance. *BBC Wildlife*, **7**, 764–7.

Redshaw, C.J., Mason, C.F., Hayes, C.R. and Roberts, R.D. (1988). Nutrient budget for a hypertrophic reservoir. *Water Res.*, **4**, 413–19.

Reimer, P. (1989). Concentrations of lead in aquatic macrophytes from Shoal Lake, Manitoba, Canada. *Environ. Pollut.*, **56**, 77–84.

Reimer, P. and Duthie, H.C. (1993). Concentrations of zinc and chromium in aquatic macrophytes from Sudbury and Muskoka regions of Ontario, Canada. *Environ. Pollut.*, **79**, 261–5.

Renberg, I. and Hedberg, T. (1982). The pH history of lakes in south-western Sweden as calculated from the subfossil diatom flora of the sediments. *Ambio*, **11**, 30–3.

Resh, V.H. and McElravy, E.P. (1993). Contemporary quantitative approaches to biomonitoring using benthic macroinvertebrates. In Rosenberg, D.M. and Resh, V.H. (eds) *Freshwater Biomonitoring and Benthic Macroinvertebrates*, pp. 159–94. Chapman and Hall, New York.

Reynolds, B., Stevens, P.A., Hughes, S., Parkinson, J.A. and Weatherley, N.S. (1995). Stream chemistry impacts of conifer harvesting in Welsh catchments. *Water, Air, Soil Pollut.*, **79**, 147–70.

Ricciardi, A., Neves, R.J. and Rasmussen, J.B. (1998). Impending extinctions of North American freshwater mussels (Unionidae) following the zebra mussel (*Dreissena polymorpha*) invasion. *J. Anim. Ecol*, **67**, 613–19.

Richardson, C.J. and Qian, S.S. (1999). Long-term phosphorus assimilative capacity in freshwater wetlands: a new paradigm for sustaining ecosystem structure and function. *Environ. Sci. Technol.*, **33**, 1545–51.

Richardson, J.S. and Kiffney, P.M. (2000). Responses of a macroinvertebrate community from a pristine, southern British Columbia, Canada, stream to metals in experimental mesocosms. *Environ. Toxicol. Chem.*, **19**, 736–43.

Ridge, I. and Barrett, P.R.F. (1992). Algal control with barley straw. *Aspects Appl. Biol.*, **29**, 457–62.

Ripl, W. (1976). Biochemical oxidation of polluted lake sediment with nitrate. A new restoration method. *Ambio*, **5**, 312–15.

Robach, F., Thiébaut, G., Trémolieres, M. and Muller, S. (1996). A reference system for continental running waters: plant communities as bioindicators of increasing eutrophication in alkaline and acidic waters in north-east France. *Hydrobiologia*, **340**, 67–76.

Robertson, L.A. and Kuenen, J.G. (1992). Nitrogen removal from water and waste. In Fry, J.C., Gadd, G.M., Herbert, R.A., Jones, C.W. and Watson-Craik, I.A. (eds) *Microbial Control of Pollution*, pp. 227–67. Cambridge University Press, Cambridge.

Rodgers-Gray, T.P., Jobling, S., Morris, S., Kelly, C., Kirby, S., Janbakhsh, A., Harries, J.E., Waldock, M.J., Sumpter, J.P. and Tyler, C.R. (2000). Long-term temporal changes in the estrogenic composition of treated sewage effluent and its biological effects on fish. *Environ. Sci. Technol.*, **34**, 1521–8.

Rodhe, W. (1969). Crystallization of eutrophication concepts in northern Europe. In *Eutrophication: Causes, Consequences, Correctives*, pp. 50–64. National Academy of Sciences, Washington, DC.

Roesijadi, G. and Robinson, W.E. (1994). Metal regulation in aquatic animals; mechanisms of uptake, accumulation and release. In Malins, D.C. and Ostrander, G.K. (eds) *Aquatic Toxicology*, pp. 387–420. Lewis, Boca Raton, FL.

Room, P.M. (1990). Ecology of a simple plant-herbivore system: biological control of *Salvinia*. *Trends Ecol. Evol.*, **5**, 74–9.

Roos, A., Greyerz, E., Olsson, M. and Sandegren, F. (2001). The otter (*Lutra lutra*) in Sweden – population trends in relation to ΣDDT and total PCB concentrations during 1968–99. *Environ. Pollut.*, **111**, 457–69.

Rose, C. (1990). *The Dirty Man of Europe*. Simon and Schuster, London.

Rosenberg, D.M. and Resh, V.H. (1982). The use of artificial Substrates in the study of freshwater macroinvertebrates. In Cairns, J. (ed.) *Artificial Substrates*, pp. 175–235. Ann Arbor Science Publishers, Ann Arbor, MI.

Rosseland, B.O. (1986). Ecological effects of acidification on tertiary consumers. Fish population responses. *Water, Air, Soil Pollut.*, **30**, 451–60.

Rosseland, B.O., Skogheim, O.K., Abrahamsen, H. and Matzow, D. (1986). Limestone slurry reduces physiological stress and increases survival of Atlantic salmon (*Salmo salar*) in an acidic Norwegian river. *Can. J. Fish. Aquat. Sci.*, **43**, 1888–93.

Round, F.E. (1991). Diatoms in river water-monitoring studies. *J. Appl. Phycol.*, **3**, 129–45.

Rowe, C.L., Kinney, O.M., Fiori, A.P. and Congdon, J.D. (1996). Oral deformities in tadpoles (*Rana catesbiana*) associated with coal ash deposition: effects on grazing ability and growth. *Freshwater Biol.*, **36**, 723–30.

Royal Commission on Environmental Pollution (2000). *Energy – The Changing Climate*. Royal Commission on Environmental Pollution, London.

Rundle, S.D., Weatherley, N.S. and Ormerod, S.J. (1995). The effects of catchment liming on the chemistry and biology of upland Welsh streams: testing model predictions. *Freshwater Biol.*, **34**, 165–75.

Rutt, G.P., Pickering, T.D. and Reynolds, N.R.M. (1993). The impact of livestock farming on Welsh streams: the development and testing of a rapid biological method for use in the assessment and control of organic pollution from farms. *Environ. Pollution*, **81**, 217–28.

Ryckman, D.P., Weseloh, D.V., Hamr, P., Fox, G.A., Collins, B., Ewins, P.J. and Norstrom, R.J. (1998). Spatial and temporal trends in organochlorine contamination and bill deformities in double-crested cormorants (*Phalacrocorax auritus*) from the Canadian Great Lakes. *Environ. Monit. Assess.*, **53**, 169–95.

Ryding, S.-O. and Rast, W. (1989). *The Control of Eutrophication of Lakes and Reservoirs*. Parthenon, Paris.

Safe, S. (1987). Determination of 2,3,7,8-TCDD toxic equivalence factors (TEFs). Support for the use of *in vitro* AHH induction assay. *Chemosphere*, **16**, 791–802.

Safe, S. (1990). Polychlorinated biphenyls (PCBs), dibenzo-*p*-dioxins (PCDDs), dibenzofurans (PCDFs) and related compounds: environmental and mechanistic considerations which support the development of toxic equivalency factors (TEFs). *Crit. Rev. Toxicol.*, **21**, 51–88.

Sanders, G., Jones, K.C., Hamilton-Taylor, J. and Dorr, H. (1992). Historical inputs of polychlorinated biphenyls and other organochlorines to a dated lacustrine sediment core in rural England. *Environ. Sci. Technol.*, **26**, 1815–21.

Sanders, J.E. (1989). PCB-pollution problem in the Upper Hudson River: from environmental disaster to 'environmental gridlock'. *NE Environ. Sci.*, **8**, 1–86.

Santillo, D., Stringer, R.L., Johnson, P.A. and Tickner, J. (1998) The Precautionary Principle: protecting against failures of scientific method and risk assessment. *Mar. Pollut. Bull.*, **36**, 939–50.

Say, P.J., Diaz, B.M. and Whitton, B.A. (1977). Influence of zinc on lotic plants. 1. Tolerance of *Hormidium* species to zinc. *Freshwater Biol.*, **7**, 357–76.

Scharenberg, W., Gramann, P. and Pfeiffer, W.H. (1994). Bioaccumulation of heavy metals and organochlorines in a lake ecosystem with special reference to bream (*Abramis brama* L.). *Sci. Total Environ.*, **155**, 187–97.

Scheffer, M. (1998). *Ecology of Shallow Lakes.* Chapman and Hall, London.

Scheffer, M., Hosper, S.H., Meijer, M.-L., Moss, B. and Jeppeson, E. (1993). Alternative equilibria in shallow lakes. *Trends Ecol. Evol.*, **8**, 275–9.

Scheider, W.A., Cox, C., Hayton, A., Hitchin, G. and Vaillancourt, A. (1998). Current status and temporal trends in concentrations of persistent toxic substances in sport fish and juvenile forage fish in the Canadian waters of the Great Lakes. *Environ. Monit. Assess.*, **53**, 57–76.

Scheuhammer, A.M. and Graham, J.E. (1999). The bioaccumulation of mercury in aquatic organisms from two similar lakes with differing pH. *Ecotoxicology*, **8**, 49–56.

Schindler, D.W. (1987a). Detecting ecosystem responses to anthropogenic stress. *Can. J. Fish. Aquat. Sci.*, **44**, 6–25.

Schindler, D.W. (1987b). Recovery of Canadian lakes from acidification. In Barth, H. (ed.) *Reversibility of Acidification*, pp. 2–13. Elsevier, London.

Schindler, D.W. (1988a). Experimental studies of chemical stressors on whole lake ecosystems. *Verh. int. Verein. Limnol.*, **23**, 11–41.

Schindler, D.W. (1988b). Effects of acid rain on freshwater ecosystems. *Science*, **239**, 149–57.

Schindler, D.W. (1997). Widespread effects of climatic warming on freshwater ecosystems in North America. *Hydrol. Proc.*, **11**, 1043–67.

Schindler, D.W. and Fee, E.J. (1973). Diurnal variation of dissolved inorganic carbon and its use in estimating primary production and CO_2 invasion in Lake 227. *J. Fish. Res. Bd. Can.*, **30**, 1501–10.

Schindler, D.W., Armstrong, F.A.J., Holmgren, S.K. and Brunskill, G.J. (1971). Eutrophication of lake 227, Experimental Lakes Area, northwestern Ontario, by addition of phosphate and nitrate. *J. Fish. Res. Bd Can.*, **28**, 1763–82.

Schindler, D.W., Kling, H., Schmidt, R.V., Prokopowich, J., Frost, V.E., Reid, R.A. and Capel, M. (1973). Eutrophication of Lake 227 by addition of phosphate and nitrate: the second, third and fourth years of enrichment, 1970, 1971 and 1972. *J. Fish. Res. Bd Can.*, **30**, 1415–40.

Schindler, D.W., Mills, K.H., Malley, D.F., Findlay, D.L., Shearer, J.A., Davies, I.J., Turner, M.A., Linsey G.A. and Cruikshank, D.R. (1985). Long-term ecosystem stress: the effects of years of experimental acidification on a small lake. *Science*, **228**, 1395–401.

Schindler, D.W., Beatty, K.G. and Fee, E.J. (1990). Effects of climatic warming on lakes of the central boreal forest. *Science*, **250**, 967–70.

Schrage, W. and Enderlein, R. (1996). Environmental impact assessment and water management; policy issues in the ECE region. *European Water Poll. Control*, **6**, 17–28.

Schriver, P., Bøgestrand, J., Jeppeson, E. and Søndergaard, M. (1996). Impact of submerged macrophytes on fish–zooplankton–phytoplankton interactions: large-scale enclosure experiments in a shallow eutrophic lake. *Freshwater Biol.*, **33**, 255–70.

Schulte-Wulwer-Leidig, A. (1995). Ecological master plan for Rhine catchment. In Harper, D.M. and Ferguson, A.J.D. (eds) *The Ecological Basis for River Management*, pp. 505–14. Wiley, Chichester.

Schultz, R. and Liess, M. (1999a). A field study of the effects of agriculturally derived insecticide input on stream macroinvertebrate dynamics. *Aquat. Toxicol.*, **46**, 155–76.

Schultz, R. and Liess, M. (1999b). Validity and ecological relevance of an active in situ bioassay using *Gammarus pulex* and *Limnephilus lunatus*. *Environ. Toxicol. Chem.*, **18**, 2243–50.

Schüpbach, M.R. (1981). Halogenierte Kohlenwasserstoffe in der Nahrung. *Proc. Int. Conf. Chem. Environ. Man.*, pp. 105–24. Gottlieb Duttweiller Institute, Zurich.

Scragg, A. (1999). *Environmental Biotechnology.* Longman, Harlow.

Scullion, J. and Edwards, R.W. (1980). The effects of coal industry pollutants on the macroinvertebrate fauna of a small river in the South Wales coalfield. *Freshwater Biol.*, **10**, 141–62.

Sears, J. (1989). A review of lead poisoning among the River Thames mute swan *Cygnus olor* population. *Wildfowl*, **40**, 151–2.

Seehausen, O., Witte, F., Katunzi, E.F., Smits, J. and Bouton, M. (1997). Patterns of the remnant cichlid fauna in southern Lake Victoria. *Conserv. Biol.*, **11**, 890–904.

Sell, N.J. (1992). *Industrial Pollution Control.* Van Nostrand Reinhold, New York.

Sepúlveda, M.S., Williams, G.E., Frederick, P.C. and Spalding, M.G. (1999). Effects of mercury on health and first-year survival of free-ranging great egrets (*Ardea albus*) from southern Florida. *Arch. Environ. Contam. Toxicol.*, **37**, 369–76.

Shales, S., Thake, B.A., Frankland, B., Khan, D.H., Hutchinson, J.D. and Mason, C.F. (1989). Biological and ecological effects of oils. In Green, J. and Trett, M. (eds) *The Fate and Effects of Oil in Freshwater*, pp. 81–171. Applied Science Publishers, London.

Shapiro, J. and Wright, D.I. (1984). Lake restoration by manipulation: Round Lake, Minnesota, the first two years. *Freshwater Biol.*, **14**, 371–83.

Sharpe, R.M. and Skakkebaek, N.E. (1993). Are oestrogens involved in falling sperm counts and disorders of the male reproductive tract? *Lancet*, **341**, 1392–5.

Sharpe, R.M., Turner, K.J. and Sumpter, J.P. (1998). Endocrine disruptors and testis development. *Environ. Health. Perspective*, **106**, A220–A221.

Sheath, R.G., Havas, M., Hellebust, J.A. and Hutchinson, T.C. (1982). Effect of long-term acidification on the algal communities of tundra ponds at the Smoking Hills, N.W.T., Canada. *Can. J. Bot.*, **60**, 58–72.

Shutes, B., Ellis, B., Revitt, M. and Bascombe, A. (1992). The use of freshwater invertebrates for the assessment of metal pollution in urban receiving waters. In Dallinger, R. and Rainbow, P.S. (eds) *Ecotoxicology of Metals in Invertebrates*, pp. 201–22. Lewis, Boca Raton, FL.

Simmons, I.G. (1996). *Changing the Face of the Earth*, 2nd edn. Blackwell, Oxford.

Simpson, K.W., Bode, R.W. and Colquhoun, J.R. (1985). The macroinvertebrate fauna of an acid-stressed headwater stream system in the Adirondack Mountains, New York. *Freshwater Biol.*, **15**, 671–81.

Simpson, V.R., Bain, M.S., Brown, R., Brown, B.F. and Lacey, R.F. (2000). A long-term study of vitamin A and polychlorinated hydrocarbon levels in otters (*Lutra lutra*) in south west England. *Environ. Pollut.*, **110**, 267–75.

Sims, J.T., Simard, R.R. and Joern, B.C. (1998). Phosphorus loss in agricultural drainage: historical perspective and current research. *J. Environ. Quality*, **27**, 277–93.

Skidmore, J.F. (1970). Respiration and osmoregulation in rainbow trout with gills damaged by zinc sulphate. *J. Exp. Biol.*, **52**, 484–94.

Sládeček, V. (1973). System of water quality from the biological point of view. *Arch. Hydrobiol.*, **7**, 1–218.

Sládeček, V. (1979). Continental systems for the assessment of river water quality. In James, A. and Evison, L. (eds) *Biological Indicators of Water Quality*, pp. 3.1–3.32. Wiley, Chichester.

Smith, A.M. (1990). The ecophysiology of epilithic diatom communities of acid lakes in Galloway, southwest Scotland. *Phil. Trans. R. Soc. London*, **327B**, 25–30.

Smith, C.D., Harper, D.M. and Barham, P.J. (1990). Engineering operations and invertebrates: linking hydrology with ecology. *Regul. Rivers*, **5**, 89–96.

Smith, H.V. (1997). Detection of *Cryptosporidium* oocysts in water and environmental concentrates. In Sutcliffe, D.W. (ed.) *The Microbiological Quality of Water*, pp. 126–38. Freshwater Biological Association, Ambleside.

Smith, M.A., Grant, L.D. and Sors, A.I. (1989). *Lead Exposure and Child Development: An International Assessment*. Kluwer Academic, Lancaster.

Smith, P. and Briggs, J. (1999). Zander – the hidden invader. *Brit. Wildlife*, **11**, 2–8.

Smith, R.V. (1993). Phosphorus and nitrogen loadings to Lough Neagh and their management. In Wood, R.B. and Smith, R.V. (eds) *Lough Neagh*, pp. 149–69. Kluwer Academic, Dordrecht.

Smith, S.R. (1994a). Effect of soil pH on availability to crops of metals in sewage sludge-treated soils. 1. Nickel, copper and zinc uptake and toxicity to ryegrass. *Environ. Pollut.*, **85**, 321–7.

Smith, S.R. (1994b). Effect of soil pH on availability to crops of metals in sewage sludge-treated soils. II. Cadmium uptake by crops and implications for human dietary intake. *Environ. Pollut.*, **86**, 5–13.

Snyder, C.D. and Hendricks, A.C. (1995). Effect of seasonally changing feeding habits on the whole-animal mercury concentrations in *Hydropsyche morosa* (Trichoptera: Hydropsychidae). *Hydrobiologia*, **229**, 115–23.

Solbé, J.F. de L.G. (1993). Freshwater fish. In Calow, P. (ed.) *Handbook of Ecotoxicology*, pp. 66–82. Blackwell, Oxford.

Søndergaard, M., Jensen, J.P. and Jeppeson, E. (1999). Internal phosphorus loading in shallow Danish lakes. *Hydrobiologia*, **409**, 145–52.

Sorensen, D.L., Eberl, S.G. and Dicksa, R.A. (1989). *Clostridium perfringens* as a point source indicator in non-point polluted streams. *Water Res.*, **23**, 191–7.

Soucek, D.J., Cherry, D.S., Currie, R.J., Latimer, H.A. and Trent, G.C. (2000a). Laboratory to field validation in an integrative assessment of an acid mine drainage-impacted watershed. *Environ. Toxicol. Chem.*, **19**, 1036–43.

Soucek, D.J., Cherry, D.S. and Trent, G.C. (2000b). Relative acute toxicity of acid mine drainage water column and sediments to *Daphnia magna* in the Puckett's Creek watershed, Virginia, USA. *Arch. Environ. Contam. Toxicol.*, **38**, 305–10.

Spalding, M.G., Bjork, R.D., Powell, G.V.N. and Sundlof, S.F. (1994). Mercury and cause of death in great white herons. *J. Wildlife Manage.*, **58**, 735–9.

Spahn, S.A. and Sherry, T.W. (1999). Cadmium and lead exposure associated with reduced growth rates, poorer fledging success of little blue herons (*Egretta caerulea*) in South Louisiana wetlands. *Arch. Environ. Contam. Toxicol.*, **37**, 377–84.

Spencer, C.N., McClelland, B.R. and Stanford, J.A. (1991). Shrimp stocking, salmon collapse, and eagle displacement. *BioScience*, **41**, 14–21.

Sprague, J.B. (1964). Lethal concentrations of copper and zinc for young Atlantic salmon. *J. Fish. Res. Bd Can.*, **21**, 17–26.

Sprague, J.B. (1970). Measurement of pollutant toxicity to fish. II. Utilizing and applying bioassay results. *Water Res.*, **4**, 3–32.

Sprague, J.B. (1971). Measurement of pollutant toxicity to fish. III. Sublethal effects and safe concentrations. *Water Res.*, **5**, 245–66.

Spry, D.J. and Wiener, J.G. (1991). Metal bioavailability and toxicity to fish in low-alkalinity lakes: a critical review. *Environ. Pollut*, **71**, 243–304.

Stallsmith, B.W., Ebersole, J.P. and Hagar, W.G. (1996). The effects of acid episodes on *Lepomis* sunfish recruitment and growth in two ponds in Massachusetts, U.S.A. *Freshwater Biol.*, **36**, 731–44.

Stansfield, J., Moss, B. and Irvine, K. (1989). The loss of submerged plants with eutrophication. III. Potential role of organochlorine pesticides: a palaeoecological study. *Freshwater Biol.*, **22**, 109–32.

Stauffer, J. (1998). *The Water Crisis. Constructing Solutions to Freshwater Pollution.* Earthscan, London.

Stegeman, J.J. and Hahn, M.E. (1994). Biochemistry and molecular biology of monooxygenases: current perspectives on forms, functions, and regulation of cytochrome P450 in aquatic species. In Malins, D.C. and Ostrander, G.K. (eds) *Aquatic Toxicology*, pp. 87–206. Lewis, Boca Raton, FL.

Steinberg, C.E. and Hartmann, H.M. (1988). Planktonic bloom-forming cyanobacteria and the eutrophication of lakes and rivers. *Freshwater Biol.*, **20**, 279–87.

Stenson, J.A.E. and Svensson, J.E. (1995). Changes in planktivore fauna and development of zooplankton after liming of acidified Lake Gårdsjön. *Water, Air, Soil Pollut.*, **85**, 979–84.

Stephenson, M. and Mackie, G.L. (1986). Lake acidification as a limiting factor in the distribution of the freshwater amphipod *Hyalella azteca*. *Can. J. Fish. Aquat. Sci.*, **43**, 288–92.

Stephenson, M. and Mackie, G.L. (1988). Multivariate analysis of correlations between environmental parameters and cadmium concentrations in *Hyalella azteca* (Crustacea: Amphipoda) from central Ontario lakes. *Can. J. Fish. Aquat. Sci.*, **45**, 1705–10.

Stoddard, J.L., Jeffries, D.S., Lükewille, A., Clair, T.A., Dillon, P.J., Driscoll, C.T., Forsius, M., Johannessen, M., Kahl, J.S., Kellogg, J.H., Kemp, A., Mannio, J., Monteith, D., Murdoch, R.S., Patrick, S., Rebsdorf, A., Skjeikvale, R.L., Stainton, M.P., Traaen, T., Van Dam, H., Webster, K.E., Wieting, J. and Wilander, A. (1999). Regional trends in aquatic recovery from acidification in North America and Europe. *Nature*, **401**, 575–8.

Stoner, J.H. and Gee, A.S. (1985). Effects of forestry on water quality and fish in Welsh rivers and lakes. *J. Inst. Water Eng. Sci.*, **39**, 27–45.

Stoner, J.H., Gee, A.S. and Wade, K.R. (1984). The effects of acidification on the ecology of streams in the upper Tywi catchment in West Wales. *Environ. Pollut. (A)*, **35**, 125–57.

Sumpter, J.P. (1998). Xenoendocrine disruptors – environmental impacts. *Toxicol. Lett.*, **103**, 337–42.

Suns, K.R., Hitchin, G.G. and Toner, D. (1993). Spatial and temporal trends of organochlorine contaminants in spottail shiners from selected sites in the Great Lakes (1975–90). *J. Great Lakes Res.*, **19**, 703–14.

Sutcliffe, D.W. and Hildrew, A.G. (1989). Invertebrate communities in acid streams. In Morris, R., Taylor, E.W., Brown, D.J.A. and Brown, J.A. (eds) *Acid Toxicity and Aquatic Animals*, pp. 13–29. Cambridge University Press, Cambridge.

Swain, W.R. (1988). Human health consequences of consumption of fish contaminated with organochlorine compounds. *Aquat. Toxicol.*, **11**, 357–77.

Swales, S. (1998). Biological monitoring of the impacts of the Ok Tedi copper mine on the fish populations in the Fly River system, Papua New Guinea. *Sci. Total Environ.*, **214**, 99–111.

Tahedl, H. and Häder, D.-P. (1999). Fast estimation of water quality using the automatic biotest Ecotox based on the movement behaviour of a freshwater flagellate. *Water Res.*, **33**, 426–32.

Taylor, E.J., Maund, S.J. and Pascoe, D. (1991). Toxicity of four common pollutants to the freshwater macroinvertebrates *Chironomus riparius* Meigen (Insecta: Diptera) and *Gammarus pulex* (L.) (Crustacea: Amphipoda). *Arch. Environ. Contam. Toxicol.*, **21**, 371–6.

Taylor, E.J., Blockwell, S.J., Maund, S.J. and Pascoe, D. (1993). Effects of lindane on the life-cycle of a freshwater macroinvertebrate *Chironomus riparius* Meigen (Insecta: Diptera). *Arch. Environ. Contam. Toxicol.*, **24**, 145–50.

Taylor, E.J., Rees, E.M. and Pascoe, D. (1994). Mortality and a drift-related response of the freshwater amphipod *Gammarus pulex* (L.) exposed to natural sediments, acidification and copper. *Aquat. Toxicol.*, **29**, 83–101.

Taylor, E.J., Underhill, K.M., Blockwell, S.J. and Pascoe, D. (1998). Haem biosynthesis in the freshwater macroinvertebrate *Gammarus pulex* (L.): effects of copper and lindane. *Water Res.*, **32**, 2202–4.

Tessier, A. and Turner, D.R. (1995). *Metal Speciation and Bioavailability in Aquatic Systems.* Wiley, Chichester.

Tessier, L., Boisvert, J.L., Vought, L.B.-M. and Lacoursiere, J.O. (2000a). Anomalies on capture nets of *Hydropsyche slossonae* larvae (Trichoptera; Hydropsychidae) following a sublethal chronic exposure to cadmium. *Environ. Pollut.*, **108**, 425–38.

Tessier, L., Boisvert, J.L., Vought, L.B.-M. and Lacoursiere, J.O. (2000b). Effects of 2,4-dichlorophenol on the net-spinning behavior of *Hydropsyche slossonae* larvae (Trichoptera; Hydropsychidae), an early warning signal of chronic toxicity. *Ecotoxicol. Environ. Safety*, **46**, 207–17.

Tessier, L., Boisvert, J.L., Vought, L.B.-M. and Lacoursiere, J.O. (2000c). Anomalies of capture nets of *Hydropsyche slossonae* larvae (Trichoptera; Hydropsychidae), a potential indicator of chronic toxicity of malathion (organophosphate insecticide). *Aquat. Toxicol.*, **50**, 125–39.

Theodorakis, C.W., Blaylock, B.G. and Shugart, L.R. (1997). Genetic ecotoxicology. 1. DNA integrity and reproduction in mosquitofish exposed *in situ* to radionuclides. *Ecotoxicology*, **6**, 205–18.

Thomas, M. (1998). Temporal changes in the movements and abundance of Thames estuary fish populations. In Attrill, M.J. (ed.) *A Rehabilitated Estuarine Ecosystem: The Environment and Ecology of the Thames Estuary*, pp. 115–39. Kluwer, Dordrecht.

Thompson, P.-A. and Rhee, G.-Y. (1994). Phytoplankton responses to eutrophication. *Ergebn. Limnol.*, **42**, 125–66.

Timmis, K.N. and Pieper, D.H. (1999). Bacteria designed for bioremediation. *Trends Biotechnol*, **17**, 201–4.

Timms, R.M. and Moss, B. (1984). Prevention of growth of potentially dense phytoplankton populations by zooplankton grazing, in the presence of zooplanktivorous fish, in a shallow wetland ecosystem. *Limnol. Oceanogr.*, **29**, 472–86.

Tipping, E., Bettney, R., Hurley, M.A., Isgren, F., James, J.B., Lawlor, A.J., Lofts, S., Rigg, E., Simon, B.M., Smith, E.J. and Woof, C. (2000). Reversal of acidification in tributaries of the River Duddon (English Lake District) between 1970 and 1998. *Environ. Pollut.*, **109**, 183–91.

Tittizer, T., Scholl, F. and Dommermuth, M. (1994). The development of the macrozoobenthos in the River Rhine in Germany during the 20th century. *Water Sci. Technol.*, **29(3)**, 21–8.

Tolba, M.R. and Holdich, D.M. (1981). The effect of water quality on the size and fecundity of *Asellus aquaticus* (Crustacea: Isopoda). *Aquat. Toxicol.*, **1**, 101–12.

Tranter, M., Davies, T.D., Wigington, P.J. and Eshleman, K.N. (1994). Episodic acidification of freshwater systems in Canada – physical and geochemical processes. *Water, Air, Soil Pollut.*, **72**, 19–39.

Trippel, E.A., Eckmann, R. and Hartmann, J. (1991). Potential effects of global warming on whitefish in Lake Constance, Germany. *Ambio*, **20**, 226–31.

Try, P.M. and Price, G.J. (1995). Sewage and industrial effluents. In Hester, R.E. and Harrison, R.M. (eds) *Waste Treatment and Disposal*, pp. 17–41. Cambridge, Royal Society of Chemistry.

Turner, B.L. and Haygarth, P.M. (1999). Phosphorus leaching under cut grassland. *Water Sci. Technol.*, **39**, 63–7.

Turnpenny, A.W.H. (1989). Field studies on fisheries in acid waters in the United Kingdom. In Morris, R., Taylor, E.W., Brown, D.J.A. and Brown, J.A. (eds) *Acid Toxicity and Aquatic Animals*, pp. 45–65. Cambridge University Press, Cambridge.

Turnpenny, A.W.H., Dempsey, C.H., Davis, M.H. and Fleming, J.M. (1988). Factors limiting fish populations in the Loch Fleet system, an acid drainage system in south-west Scotland. *J. Fish Biol.*, **32**, 101–18.

Turnpenny, A.W.H., Fleming, J.M. and Wood, R. (1995). The brown trout population at Loch Fleet eight years after liming. *Chem. Ecol.*, **9**, 185–97.

Tüzün, I. and Mason, C.F. (1996). Eutrophication and its control by biomanipulation: an enclosure experiment. *Hydrobiologia*, **331**, 79–95.

Twitchen, J.B. (1990). The physiological bases of resistance to low pH among aquatic insect larvae. In Mason, J.B. (ed.) *The Surface Waters Acidification Programme*, pp. 413–19. Cambridge University Press, Cambridge.

Twitchen, I.B. and Eddy, F.B. (1994). Sublethal effects of ammonia on freshwater fish. In Müller, R. and Lloyd, R. (eds) *Sublethal and Chronic Effects of Pollutants on Freshwater Fish*, pp. 135–47. Fishing News Books, Oxford.

Tyler, C.R., Jobling, S. and Sumpter, J.P. (1998). Endocrine disruption in wildlife: a critical review of the evidence. *Crit. Rev. Toxicol.*, **28**, 319–61.

Tyler, S.J. and Ormerod, S.J. (1992). A review of the likely causal pathways relating the reduced density of breeding dippers *Cinclus cinclus* to the acidification of upland streams. *Environ. Pollut.*, **78**, 49–55.

Tyler-Jones, R., Beattie, R.C. and Aston, R.J. (1989). The effects of acid water and aluminium on the embryonic development of the common frog, *Rana temporaria*. *J. Zool., Lond.*, **219**, 355–72.

Vandermeulen, J.H. (1987). Toxicity and sublethal effects of petroleum hydrocarbons in freshwater biota. In Vandermeulen, J.H. and Hrudey, S.E. (eds) *Oil in freshwater: chemistry, biology, countermeasure technology*, pp. 267–303. Pergamon, New York.

Vandermeulen, J.H. and Hrudey, S.E. (eds) (1987). *Oil in freshwater: chemistry, biology and countermeasure technology*. Pergamon, New York.

Vanderpoorten, A. (1999). Aquatic bryophytes for a spatio-temporal monitoring of the water pollution of the rivers Meuse and Sambre (Belgium). *Environ. Pollut.*, **104**, 401–10.

Van Hattum, B., De Voogt, P., Van Den Bosch, L., Van Straalen, N.M. and Joosse, E.N.G. (1989). Bioaccumulation of cadmium by the freshwater isopod *Asellus aquaticus* (L.) from aqueous and dietary sources. *Environ. Pollution*, **62**, 129–51.

Van Hattum, B., Timmermans, K.R. and Govers, H.A. (1991). Abiotic and biotic factors influencing *in situ* trace metal levels in macroinvertebrates in freshwater ecosystems. *Environ. Toxicol. Chem.*, **10**, 275–92.

Van Hattum, B., Korthals, G., Van Straalen, N.M., Govers, H.A.J. and Joosse, E.N.G. (1993). Accumulation patterns of trace metals in freshwater isopods in sediment bioassays – influence of substrate characteristics, temperature and pH. *Water Res.*, **27**, 669–84.

Van Leeuwen, F.X.R., Feeley, M., Schrenk, D., Larsen, J.C., Farland, W. and Younes, M. (2000). Dioxins: WHO's tolerable daily intake (TDI) revisited. *Chemosphere*, **40**, 1095–1101.

Van Liere, L., Parma, S., Mur, L.R., Leentwaar, P. and Engelen, G.B. (1984). Loosdrecht Lakes restoration project, an introduction. *Verh. int. verein. Limnol.*, **22**, 829–34.

Van Rijen, J.P.M. (1998). Practical approaches for nature development: let nature do its own thing again. In De Waal, L., Large, A.R.G. and Wade, P.M. (eds), *Rehabilitation of rivers: principles and implementation*, pp. 113–30. Wiley, Chichester.

Vaughan, N., Jones, G. and Harris, S. (1996). Effects of sewage effluent on the activity of bats (Chiroptera: Vespertilionidae) foraging along rivers. *Biol. Conserv.*, **78**, 337–43.

Veith, G.D., Defoe, D.L. and Bergstedt, B.V. (1979). Measuring and estimating the bioconcentration factor of chemicals in fish. *J. Fish. Res. Bd Can.*, **36**, 1040–8.

Ventura, M. and Harper, D. (1996). The impacts of acid precipitation mediated by geology and forestry upon upland stream invertebrate communities. *Arch. Hydrobiol.*, **138**, 161–73.

Vesely, J. (1994). Effects of acidification on trace metal transport in fresh waters. In Steinberg, C.E.W. and Wright, R.F. (eds) *Acidification of freshwater ecosystems: implications for the future*, pp. 141–51. Wiley, Chichester.

Vickery, J.A. (1991). Breeding density of dippers *Cinclus cinclus*, grey wagtails *Motacilla cinerea* and common sandpipers *Actitis hypoleucos* in relation to the acidity of streams in south-west Scotland. *Ibis*, **133**, 18–23.

Viessman, W. and Hammer, R.J. (1993). *Water supply and pollution control*, 5th edition Harper Collins, New York.

Vincent, W.F. (1989). Cyanobacterial growth and dominance in two eutrophic lakes: a review and synthesis. *Arch. Hydrobiol.*, **32**, 239–54.

Visser, P.M., Ibelings, B.W., Veer, B. Van Der, Koedood, J. and Mur, L.C. (1996). Artificial mixing prevents nuisance blooms of the cyanobacterium in Lake Nieuwe Meer, the Netherlands. *Freshwater Biol.*, **36**, 435–50.

Vollenweider, R.A. (1969). Möglichkeite und Grenzen elementarer Modelle der Stoffbilanz von See. *Arch. Hydrobiol.*, **66**, 1–36.

Vollenweider, R.A. (1975). Input–output models with special reference to the phosphorus loading concept in limnology. *Schweiz. Z. Hydrol.*, **37**, 53–84.

Von Castein, H., Li, Y., Timmis, K.N., Deckwer, W.D. and Wagner-Dobler, I. (1999). Removal of mercury from chloralkali electrolysis wastewater by a mercury-resistant *Pseudomonas putida* strain. *Appl. Environ. Microbiol.*, **65**, 5279–84.

Vuori, K-M. (1996). Acid-induced toxicity of aluminium to three species of filter feeding caddis larvae (Trichoptera, Arctopsychidae and Hydropsychidae). *Freshwater Biol.*, **35**, 179–88.

Vuorinen, M., Vourinen, P.J., Rask, M. and Suomela, J. (1994). The sensitivity to acidity and aluminium of newly-hatched perch (*Perca fluviatilis*) originating from strains from four lakes with different degrees of acidity. In Müller, R. and Lloyd, R. (eds) *Sublethal and chronic effects of pollutants on freshwater fish*, pp. 273–82. Fishing News Books, Oxford.

Vuorinen, P.J. and Vuorinen, M. (1992). Acidification in Finland: a review of studies of fish physiology and toxicology. *Finnish Fish. Res.*, **13**, 119–32.

Wade, K.R., Ormerod, S.J. and Gee, A.S. (1989). Classification and ordination of macroinvertebrate assemblages to predict stream acidity in upland Wales. *Hydrobiologia*, **171**, 59–78.

Waite, S. (2000). *Statistical ecology in practice.* Prentice Hall, Harlow.

Wakao, R., Tachibana, H., Tanaka, Y., Sukarai, Y. and Shiota, H. (1985). Morphological and physiological characteristics of streamers in acid mine drainage waters from a pyrite mine. *J. Gen. Appl. Microbiol.*, **31**, 17–28.

Walesh, S.G. (1989). *Urban surface water management.* Wiley, Chichester.

Walker, C.H., Hopkin, S.P., Sibly, R.M. and Peakall, D.B. (1996). *Principles of ecotoxicology.* Taylor & Francis, London.

Walter, R., Macht, W., Dürkop, J., Hecht, R., Hornig, U. and Schulze, P. (1989). Virus levels in river waters. *Water Res.*, **23**, 133–8.

Walley, W.J. and Hawkes, H.A. (1996). A computer-based reappraisal of the Biological Monitoring Working Party scores using data from the 1990 river quality survey of England and Wales. *Water Res.*, **30**, 2086–94.

Walley, W.J. and Hawkes, H.A. (1997). A computer-based development of the Biological Monitoring Working Party score system incorporating abundance rating, site type and indicator value. *Water Res.*, **31**, 201–10.

Wang, W. (1990). Literature review on duckweed toxicity testing. *Environ. Res.*, **52**, 7–22.

Ward, D. (1994). Management of lowland wet grassland for breeding waders. *Brit. Wildlife*, **6**, 89–98.

Warn, A.E. and Page, C. (1984). Estimating the effect of water quality on surface water supplies. *Water Res.*, **18**, 167–72.

Warren, R.D., Waters, D.A., Altringham, J.D. and Bullock, D.J. (2000). The distribution of Daubenton's bats (*Myotis daubentonii*) and pipistrelle bats (*Pipistrellus pipistrellus*) (Vespertilionidae) in relation to small-scale variation in riverine habitats. *Biol. Conserv.*, **92**, 85–91.

Watkins, J. and Jian, X. (1997). Cultural methods for the detection of microorganisms: recent advances and successes. In Sutcliffe, D.W. (ed.) *The microbiological quality of water*, pp. 19–27. Freshwater Biological Association, Ambleside.

Weatherley, N.S. and Ormerod, S.J. (1992). The biological response of acidic streams to catchment liming compared to the changes predicted from stream chemistry. *J. Environ. Manage.*, **34**, 105–15.

Weatherley, N.S., Ormerod, S.J., Thomas, S.P. and Edwards, R.W. (1988). The response of macroinvertebrates to experimental episodes of low pH with different forms of aluminium, during a natural spate. *Hydrobiologia*, **169**, 225–32.

Weatherley, N.S., Thomas, S.P. and Ormerod, S.J. (1989). Chemical and biological effects of acid, aluminium and lime additions to a Welsh hill-stream. *Environ. Pollut.*, **56**, 283–97.

Weatherley, N.S., McCahon, C.P., Pascoe, D. and Ormerod, S.J. (1990). Ecotoxicological studies of acidity in Welsh streams. In Edwards, R.W., Gee, A.S. and Stoner, J.H. (eds) *Acid waters in Wales*, pp. 159–72. Kluwer, Dordrecht.

Weber, D.N. and Spieler, R.E. (1994). Behavioural mechanisms of metal toxicity in fishes. In Malins, D.C. and Ostrander, G.K. (eds) *Aquatic toxicology*, pp. 421–67. Lewis, Boca Raton, FL.

Wehr, J.D., Empain, A., Mouvet, C., Say, P.J. and Whitton, B.A. (1983). Methods for processing aquatic mosses used as monitors of heavy metals. *Water Res.*, **17**, 985–92.

Welch, E.B. and Schrieve, G.D. (1994). Alum treatment effectiveness and longevity in shallow lakes. *Hydrobiologia*, **275/276**, 423–31.

Welch, I.M., Barrett, P.R.F., Gibson, M.T. and Ridge, I. (1990). Barley straw as an inhibitor of algal growth I: studies in the Chesterfield Canal. *J. Appl. Phycol.*, **2**, 231–9.

Werritty, A. (1997). Enhancing the quality of freshwater resources: the role of integrated catchment management. In Boon, P.J. and Howell, D.L. (eds) *Freshwater quality: defining the indefinable?*, pp. 489–505. The Stationary Office, Edinburgh.

Westlake, G.F. and Van der Schalie, W.H. (1977). Evaluation of an automated biological monitoring system at an industrial site. In Cairns, J., Dickson, K.L. and Westlake, G.F. (eds) *Biological monitoring of water and effluent quality*, pp. 30–7. American Society for Testing and Materials, Philadelphia, PA.

Wharfe, J. and Heber, M. (1998). Toxicity assessment: its role in regulation. In Douben, P.E.T. (ed.) *Pollution risk assessment and management*, pp. 311–30. Wiley, Chichester.

Whitehead, P.G., Neal, C. and Neale, R. (1987). Modelling stream acidity in U.K. catchments. In Barth, H. (ed.) *Reversibility of acidification*, pp. 126–41. Elsevier, London.

Whitehouse, P., Van Dijk, P.A.H., Delaney, P.J., Roddie, B.D., Redshaw, C.J. and Turner, C. (1996). The precision of aquatic toxicity tests: Its implications for the control of effluents by direct toxicity testing. In Tapp, J.F., Wharfe, J.R. and Hunt, S.M. (eds) *Toxic impacts of wastes on the aquatic environment*, pp. 44–53. Royal Society of Chemistry, Cambridge.

Whitehurst, I.T. (1991). The *Gammarus: Asellus* ratio as an index of organic pollution. *Water Res.*, **25**, 333–9.

Whitton, B.A. and Diaz, B.M. (1981). Influence of environmental factors on photosynthetic species in highly acidic waters. *Verh. int. verein. Limnol.*, **21**, 1459–65.

Whitton, B.A. and Shehata, F.H.A. (1982). Influence of cobalt, nickel, copper and cadmium on the blue-green alga *Anacystis nidulans*. *Environ. Pollution A*, **27**, 275–81.

Whitton, B.A., Burrows, I.G. and Kelly, M.G. (1989). Use of *Cladophora glomerata* to monitor heavy metals in rivers. *J. Appl. Phycol.*, **1**, 293–9.

Wickham, P., Van De Walle, E and Planas, D. (1987). Comparative effects of mine wastes on the benthos of an acid and an alkaline pool. *Environ. Pollut.*, **44**, 83–9.

Wiener, J.G., Fitzgerald, W.F., Watras, C.J. and Rada, R.G. (1990). Partitioning and bioavailability of mercury in an experimentally acidified Wisconsin lake. *Environ. Toxicol. Chem.*, **9**, 909–18.

Wilkinson, J., Martin, J., Boon, P.J. and Holmes, N.T.H. (1998). Convergence of field survey protocols for SERCON (System for Evaluating Rivers for Conservation) and RHS (River Habitat Survey). *Aquat. Conserv.*, **8**, 579–96.

Williams, D.D. and Feltmate, B.W. (1992). *Aquatic insects*. C.A.B. Interational, Wallingford.

Williams, J.H., Kingham, H.G., Cooper, B.J. and Eagle, D.J. (1977). Growth regulator injury to tomatoes in Essex, England. *Environ. Pollut.*, **12**, 149–66.

Williams, P., Biggs, J., Corfield, A., Fox, G., Walker, D. and Whitfield, M. (1997). Designing new ponds for wildlife. *Brit. Wildlife*, **8**, 137–50.

Williams, W.D. and Aladin, N.V. (1991). The Aral sea: recent limnological changes and their conservation significance. *Aquat. Conserv.*, **1**, 3–23.

Williamson, C.E., Zagarese, H.E., Schulze, P.C., Hargreaves, P.R. and Seva, J. (1994). The impact of short-term exposure to UV-B radiation on zooplankton communities in north temperate lakes. *J. Plankton Res.*, **16**, 205–18.

Williamson, M. (1996). *Biological invasions*. Chapman and Hall, London.

Wilson, R.S. (1994). Monitoring the effect of sewage effluent on the Oxford Canal using chironomid pupal exuviae. *J. Inst. Water Environ. Manage.*, **8**, 171–82.

Wilson, R.S. and McGill, J.D. (1977). A new method of monitoring water quality in a stream receiving sewage effluent, using chironomid pupal exuviae. *Water Res.*, **11**, 959–62.

Winner, R.W., Van Dyke, J.S., Caris, N. and Farrel, M.P. (1975). Response of the macro-invertebrate fauna to a copper gradient in an experimentally polluted stream. *Verh. int. verein. Limnol.*, **19**, 2121–7.

Winterbourn, M.J., McDiffet, W.F. and Eppley, S.J. (2000). Aluminium and iron burdens of aquatic biota in New Zealand streams contaminated by acid mine drainage: effects of trophic level. *Sci. Total Environ.*, **254**, 45–54.

Wissel, B. and Benndorf, J. (1998). Contrasting effects of the invertebrate predator *Chaoborus obscuripes* and planktivorous fish on plankton communities of a long term bio-manipulation experiment. *Arch. Hydrobiol.*, **143**, 129–46.

Witte, F., Goldschmit, T., Goudswaard, P.C., Ligtvoet, W., Van Oijen, M.J.P. and Wanink, J.H. (1992). Species extinctions and concominant ecological changes in Lake Victoria. *Neth. J. Zool.*, **42**, 214–32.

Woin, P. (1998). Short- and long-term effects of the pyrethroid insecticide fenvalerate on an invertebrate pond community. *Ecotox. Environ. Safety*, **41**, 137–146.

Woltering, D.M. (1984). The growth response in fish chronic and early life stages: a critical review. *Aquat. Toxicol.*, **5**, 1–21.

Wood, C.M. and McDonald, D.G. (eds) (1997). *Global warming: implications for freshwater and marine fish*. Cambridge University Press, Cambridge.

Wood, L.B. (1982). *The restoration of the tidal Thames*. Hilger, Bristol.

Wood, P.J. and Petts, G.E. (1994). Low flows and recovery of macroinvertebrates in a small regulated chalk stream. *Regul. Rivers*, **9**, 303–16.

Worrall, P., Peberdy, K.J. and Millett, M.C. (1997). Constructed wetlands and nature conservation. *Water Sci. Tech.*, **35**, 205–13.

Wortley, J.S. and Phillips, G.L. (1987). Fish mortalities and *Prymnesium* in the Norfolk Broads. In Wortley, J.S. (ed.) *Proc. 18th Study Course of Inst. Fish. Management, Cambridge*, pp. 152–62.

Wren, C.D. (1991) Cause–effect linkages between chemicals and populations of mink (*Mustela vision*) and otter (*Lutra canadensis*) in the Great Lakes basin. *J. Toxicol. Environ. Health*, **33**, 549–85.

Wren, C.D. and MaCrimmon, H.R. (1986). Comparative bioaccumulation of mercury in two adjacent freshwater ecosystems. *Water Res.*, **20**, 763–9.

Wren, C.D. and Stephenson, G.L. (1991). The effect of acidification and toxicity of metals to freshwater invertebrates. *Environ. Pollut.*, **71**, 205–41.

Wren, C.D., Harris, S. and Harttrup, N. (1995). Ecotoxicology of mercury and cadmium. In Hoffman, D.J., Rattner, B.A., Burton, G.A. and Cairns, J. (eds) *Handbook of ecotoxicology*, pp. 392–424. Lewis, Boca Raton, FL.

Wright, J.F. (2000). An introduction to RIVPACS. In Wright, J.F., Sutcliffe, D.W. and Furse, M.T. (eds) *Assessing the biological quality of fresh waters*, pp. 1–24. Freshwater Biological Association, Ambleside.

Wright, J.F., Armitage, P.D., Furse, M.T. and Moss, D. (1989). Prediction of invertebrate communities using stream measurements. *Regul. Rivers*, **4**, 147–55.

Wright, J.F., Furse, M.T. and Armitage, P.D. (1993). RIVPACS – a technique for evaluating the biological quality of rivers in the U.K. *European Water Pollut. Control*, **3(4)**, 15–25.

Wright, J.F., Furse, M.T. and Armitage, P.D. (1994). Use of macroinvertebrate communities to detect environmental stress in running waters. In Sutcliffe, D.W. (ed.) *Water quality stress indicators in marine and freshwater systems: linking levels of organization*, pp. 15–34. Freshwater Biological Association, Ambleside.

Wright, J.F., Sutcliffe, D.W. and Furse, M.T. (eds) (2000). *Assessing the biological quality of fresh waters*. Freshwater Biological Association, Ambleside.

Wright, P.J. and Tillitt, D.E. (1999). Embryotoxicity of Great Lakes lake trout extract to developing rainbow trout. *Aquat. Toxicol.*, **47**, 77–92.

Wright, R.F. (1985). Chemistry of Lake Hovvatn, Norway, following liming and reacidification. *Can. J. Fish. Aquat. Sci.*, **42**, 1103–13.

Wright, R.F., Lotse, E. and Semb, A. (1994). Experimental acidification of Alpine catchments at Sogndal, Norway: results after eight years. *Water Air Soil Pollut.*, **72**, 297–315.

Wright, R.M. and Phillips, V.E. (1992). Changes in the aquatic vegetation of two gravel pit lakes after reducing the fish population density. *Aquat. Bot.*, **43**, 43–9.

Xu, Q. and Pascoe, D. (1993). The bioconcentration of zinc by *Gammarus pulex* (L.) and the application of a kinetic model to determine bioconcentration factors. *Water Res.*, **27**, 1683–8.

Yamashita, N., Tanabe, S., Ludwig, J.P., Kurita, H., Ludwig, M.E. and Tatsukawa, R. (1993). Embryonic abnormalities and organochlorine contamination in double-crested cormorants (*Phalacrocorax auritus*) and Caspian terns (*Hydroprogne caspia*) from the upper Great Lakes in 1988. *Environ. Pollut.*, **79**, 163–73.

Zambrano, L. and Hinojosa, D. (1999). Direct and indirect effects of carp (*Cyprinus carpio* L.) on macrophytes and benthic communities in experimental shallow ponds in central Mexico. *Hydrobiologia*, **408/409**, 131–8.

Zamora-Muñoz, C. Sainz-Cantero, C.E., Sanchez-Ortega, A. and Alba-Tercedor, J. (1995). Are biological indices BMWP and ASPT and their significance regarding water quality seasonally dependent? Factors explaining their variations. *Water Res.*, **29**, 285–90.

Zillioux, E.J., Porcella, D.B. and Benoit, J.M. (1993). Mercury cycling and effects in freshwater wetland ecosystems. *Environ. Toxicol. Chem.*, **12**, 2245–64.

INDEX